T0214891

Low-Power High-Resolution Analog
to Digital Converters

ANALOG CIRCUITS AND SIGNAL PROCESSING

Series Editors: Mohammed Ismail
Mohamad Sawan

For other titles published in this series; go to
http://www.springer.com/series/7381

Amir Zjajo · José Pineda de Gyvez

Low-Power High-Resolution Analog to Digital Converters

Design, Test and Calibration

 Springer

Amir Zjajo
Delft University of Technology
Delft, The Netherlands
azjajo@gmail.com

José Pineda de Gyvez
Eindhoven University of Technology,
and NXP Semiconductors
Eindhoven, The Netherlands
j.pineda.de.gyvez@tue.nl

ISBN 978-94-024-0530-9   ISBN 978-90-481-9725-5 (eBook)
DOI 10.1007/978-90-481-9725-5
Springer Dordrecht New York Heidelberg London

Springer is part of Springer Science+Business Media (www.springer.com)

"To my family"

Amir Zjajo

"To the free spirited"

José Pineda de Gyvez

Contents

Foreword

Exploring power efficiency under low supply voltages in deep-submicron CMOS technology is a major challenge of mixed-signal IC researchers. Low voltage and parametric variability seem, at a first glance, incompatible to high performance and robust analog components. In addition, testing and debugging issues grow in difficulty as manufacturing parameters become harder to be controlled and failure mechanisms become complex failure modes. Luckily, innovations have been made to face these challenges, providing low-power circuit and calibration techniques, new mixed-signal testing paradigms, and error detection and isolation processes and tools.

In this book, authors outline valuable examples of these innovations and give us the opportunity to see their benefits when applied to the development of high-resolution Analog to Digital Converters. More specifically, the reader of the book can find main contributions in: the design of multi-step ADCs using time-interleaved signal processing and calibration, the proposal and implementation of a DfT technique for full observability and controllability of these converters, the methodology and design of sensor network to allow identification of process parameter variations, the proposal of algorithms to estimate process variations using small samples sizes, and the development of test pattern generator for wafer level test.

Behind the merit of this book is the industrial expertise and academic knowledge of the authors. Both have been with Philips Research Laboratories and Corporate Research of NXP Semiconductors and in university institutions, and are well-known researchers in the fields of low power design and testing. Several of their books and journal publications are currently indispensable references for understanding design, testing and debugging in deep-submicron technologies. It is my hope that readers enjoy the book as I done myself.

Adoración Rueda
Professor at the University of Sevilla and researcher at IMSE-CNM

Abbreviations

A/D	Analog to Digital
ADC	Analog to Digital Converter
ADSL	Asynchronous DSL
ATE	Automatic Test Equipment
ATPG	Automatic Test Pattern Generator
BIST	Built-In Self-Test
CAD	Computer Aided Design
CDMA	Code Division Multiple Access
CMFB	Common-Mode Feedback
CMOS	Complementary MOS
CMRR	Common-Mode Rejection Ratio
CPU	Central Processing Unit
D/A	Digital to Analog
DAC	Digital to Analog Converter
DAE	Differential Algebraic Equations
DEM	Dynamic Element Matching
DFT	Discrete Fourier Transform
DfT	Design for Testability
DIBL	Drain-Induced Barrier Lowering
DLPM	Die-Level Process Monitor
DMT	Discrete Multi Tone
DNL	Differential Non-Linearity
DR	Dynamic Range
DSL	Digital Subscriber Line
DSP	Digital Signal Processor
DTFT	Discrete Time Fourier Transform
DUT	Device under Test
EM	Expectation-Maximization
ENOB	Effective Number of Bits
ERBW	Effective Resolution Bandwidth
ESSCIRC	European Solid-State Circuit Conference
FFT	Fast Fourier Transform
FoM	Figure of Merit

FPGA	Field Programmable Gate Array
GBW	Gain-Bandwidth Product
GSM	Global System for Mobile Communication
IC	Integrated Circuit
IF	Intermediate Frequency
INL	Integral Non-Linearity
IP	Intellectual Property
ISDN	Integrated Services Digital Network
ISSCC	International Solid-State Circuit Conference
ITDFT	Inverse Time Discrete Fourier Transform
KCL	Kirchhoff' Current Law
LMS	Least Mean Square
LSB	Least Significant Bit
ML	Maximum Likelihood
MNA	Modified Nodal Analysis
MOS	Metal Oxide Semiconductor
MOSFET	Metal Oxide Semiconductor Field Emitter Transistor
MSE	Mean Square Error
MSB	Most Significant Bit
NMOS	Negative doped MOS
OFDM	Orthogonal Frequency Division Multiplex
OTA	Operational Transconductance Amplifier
PCB	Printed Circuit Board
PCM	Process Control Monitoring
PDF	Probability Density Function
PGA	Programmable Gain Amplifier
PLL	Phase Locked Loop
PMOS	Positive Doped MOS
PSK	Phase Shift Keying
PSRR	Power Supply Rejection Ratio
PTAT	Proportional to Absolute Temperature
RF	Radio Frequency
RSD	Redundant Sign Digit
S/H	Sample and Hold
SDM	Steepest Descent Method
SC	Switched Capacitor
SEIR	Stimulus Error Identification and Removal
SFDR	Spurious Free Dynamic Range
SINAD	Signal to Noise and Distortion
SNR	Signal to Noise Ratio
SNDR	Signal to Noise plus Distortion Ratio
SoC	System on Chip
SR	Slew Rate
SVM	Support Vector Machine

THD   Total Harmonic Distortion
UMTS   Universal Mobile Telecommunication System
VGA   Variable Gain Amplifier
VLSI   Very Large-Scale Integrated Circuit
WCDMA   Wideband Code Division Multiple Access
WLAN   Wireless Local Area Network
xDSL   HDSL, ADSL, VDSL, ...

Symbols

a	Elements of the incidence matrix A
A	Amplitude, area, amplifier voltage gain, incidence matrix
A_f	Voltage gain of feedback amplifier
A_0	Open loop dc gain
b	Number of circuit branches
B_i	Number of output codes
B	Bit, effective stage resolution
B_n	Noise bandwidth
c_i	class to which the data x_i from the input vector belongs
c_{xy}	Process correction factors depending upon the process maturity
$c_{h(i)}$	Highest achieved normalized fault coverage
C^*	Neyman-Pearson Critical region
C	Capacitance, covariance matrix
C_C	Compensation capacitance, cumulative coverage
C_{eff}	Effective capacitance
C_F	Feedback capacitance
C_G	Gate capacitance, input capacitance of the operational amplifier
C_{GS}	Gate-Source capacitance
C_H	Hold capacitance
C_{in}	Input capacitance
C_L	Load capacitance
C_{out}	Parasitic output capacitance
C_{ox}	Gate-oxide capacitance
C_{par}	Parasitic capacitance
C_{tot}	Total load capacitance
C_Q	Function of the deterministic initial solution
$C_{\Xi\Xi}$	Autocorrelation matrix
C	Symmetrical covariance matrix
$CH[]$	Cumulative histogram
d_i	Location of transistor i on the die with respect to a point of origin
D_i	Multiplier of reference voltage
D_{out}	Digital output
D_T	Total number of devices

e	Noise, error, scaling parameter of transistor current
e_q	Quantization error
e^2	Noise power
E_{conv}	Energy per conversion step
f_{clk}	Clock frequency
f_{in}	Input frequency
$f_{p,n}(d_i)$	Eigenfunctions of the covariance matrix
f_S	Sampling frequency
f_{sig}	Signal frequency
f_{spur}	Frequency of spurious tone
f_T	Transit frequency
F_F	Folding factor
F_Q	Function of the deterministic initial solution
g	Conductance
G_i	Interstage gain
G_k	Fourier series coefficient of gain mismatch
G_m	Transconductance
i	Index, circuit node, transistor on the die
I	Current
I_{amp}	Total amplifier current consumption
I_D	Drain current
I_{DD}	Power supply current
I_{ref}	Reference current
j	Index, circuit branch
J_0	Jacobian of the initial data z_0 evaluated at p_i
k	Boltzmann's coefficient, error correction coefficient, index
K	Amplifier current gain, gain error correction coefficient
$l()$	Likelihood function
L	Channel length
L_R	Length of the measurement record
$L(\theta/T_X)$	Log-likelihood of parameter θ with respect to input set T_X
m	Number of different stage resolutions, index
M	Number of terms
n	Index, number of circuit nodes, number of faults in a list
N	Number of bits, number of parallel channels, noise power
$N_{aperture}$	Aperture jitter limited resolution
P	Power
p	Process parameter
$p(d_i,\theta)$	Stochastic process corresponding to process parameter p
$p_{X/\Theta}(x/\theta)$	Gaussian mixture model
p^*	Process parameter deviations from their corresponding nominal values
p_1	Dominant pole of amplifier
p_2	Non-dominant pole of amplifier
q	Channel charge, circuit nodes, index

.

Q	Quality factor
Q_i	Number of quantization steps, cumulative probability
$Q(x)$	Normal accumulation probability function
$Q(\theta/\theta^{(t)})$	Auxillary function in EM algorithm
r	Resolution, resistance, circuit nodes
R	Resistance
r_{ds}	Output resistance of a transistor
R_{eff}	Effective thermal resistance
R_{on}	Switch on-resistance
r_{out}	Amplifier output resistance
R_{ref}	Reference value (current or voltage)
s	Scaling parameter of transistor size, observed converter stage
t	Time
T	Absolute temperature, sampling period, transpose, test, test stimuli
t_{ox}	Oxide thickness
t_S	Sampling time
v_f	Fractional part of the analog input signal
UB_i	Upperbound of the ith level
V	Voltage
V_{DD}	Positive supply voltage
V_{DS}	Drain-source voltage
$V_{DS,SAT}$	Drain-source saturation voltage
V_{FS}	Full-scale voltage
V_{GS}	Gate-source voltage
V_{be}	Base-emitter voltage
V_{in}	Input voltage
$V_{[k]}$	Fourier series coefficient of offset mismatch
V_{LSB}	Voltage corresponding to the least significant bit
V_{margin}	Safety margin of drain-source saturation voltage
V_{off}	Offset voltage
V_{ped}	Pedestal voltage
V_{res}	Residue voltage
V_T	Threshold voltage
w	Normal vector perpendicular to the hyperplane, weight
w_i	Cost of applying test stimuli performing test number i
W	Channel width, parameter vector, loss function
W^*, L^*	Geometrical deformation due to manufacturing variations
x_i	Vectors of observations
$x(t)$	Analog input signal
X	Input
y_0	Arbitrary initial state of the circuit
$y[k]$	Output digital signal
Y	Output, yield
z_0	Nominal voltages and currents

$z_{(1-\alpha)}$	$(1-\alpha)$-quantile of the standard normal distribution Z
$z[k]$	Reconstructed output signal
α	Neyman-Pearson significance level, weight vector of the training set
β	Feedback factor, transistor current gain
γ	Noise excess factor, measurement correction factor, reference errors
δ	Relative mismatch
δ_{ramp}	Slop of the ramp signal for given full-scale-range V_{FS}
ε	Error
ζ	Samples per code, forgetting factor
η	Distance based weight term, stage gain errors
θ	Die, unknown parameter vector
$_{p,n}$	Eigenvalues of the covariance matrix
κ	Converter transition code
λ	Threshold of significance level α, decision stage offset error
λ_κ	Central value of the transition band
μ	Carrier mobility, mean value, iteration step size
ν	Fitting parameter estimated from the extracted data
ξ_i	measure the degree of misclassification of the data x_i
$\xi_n(\theta)$	Vector of zero-mean uncorrelated Gaussian random variables
ρ	Correlation parameter reflecting the spatial scale of clustering
p	Random vector accounting for device tolerances
σ	Standard deviation
σ_a	Gain mismatch standard deviation
σ_b	Bandwidth mismatch standard deviation
σ_d	Offset mismatch standard deviation
σ_r	Time mismatch standard deviation
τ	Time constant
ϕ	Flux stored in inductors
ϕ	Clock phase
χ	Circuit dependent proportionality factor
ω_S	Dominant pole frequency, angular sampling frequency
ω_{GBW}	Angular gain-bandwidth frequency
$r_f[.]$	Probability function
Δ	Relative deviation
Δ_{bi}	Bandwidth error parameter in ith channel
Δ_{gi}	Gain error parameter in ith channel
Δ_{oi}	Offset error parameter in ith channel
Δ_{ti}	Time error parameter in ith channel
Λ	Linearity of the ramp
Ξ	Quasi-static fault model
Ξ_r	Boundaries of quasi-static node voltage
Ω	Sample space of the test statistics

Chapter 1
Introduction

1.1 A/D Conversion Systems

Analog-to-digital (A/D) conversion and digital-to-analog (D/A) conversion lie at the heart of most modern signal processing systems where digital circuitry performs the bulk of the complex signal manipulation. As digital signal processing (DSP) integrated circuits become increasingly sophisticated and attain higher operating speeds more processing functions are performed in the digital domain. Driven by the enhanced capability of DSP circuits, A/D converters (ADCs) must operate at ever-increasing frequencies while maintaining accuracy previously obtainable at only moderate speeds. This trend has several motivations and poses important consequences for analog circuit design. The motivations for processing most signals digitally are manifold: digital circuits are much less expensive to design, test, and manufacture than their analog counterparts; many signal processing operations are more easily performed digitally; digital implementations offer flexibility through programmability; and digital circuitry exhibits superior dynamic range, thereby better preserving signal fidelity. As a consequence of the aforementioned advantages accrued by DSP, fewer and fewer operations benefit from analog solutions. Since A/D conversion generally requires more power and circuit complexity than D/A conversion to achieve a given speed and resolution, ADCs frequently limit performance in signal processing systems. This fact underscores the second consequence of enhanced DSP performance on the role of analog circuit design. That is, since A/D conversion limits overall system performance, development of improved A/D conversion algorithms and circuitry represents an extremely important area of research for the foreseeable future.

Propelling the great venture and unprecedented success of digital techniques, the CMOS technology has emerged and dominated the mainstream silicon IC industry in the last few decades. As the lithography technology improved, the MOS device dimensions have reduced its minimum feature size over the last 40 years and greatly impacted the performance of digital integrated circuits. During the course of pursuing a higher level of system integration and lower cost, the economics has driven

A. Zjajo and J. Pineda de Gyvez, *Low-Power High-Resolution Analog to Digital Converters*, Analog Circuits and Signal Processing, DOI 10.1007/978-90-481-9725-5_1, © Springer Science+Business Media B.V. 2011

technology to seek solutions to integrate analog and digital functionalities on a single die using the same or compatible fabrication processes. With the inexorable scaling of the MOS transistors, the raw device speed takes great leaps over time, measured by the exponential increase of the transit frequency f_T – the frequency where a transistor still yields a current gain of unity. The advancement of technology culminated in a dramatic performance improvement of CMOS analog circuits, opening an avenue to achieve system integration using a pure CMOS technology. Process enhancements, such as the triple-well option, even helped to reduce the noise crosstalk problem – one of the major practical limitations of sharing the substrate of precision analog circuits with noisy digital logic gates. As CMOS integrated circuits are moving into unprecedented operating frequencies and accomplishing unprecedented integration levels (Fig. 1.1), potential problems associated with device scaling – the short-channel effects – are also looming large as technology strides into the deep-submicron regime. Besides that it is costly to add sophisticated process options to control these side effects, the compact device modeling of short-channel transistors has become a major challenge for device physicists. In addition, the loss of certain device characteristics, such as the square-law I–V relationship, adversely affects the portability of the circuits designed in an older generation of technology. Smaller transistors also exhibit relatively larger statistical variations of many device parameters (i.e., doping density, oxide thickness, threshold voltage, etc.). The resultant large spread of the device characteristics also causes severe yield problems for both analog and digital circuits.

The analog-to-digital interface circuit in highly integrated CMOS wireless transceivers exhibits keen sensitivity to technology scaling. The trend toward more digital signal-processing for multi-standard agility in receiver designs has recently created a great demand for low-power, low-voltage analog-to-digital converters (ADCs) that can be realized in a mainstream deep-submicron CMOS technology. Intended for embedded applications, the specifications of such converters

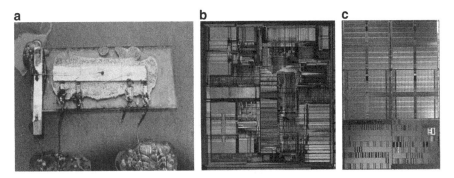

Fig. 1.1 (a) *Left*: first working integrated circuit, 1958 (Copyright© Texas Instruments: source – www.ti.com–public domain); (b) *Middle*: revolutionary Intel Pentium processor fabricated in 0.8 μm technology containing 3.1 million transistors, 1993; (c) *Right*: Intel's 45-nm test chip containing over a billion transistors, 2008 (Copyright© Intel Corporation: source – www.intel.com–public domain)

emphasize high dynamic range and low spurious spectral performance. In a CMOS radio SoC, regardless of whether frequency translation is accomplished with a single conversion, e.g., the direct-conversion and low-IF architecture, or a wideband-IF double conversion and low-IF architecture, or a wideband-IF double conversion, the lack of high-Q on-chip IF channel-select filters inevitably leads to a large dynamic range imposed on the baseband circuits in the presence of in-band blockers (strong adjacent channel interference signals). For example, the worst-case blocking specifications of some wireless standards, such as GSM, dictate a conversion linearity of twelve bits or more to avoid losing a weak received signal due to distortion artifacts. Recent works also underline the trend toward the IF digitizing architecture to enhance programmability and to achieve a more digital receiver. However, advancing the digitizing interface toward the antenna exacerbates the existing dynamic range problem, as it also requires a high oversampling ratio. To achieve high linearity, high dynamic range, and high sampling speed simultaneously under low supply voltages in deep-submicron CMOS technology with low power consumption has thus far been conceived of as extremely challenging.

A typical analog-to-digital and digital-to-analog interface in a digital system is depicted in Fig. 1.2. The majority of signals encountered in nature are continuous in both time and amplitude. The analog-to-digital converter (ADC) converts analog signals to discrete time digitally coded form for digital processing and transmission. Usually analog-to-digital conversion includes two major processes: sampling and quantization, as shown in the sub-blocks in the figure. The bandwidth of the analog input signal is limited to the Nyquist frequency by means of an anti-alias filter. This band-limited signal is then fed to the sample-and-hold stage that performs the first step towards a digital signal. The input signal that has been continuous in time is converted to a discrete-time signal. The amplitude, however, remains continuous after the sample-and-hold stage. The quantization of the amplitude is performed by the A/D converter. While during the conversion of a continuous-time signal to a discrete-time signal no information is lost if the signal is band-limited to the Nyquist frequency, the amplitude quantization that maps a continuous signal to a finite number of discrete values inevitably causes a decrease of information such that a complete reconstruction of the signal is not possible any more. The A/D converter thus limits the accuracy and the dynamic range of the entire system; the design of the converter must therefore be carried out with special care.

The output of the A/D converter consists of digital data that represents the analog input signal apart from the limitations mentioned above. A digital system processes this data. This system can be as simple as a single digital filter, but can also be as complex as e.g., a digital telephone system. In either case the resulting

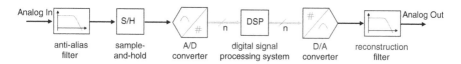

Fig. 1.2 A typical digital system with analog input and output quantities

output is also digital data that has to be converted back to the analog domain, which is performed by the digital-to-analog (D/A) converter, whose output is an analog, but still discrete-time signal. A reconstruction filter finally creates the continuous-time output signal. Although the entire system seems to be very complex, it has a number of advantages over purely analog signal processing. The most obvious advantage is the flexibility of the system. As mentioned above, the digital signal processing block can perform any function, from very simple to very complex. Many tasks that are complicated to achieve in the analog domain, such as storing large amounts of data, are comparatively easy in the digital domain. Moreover, the entire system can be made programmable or even adaptive to the current situation. In an analog signal processing system, the addition of functionality often degrades the signal. Any filtering stage, e.g., that is added to the signal path also adds noise to

Table 1.1 A/D converter's full application scope

Medical imaging	Antenna array position	Portable instrumentation (battery)
Positron emission tomography	IF sampling	Handheld oscilloscope
MRI receivers	CDMA2k, WCDMA, TD-SCDMA	Digital oscilloscope
Nondestructive ultrasound	IS95, CDMA-One, IMT2000	Spectrum analyzers
Ultrasound	BS infrastructure	Communications instrumentation
Ultrasound beam forming system	AMPS, IS136, (W)-CDMA, GSM	Instrumentation
X-ray imaging	Direct conversion	Radar, infrared imaging
Medical scan converters	Digital receiver single channel	Radar, sonar and satellite subsystems
Optical networking	Communication subsystem	Power-sensitive military applications
Broadband access	Wideband carrier frequency system	Astronomy
Broadband LAN	Point to point radio	Flat panel displays
Communications (modems)	GPS anti-jamming receiver	Projection systems
Powerline networking	MMDS base station	CCD imaging
Home phone networking	Wireless local loop (WLL)	Set-top boxes
Wireless local loop, fixed access	I & Q communications	Vsat terminal/receiver
WLAN	DSP front-end	Multimedia
VDSL, XDSL & HPNA	Tape drives	Film scanners
Power amplifier linearization	Phased array receivers	Data acquisition
Broadband wireless	Secure communications	Bill validation
Quadrature radio receivers	Digital receivers	Motor control
Cable reverse path	Antenna array processing	Industrial process control
Communications receivers	Antenna array processing	Optical sensor
Diversity radio receivers	Digital receivers	Cable head-end systems
Viterbi decoders	Video imaging	Test and measurement equipment

the signal and thus reduces the dynamic range. In a digital system, however, the losses are small once the analog signal is converted to the digital domain, no matter how complex the operations are that are performed with the digital data.

A/D converters are widely used in numerous applications (Table 1.1). Recently, the applications for A/D converters have expanded widely as many electronic systems that used to be entirely analog have been implemented using digital electronics. Examples of such applications include digital telephone transmission, cordless phones, transportation, and medical imaging. Consumer products, such as high-fidelity audio and image processing, require very high resolution, while advanced radar systems and satellite communications with ultra-wide-bandwidth require very high sampling rates (above 1 GHz). Advanced radar, surveillance, and intelligence systems, which demand even higher frequency and wider bandwidth, would benefit significantly from high resolution A/D converters having broad bandwidths.

1.2   Remarks on Curent Design and Debugging Practice

Reduction of the power dissipation associated with high speed sampling and quantization is a major problem in many applications, including portable video devices such as camcorders, cellular phones, personal communication devices such as wireless LAN transceivers, in the read channels of magnetic storage devices using digital data detection, and many others. With the rapid growth of internet and information-on-demand, handheld wireless terminals are becoming increasingly popular. With limited energy in a reasonable size battery this level of power consumption may not be suitable and further power reduction is essential for power-optimized A/D interfaces. The trend of increasing integration level for integrated circuits has forced the A/D converter interface to reside on the same silicon in complex mixed-signal ICs containing mostly digital blocks for DSP and control. The use of the same supply voltage for both analog and digital circuits can give advantages in reducing the overall system cost by eliminating the need of generating multiple supply voltages with *dc-dc* converters. However, specifications of the converters in various applications, such as communication applications emphasize high dynamic range and low spurious spectral performance. It is nontrivial to achieve this level of linearity in a monolithic environment where post-fabrication component trimming or calibration is cumbersome to implement for certain applications or/and for cost and manufacturability reasons. Another hurdle to achieve full system integration stems from the power efficiency of the A/D interface circuits supplied by a low voltage dictated by the gate-oxide reliability of the deeply scaled digital CMOS devices.

Similarly, the integrated circuits (ICs) manufacturing process in itself is neither deterministic nor fully controllable. Microscopic particles present in the manufacturing environment and slight variations in the parameters of manufacturing steps can all lead to the geometrical and electrical properties of an IC to deviate from those generated at the end of the design process. Those defects can cause various types of malfunctioning, depending on the IC topology and the nature of the defect. Silicon wafers produced in a semiconductor fabrication facility routinely go through electrical and optical measurements to determine how well the electrical parameters fit within the allowed limits. These measurements are augmented with process control monitoring (PCM) data. The information obtained is then used in the Fab to decide if some wafer process layers need to be re-worked and if the devices should be tested by a special characterization at the back end of the line to make certain that their electrical operating values meet the a priori specifications, e.g. temperature range, durability, speed, etc.

Various models have been constructed for estimating the device yield of a wafer – usually based on the die size, process linewidth, and particle accumulation. The yield is determined by the outcome of the wafer probing (electrical testing), carried out before wafer dicing. The functional testing of mixed-signal devices is very thorough, and much information can be acquired about circuit failure blocks and mechanisms based on these test results. The simplest form of yield information is the aggregate pass/fail statistics of the device, where the yield is usually expressed as a percentage of good dice per all dice on the wafer to make process and product comparisons easier. In principle, yield loss can be caused by several factors, e.g., wafer defects and contamination, IC manufacturing process defects and contamination, process variations, packaging problems, and design errors or inconsiderate design implementations or methods. Constant testing in various stages is of utmost importance for minimizing costs and improving quality.

Early discovering the presence of a defect chip is therefore most desirable as the cost of detecting a bad component in a manufactured part increases tenfold at each level of assembly. In the semiconductor industry, although the cost to fabricate a transistor has fallen dramatically, at the same time, the cost of testing each transistor has remained relatively stable. As a result, it is expected that testing a transistor in the near future (around 2012) will cost the same amount of money as manufacturing it. In general, for a fault-free die, the IC test cost is given by the costs of running the tester per unit time multiplied by the total test time per IC. However, for a given industrial test facility, consisting of ATE and prober or handler, and a given test yield, the test cost parameters are the test time of a fault-free product and the average test time of a faulty product. For the average cost of testing a die, the parameters test yield and average test time of a faulty die must also be taken into account. The test time of a fault-free die is the total time the complete test program takes to measure and process measurement data. The fact that the test time of a faulty die plays also a role implies that the tests should be ordered according to their success in detecting defects.

It should also be noted that wafer loading or package handling times are also important parameters for test time. In the situations where the handling times are comparable with the test time, decreasing the test length will obviously not be sufficient. In these cases alternative solutions such as multisite testing have to be applied, in order to make use of (a part of) the handling time to test another IC. The remaining factor, the test yield, depends on the process yield and the presence of test and measurement errors.

It becomes clear from the discussion of cost and quality issues of an IC that having tests with high fault coverage can improve both the quality and the cost figures. With the increasing demand for low defect levels, it is imperative that the effects of all possible process parameter variations be modeled and tested for. These defects are highly dependent on the type of process, and their effect on the overall circuit behavior depends on the design's process tolerance. The list of possible faults for an integrated circuit is in fact infinite, but by considering only the most likely faults, a finite set can be created. To develop realistic fault model various types of failures, their causes and effects should be considered. Parametric variations (measured in terms of their standard deviation σ) arise due to statistical fluctuations in process parameters such as oxide thickness, doping, line width, and mask misalignment. The effect varies from complete malfunction of the circuit (catastrophic), to marginally out-of-specification performance of some circuit parameters (e.g. gain, linearity), to performance that may even lie within given specifications but poses a reliability risk. Typically, process monitor circuits on each wafer are tested to ensure that all key process parameters are well within specifications so that the as yet untested chips on the wafer can be assumed to be free of excessive process variation. Nevertheless, process variations local to a chip (or to a circuit within a chip) might be out of specification.

Table 1.2 shows all-comprising list of defects and faults. The rows correspond to the manufacturing process defect that caused the fault, ranging from global process variation to local variation and shorts and opens. The columns correspond to the impact

Table 1.2 Categories of defects and faults

Defect (cause)	Fault (effect)		
	Within specification limits	Parametric fail	Catastrophic fail
Process parameters within specification limits	Defect-free and fault-free	Specific design cannot account for all possible parameter combinations for all possible conditions	Oversight in design, example: insufficient phase margin for combination of process parameters
Process parameter out of specification limits	Reliability risk, example: inter-metal dielectric that is too thin	Classic parametric faults and soft faults	Example: a low V_T causing a transistor to not turn off
Shorts and opens	Occur due to unspecified performances, example: a short circuit that causes excessive current while circuit remains within specification	Marginally fail a specification, example: a short circuit in flash A/D converter resistor chain causing a few incorrect output codes	Classic catastrophic and hard faults

of the defect on the performance of the circuit under test, ranging from parametric failure to catastrophic failure. The most likely parametric faults that are also difficult to test are of most significance. A test set is having 100% fault coverage if a circuit, which passes the test, meets all performance specifications at all operating conditions. Defects, which cause no specification failure, but may reduce reliability can be distinguished in three failure regimes: early, where the products show a high, but decreasing failure rate as a function of time until the failure rate stabilizes, random failure period and wear-out period, where the failure rate increases again when end-of-life of the products is reached. The nature of the failures in the three periods is generally very different. The majority of the failures in the early failure period are caused by manufacturing defects like near opens and shorts in metal lines, weak spots in isolating dielectrics or poorly bonded bondwires in the package. In the random failure period many different rootcauses occur but failures related to specific events like lightning, load dump spikes occurring during disconnection of car batteries or other overstress situations are most notable. Failures in the wear-out period are related to intrinsic properties of the materials and devices used in the product in combination with the product use conditions like temperature, voltage and currents including their time dependence. Examples of wear-out failure mechanisms are electromigration (gate) oxide breakdown, hot carrier degradation and mobile ion contamination. Reliability engineering, which deals with on one hand systematically reducing the infant mortality and random failures and on the other hand keeping the wear-out phase beyond practical duration is beyond the scope of this book and it will not be further discussed.

Another test-related factor that contributes indirectly to the cost of an IC is the area contribution of the built-in self-test (BIST) and design for testability (DfT) circuitry. In fact, adding BIST and DfT in an efficient way can help decrease the test development and debugging costs, thereby compensating for at least a part of the additional area cost. Finally, on-line debugging of tests has also often been pointed out as a cost factor, because of the high costs of operating the expensive ATE. Some ATE manufacturers supply debugging software for off-line debugging, but the debugging of the device under test (DUT) is still performed for a large part on-line. Standardization in terms of tester and interface board models have not been achieved yet for mixed-signal circuits.

Historically, digital and analog testing have developed at very different paces, causing analog test methodology to be in a far earlier stage today than its digital counterpart. Computer aided design (CAD) tools for automatic test generation and test circuitry insertion are available already since 2 decades for digital circuits. The main reason for this is the ease of formulating the test generation as a mathematical problem due to the discrete signal and time values. The distinction between what does and what does not work is evident for digital circuitry. For analog designs, the definition of fault-free and faulty circuits is much more a matter of specification thresholds and sensitivity of application than a sharp distinction as in the case of digital circuits. In analog signal processing circuits there are not only two choices of signal values to choose from, but in principle an infinite number of signal values are possible. Similarly, the time variation properties of analog signals bring an extra dimension to the problem. Additionally, the propagation of fault effects to the output is not possible in the digital sense, since the fault effect propagates in all directions and the calculation of this propagation pattern becomes therefore much more complex than in the digital case. The information that a fault is present at a certain node does not readily comprise the signal value information for that node, making time consuming calculations of signal values necessary. Nonlinearity, loading between circuit blocks, presence of energy-storing components and parasitics further complicate these calculations.

1.3 Motivation

With the fast advancement of CMOS fabrication technology, more and more signal-processing functions are implemented in the digital domain for a lower cost, lower power consumption, higher yield, and higher re-configurability. This has recently generated a great demand for low-power, low-voltage A/D converters that can be realized in a mainstream deep-submicron CMOS technology. However, the discrepancies between lithography wavelengths and circuit feature sizes are increasing. Lower power supply voltages significantly reduce noise margins and increase variations in process, device and design parameters. Consequently, it is steadily more difficult to control the fabrication process precisely enough to maintain uniformity. The inherent randomness of materials used in fabrication at nanoscopic scales means that performance will be increasingly variable, not only from die-to-die but also within each individual die. Parametric variability will be compounded by degradation in nanoscale integrated circuits resulting in instability of parameters over time, eventually leading to the development of faults. Process variation cannot be solved by improving manufacturing tolerances; variability must be reduced by new device technology or managed by design in order for scaling to continue. Similarly, within-die performance variation also imposes new challenges for test methods.

In an attempt to addresses these issues, this book specifically focuses on: (i) improving the power efficiency for the high-resolution, high-speed, and low spurious spectral A/D conversion performance by exploring the potential of low-voltage analog design and calibration techniques, respectively, and (ii) development of circuit techniques and algorithms to enhance testing and debugging potential to detect errors dynamically, to isolate and confine faults, and to recover errors continuously. This will become increasingly important as devices experience parametric degradation over time, requiring run-time reconfiguration.

1.4 Organization

Chapter 2 of this book reviews high-speed, high resolution A/D converter architectures and discusses the design challenges for analog circuits and the design choices for converter's common building blocks in deep-submicron CMOS technology. On the basis of this assessment, a multi-step A/D converter is selected to explore the prominent issue of power efficiency under low supply voltages. Chapter 3 describes the multi-step A/D converter architecture in more detail. Key design techniques for each stage are highlighted and the details of the circuit implementation presented. In particular, error sources are identified and circuit techniques to lessen their impact are shown. The chapter ends with the summary of the prototype experimental results. Chapter 4 firstly focuses on novel computer-aided and design-for-test circuit technique to augment analysis of random process variations on converter's performance. The chapter further presents the continuous-time on-chip waveform generator and a method for the built-in characterization of the converter parameters. The chapter closes with a novel statistical simulation tool for evaluation of IC designs affected by process variations and circuit noise. The effectiveness of the approach was evaluated on the continuous-time on-chip wave-form generator as a representative example. Development of circuit techniques and algorithms to enhance debugging prospectives are presented in Chapter 5. Initially, an approach to monitor die-level process variations to allow the estimation of selected performance figures and in certain cases guide the test is introduced. The chapter further continues with discussions on how to guide the test with the information obtained through monitoring process variations and how to estimate the selected performance figures. The chapter closes with for diagnostic analysis of static errors based on the steepest-descent method. In this approach, the most common errors are identified and modeled before the model is applied to estimate adaptive filtering algorithm look-up table for error estimation and fault isolation. Finally, debugging – calibration relationship is discussed and experimental results given. In Chapter 6 the main conclusions are summarized and recommendations for further research are presented.

Chapter 2
Analog to Digital Conversion

2.1 High-Speed High-Resolution A/D Converter Architectural Choices

Since the existence of digital signal processing, A/D converters have been playing a very important role to interface analog and digital worlds. They perform the digitalization of analog signals at a fixed time period, which is generally specified by the application. The A/D conversion process involves sampling the applied analog input signal and quantizing it to its digital representation by comparing it to reference voltages before further signal processing in subsequent digital systems. Depending on how these functions are combined, different A/D converter architectures can be implemented with different requirements on each function. To implement power-optimized A/D converter functions, it is important to understand the performance limitations of each function before discussing system issues. In this section, the concept of the basic A/D conversion process and the fundamental limitation to the power dissipation of each key building block are presented.

2.1.1 Multi-Step A/D Converters

Parallel (Flash) A/D conversion is by far the fastest and conceptually simplest conversion process [1–21], where an analog input is applied to one side of a comparator circuit and the other side is connected to the proper level of reference from zero to full scale. The threshold levels are usually generated by resistively dividing one or more references into a series of equally spaced voltages, which are applied to one input of each comparator. For n-bit resolution, $2^n - 1$ comparators simultaneously evaluate the analog input and generate the digital output as a thermometer code. Since flash converter needs only one clock cycle per conversion, it is often the fastest converter. On the other hand, the resolution of flash ADCs is limited by circuit complexity, high power dissipation, and comparator and reference mismatch. Its complexity grows

A. Zjajo and J. Pineda de Gyvez, *Low-Power High-Resolution Analog to Digital Converters*, Analog Circuits and Signal Processing, DOI 10.1007/978-90-481-9725-5_2, © Springer Science+Business Media B.V. 2011

exponentially as the resolution bit increases. Consequently, the power dissipation and the chip area increase exponentially with the resolution. The component-matching requirements also double for every additional bit, which limits the useful resolution of a flash converter to 8–10 bits. The impact of various detrimental effects on flash A/D converter design will be discussed further in Section 3.4.1.

To reduce hardware complexity, power dissipation and die area, and to increase the resolution but to maintain high conversion rates, flash converters can be extended to a two-step/multi-step [22–39] or sub-ranging architecture [40–53] (also called series-parallel converter). Conceptually, these types of converters need $m \times 2^n$ instead of $2^{m \times n}$ comparators for a full flash implementation assuming $n_1, n_2, \ldots, n_m$ are all equal to n. However, the conversion in sub-range, two-step/multi-step ADC does not occur instantaneously like a flash ADC, and the input has to be held constant until the sub-quantizer finishes its conversion. Therefore, a sample-and-hold circuit is required to improve performance. The conversion process is split into two steps as shown in Fig. 2.1. The first A/D sub-converter performs a coarse conversion of the input signal. A D/A converter is used to convert the digital output of the A/D sub-converter back into the analog domain. The output of the D/A converter is then subtracted from the analog input. The resulting signal, called the residue, is amplified and fed into a second A/D sub-converter which takes over the fine conversion to full resolution of the converter. The amplification between the two stages is not strictly necessary but is carried out nevertheless in most of the cases. With the help of this amplifying stage, the second A/D sub-converter can work with the same signal levels as the first one, and therefore has the same accuracy requirements. At the end of the conversion the digital outputs of both A/D sub-converters are summed up.

By using concurrent processing, the throughput of this architecture can sustain the same rate as a flash A/D converter. However, the converted outputs have a latency of two clock cycles due to the extra stage to reduce the number of precision comparators. If the system can tolerate the latency of the converted signal, a two-step converter is a lower power, smaller area alternative.

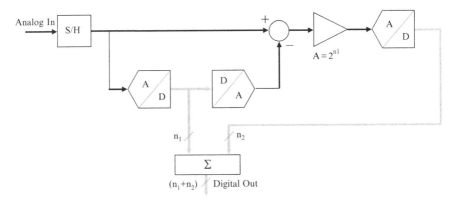

Fig. 2.1 Two-step A/D converter

2.1.2 Pipeline A/D Converters

The two-step architecture is equipped with a sample-and-hold (S/H) circuit in front of the converter (Fig. 2.1). This additional circuit is necessary because the input signal has to be kept constant until the entire conversion (coarse and fine) is completed. By adding a second S/H circuit between the two converter stages, the conversion speed of the two-step A/D converter can be significantly increased (Fig. 2.2). In a first clock cycle the input sample-and-hold circuit samples the analog input signal and holds the value until the first stage has finished its operation and the outputs of the subtraction circuit and the amplifier have settled. In the next clock cycle, the S/H circuit between the two stages holds the value of the amplified residue. Therefore, the second stage is able to operate on that residue independently of the first stage, which in turn can convert a new, more recent sample. The maximum sampling frequency of the pipelined two-step converter is determined by the settling time of the first stage only due to the independent operation of the two stages. To generate the digital output for one sample, the output of the first stage has to be delayed by one clock cycle by means of a shift register (SR) (Fig. 2.2). Although the sampling speed is increased by the pipelined operation, the delay between the sampling of the analog input and the output of the corresponding digital value is still two clock cycles. For most applications, however, latency does not play any role, only conversion speed is important. In all signal processing and telecommunications applications, the main delay is caused by digital signal processing, so a latency of even more than two clock cycles is not critical.

The architecture as described above is not limited to two stages. Because the inter-stage sample-and-hold circuit decouples the individual stages, there is no difference in conversion speed whether one single stage or an arbitrary number of stages follow the first one. This leads to the general pipelined A/D converter architecture, as depicted in Fig. 2.3 [54–89]. Each stage consists of an S/H, an N-bit flash A/D converter, a reconstruction D/A converter, a subtracter, and a residue amplifier. The

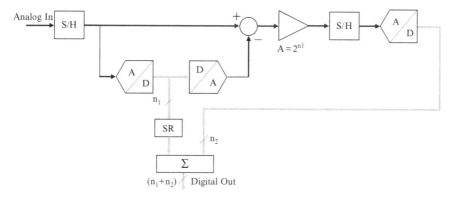

Fig. 2.2 Two-Step converter with an additional sample-and-hold circuit and a shift register (SR) to line up the stage output in time

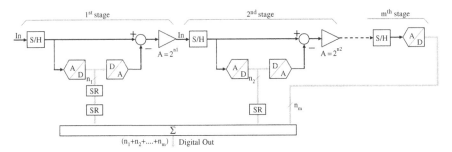

Fig. 2.3 Multi-stage pipeline A/D converter architecture

conversion mechanism is similar to that of sub-ranging conversion in each stage. Now the amplified residue is sampled by the next S/H, instead of being fed to the following stage.

All the n-bit digital outputs emerging from the quantizer are combined as a final code by using the proper number of delay registers, combination logic and digital error correction logic. Although this operation produces a latency corresponding to the sub-conversion stage before generating a valid output code, the conversion rate is determined by each stage's conversion time, which is dependent on the reconstruction D/A converter and residue amplifier settling time. The multi-stage pipeline structure combines the advantages of high throughput by flash converters with the low complexity, power dissipation, and input capacitance of sub-ranging/multi-step converters. The advantage of the pipelined A/D converter architecture over the two-step converter is the freedom in the choice of number of bits per stage. In principle, any number of bits per stage is possible, down to one single bit. It is even possible to implement a non-integer number of bits such as 1.5 bit per stage by omitting the top comparator of the flash A/D sub-converter used in the individual stages [59]. It is not necessary, although common, that the number of bits per stage is identical throughout the pipeline, but can be chosen individually for each stage [65–69]. The only real disadvantage of the pipelined architecture is the increased latency. For an A/D converter with m stages, the latency is m clock cycles. For architectures with a small number of bits per stage, the latency can thus be 10–14 clock cycles or even more.

2.1.3 Parallel Pipelined A/D Converters

The throughput rate can be increased further by using a parallel architecture [90–106] in a time-interleaved manner as shown in Fig. 2.4. The first converter channel processes the first input sample, the second converter channel the next one and so on until, after the last converter channel has processed its respective sample, the first converter has its turn again (see Section 3.1 for extensive discussion on timing-related issues in time-interleaved systems).

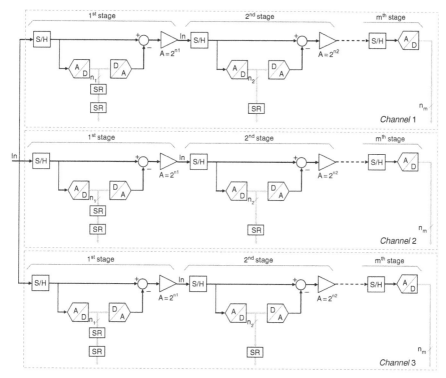

Fig. 2.4 Parallel pipeline A/D converter architecture

The individual A/D converters therefore operate on a much lower sampling rate than the entire converter, with the reduction in conversion speed for each individual converter equal to the number of A/D converters in parallel. The only building block that sees the full input signal bandwidth of the composite converter is the sample-and-hold circuit of each A/D converter. Theoretically, the conversion rate can be increased by the number of parallel paths, at the cost of a linear increase in power consumption and large silicon area requirement. A second problem associated with parallel A/D converters is path mismatch. During operation, the input signal has to pass different paths from the input to the digital output. If all A/D converters in parallel are identical, these paths are also identical. However, if offset, gain, bandwidth or time mismatch occur between the individual converters, the path for the input signal changes each time it is switched from one converter to another. This behavior gives rise to fixed-pattern noise at the output of the composite A/D converter which can be detected as spurious harmonics in the frequency domain [90]. How these errors are seen in the spectrum of the sampled signal will be discussed in conjunction with the time-interleaved S/H in Section 3.2. The parallel architecture is advantageous when high sampling rates are necessary which are difficult to achieve with single A/D converter. Although the architecture is straight-forward, parallel A/D converters usually are not the best compromise when it comes to increasing the conversion rate of medium speed converters. For the A/D converter family described in this book, it has therefore been decided in favor of two-step/multi-step converter to obtain higher speed.

2.1.4 A/D Converters Realization Comparison

In this section, a number of recently published high resolution analog to digital converters are compared. Tables 2.1, 2.2, 2.3, and 2.4 show the FoM of the realized A/D converters from 1998 until 2008 distributed over the categories: flash, two-step/multi-step/subranging, pipeline and parallel pipeline. Normalizing the dissipated power P to the effective resolution ENOB and to the effective resolution bandwidth ERBW, the figure of merit, FoM $= P/(2^{ENOB} \times 2$ ERBW$))$ [107], is a measure for the required power per achieved resolution per conversion. Today, the state-of-the-art FoM for Nyquist A/D converters is around 1 pJ/conversion.

From Table 2.1 it can be seen that the flash architecture is (barely or) not used at all for accuracies above 6 bits due to the large intrinsic capacitance required. The most prominent drawback of flash A/D converter is the fact that the number of comparators grows exponentially with the number of bits. Increasing the quantity of the comparators also increases the area of the circuit, as well as the power consumption. Other issues limiting the resolution and speed include nonlinear input capacitance, location-dependent reference node time constants, incoherent timing of comparators laid out over a large area, and comparator offsets. To lessen the impact of mismatch in the resistor reference ladder and the unequal input offset voltage of the comparators on the linearity of ADC, several schemes, such as inserting a preamplifier [1] in front of the latch, adding a chopper amplifier [2] and auto-zero scheme to sample an offset in the capacitor in front of the latch or digital background calibration [18] have been developed. Alternatively, the offsets and input capacitance can be reduced by means of distributed pre-amplification combined with averaging [108, 109] and possibly also with interpolation [110]. As thin oxide improves the matching property of transistors, smaller devices can be

Table 2.1 Table of realized flash A/D converters

Reference	N	ENOB	f_S (MS/s)	ERBW (MHz)	P (mW)	FoM (pJ)
[8]	6	5.8	500	160	225	12.6
[9]	6	5.2	500	250	330	17.3
[11]	6	5.2	1300	750	545	11.4
[13]	6	5.2	22	11	0.48	0.6
[14]	6	5.3	1300	600	600	12.7
[15]	6	5.4	2000	1000	310	3.5
[16]	6	5.7	1200	600	135	2.2
[17]	6	5.2	4000	1000	990	13.5
[19]	5	4.7	1000	200	46	1.8
[21]	5	4.0	5000	2500	102	1.3

Table 2.2 Table of realized two-step/multi-step/subranging A/D converters

Reference	N	ENOB	f_s (MS/s)	ERBW (MHz)	P (mW)	FoM (pJ)
[30]	10	9.1	25	12.5	195	14.2
[31]	14	12	100	50	1250	3.1
[32]	12	10.3	50	25	850	13.5
[33]	13	11.1	40	15	800	12.1
[34]	12	10.3	54	4	295	29.2
[35]	10	9.1	160	10	190	17.3
Ch 3 [37]	**12**	**10.5**	**60**	**30**	**100**	**1.1**
[38]	15	10.8	20	2	140	19.7
[50]	12	10.1	40	20	30	0.7

used in newer technology generations to achieve the same matching accuracy; this fact has been exploited by many recent works of flash-type converters to improve the energy efficiency of the conversion.

The realized FoM of the two-step/multi-step or subranging architectures is relatively constant for different values of the ENOB. In Table 2.2 it was shown that the FoM of a two-step/multi-step or subranging converter increases less rapidly than the noise-limited architectures such as multi-stage pipeline. Although the number of comparators is greatly reduced from the flash architecture, path matching is a problem and in some cases the input bandwidth is limited to relatively low frequency compared to the conversion rate [23, 24, 26]. Fine comparators accuracy requirements can be relaxed by including an inter-stage gain amplifier to amplify the signal for the fine comparator bank [22, 25]. While both, D/A converter and input of the residue amplifier require full resolution requirement, the coarse A/D converter section requirement can be relaxed with the digital error correction. If one stage subtracts a smaller reference than it nominally should due to the comparator offset, the subsequent stages compensate for this by subtracting larger references. This widely employed error correction method is referred to as redundant signed digit (RSD) correction, which was firstly developed for algorithmic ADCs in [111,112] and later utilized in pipelined ADC [59]. Other related methods have also been used [54].

The redundancy allows for quantization errors as far as the residue stays in the input range of the next stage. The errors can be static or dynamic; it is only essential that the bits going to the correction logic circuitry match those which are D/A converted and used in residue formation. The same correction method can easily be expanded to larger resolution stages as well. As a minimum, one extra quantization level is required [113], but for maximum error tolerance the nominal number of comparators has to be doubled. The level of error tolerance on the coarse A/D converter section depends on how much digital error correction range the fine A/D converter section can provide. The correction range varies from ± 3 LSBs in [24] to a much larger value in [22, 25] with an S/H inter-stage amplifier.

If over/under-range protection is used, the offset requirements for the coarse converter can be greatly relaxed; but the fine one shares similar matching concerns as the flash architecture. Interpolation can be applied as well to reduce the number of preamps and their sizes. A balanced design can often achieve energy efficiency per conversion close to that of the pipeline converters. Since the two-step/multi-step architecture is matching limited, calibration can be applied to reduce the intrinsic capacitance. Two approaches can be taken to calibrate out the errors: mixed signal or fully digital. In mixed signal calibration, the erroneous component values are measured from the digital output and adjusted closer to their nominal ones [34]. The correction is applied to the analog signal path and thus requires extra analog circuitry. In the fully digital approach the component values are not adjusted [114]; however, the accuracy of this method depends on the accuracy of the measurement. In the sub-ranging converter the absence of a residue amplifier places stringent offset and noise requirements on the second quantizer, which can be overcome at modest power dissipation through the use of auto-zeroing [48], averaging [50] or background offset calibration [49]. The utilization of time-interleaved second quantizers increases the effective sampling rate [46].

The FoM of the pipeline converter increases as a function of the realized ENOB (Table 2.3). The architecture has evolved by making use of the strengths of the switched capacitor technique, which provides very accurate and linear analog amplification and summation operations in the discrete time domain. When the input is a rapidly changing signal, the relative timing of the first stage S/H circuit and the sub-ADC is critical and often relaxed with a front-end S/H circuit. Consecutive stages operate in opposite clock phases and as a result one sample traverses two stages in one clock cycle. So, the latency in clock cycles is typically half the number of stages plus one, which is required for digital error correction. For feedback purposes, where low latency is essential, a coarse result can be taken after the first couple of stages. The different bits of a sample become ready at different times. Thus, digital delay lines are needed for aligning the bits.

Several techniques for achieving resolutions higher than what is permitted by matching have been developed; the reference feedforward technique [55] and commutated feedback capacitor switching [56] improve the differential non-linearity, but do not affect the integral non-linearity. In 1-bit/stage architecture the

Table 2.3 Table of realized pipelined A/D converters

Reference	N	ENOB	f_S (MS/s)	ERBW (MHz)	P (mW)	FoM (pJ)
[66]	14	12	20	10	720	8.8
[72]	14	12.5	2.5	1.2	145	5.0
[73]	12	10.5	100	50	390	2.2
[74]	10	9.3	80	40	69	1.4
[75]	12	10.4	110	10	97	3.6
[76]	10	9.4	100	50	67	1.0
[78]	10	8.6	200	90	55	0.8
[80]	14	13.1	125	60	1850	1.7
[84]	11	10.5	45	5	81	5.6

capacitive error averaging technique, which has previously been used in algorithmic ADCs [115] can be used [66]. With it, a virtually capacitor ratio-independent gain-of-two stage can be realized. The technique, however, requires two opamps per stage (a modification proposed in [70]) and needs at least one extra clock phase. Pipeline architecture has been found very suitable for calibration [57, 62]. The number of components to be calibrated is sufficiently small, since only the errors in the first few stages are significant as a result of the fact that, when referred to the input, the errors in the latter stages are attenuated by the preceding gain. Furthermore, no extra A/D converter is necessarily required for measuring the calibration coefficients, since the back-end stages can be used for measuring the stages in front of them.

Similar to a subranging converter, over/under-range protection is necessary in a pipeline A/D converter. Since the comparator offset specs are substantially relaxed due to a low stage resolution and over/under-range protection, comparator design in pipeline A/D converters is far simpler than that of the flash ones, and usually does not impose limitation on the overall conversion speed or precision. It is how fast and how accurate the residue signals can be produced and sampled that determines the performance of a pipeline converter, especially for the first stage that demands the highest precision. Negative feedback is conventionally employed to stabilize the voltage gain and to broaden the amplifier bandwidth. It is expected that with technology advancement, the accompanying short-channel effects will pose serious challenges to realizing high open-loop gain, low noise, and low power consumption simultaneously at significantly reduced supply voltages. The tradeoff between speed, dynamic range, and precision will eventually place a fundamental limit on the resolution of pipeline converters attainable in ultra-deep-submicron CMOS technologies.

The realized FoM of the parallel pipeline architectures is severely limited by required power (Table 2.4). Up to a certain resolution, component matching is satisfactory enough and the errors originating from channel mismatch can be kept to a tolerable level with careful design. High-resolution time-interleaved A/D converters, however, without exception, use different techniques to suppress errors. The offset can be rather easily calibrated using a mixed signal [116] or all-digital circuitry [91]. Calibrating the gain mismatch is also possible, but requires more complex circuitry than offset calibration [95, 96]. The timing skew may originate

Table 2.4 Table of realized parallel-pipelined A/D converters

Reference	N	ENOB	f_S (MS/s)	ERBW (MHz)	P (mW)	FoM (pJ)
[94]	8	7.6	75	35	75	5.5
[95]	10	9.4	40	20	650	24.1
[96]	10	9.5	40	10	565	39.0
[98]	10	9.4	120	20	234	8.7
[101]	10	9.7	120	2	75	22.5
[103]	10	9.4	200	60	104	1.3
[104]	8	7.8	150	75	71	2.1
[105]	11	9.4	800	400	350	0.65
[106]	15	12.3	125	60	909	1.5

from the circuit generating the clock signals for different channels or it may be due to different propagation delays to the sampling circuits. Skew can be most easily avoided by using a full-speed front-end sample-and-hold circuit [117]. The A/D converter channels resample the output of the S/H when it is in a steady state, and so the timing of the channels is not critical. However, the S/H circuit has to be very fast, since it operates at full speed.

2.2 Notes on Low Voltage A/D Converter Design

Explosive growth in wireless and wireline communications is the dominant driver for high-resolution, high-speed, low power, and low cost integrated A/D converters. From an integration point of view the analog electronics must be realized on the same die as the digital core and consequently must cope with the CMOS evolution dictated by the digital circuit. Technology scaling offers significantly lowering of the cost of digital logic and memory, and there is a great incentive to implement high-volume baseband signal processing in the most advanced process technology available. Concurrently, there is an increased interest in using transistors with minimum channel length (Fig. 2.5a) and minimum oxide thickness to implement

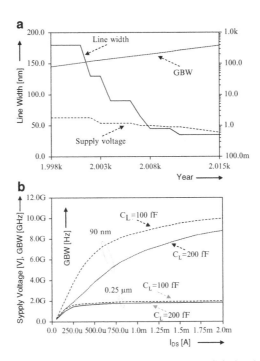

Fig. 2.5 (a) Trend of analog features in CMOS technologies. (b) Gain-bandwidth product versus drain current in two technological nodes

analog functions, because the improved device transition frequency, f_T, allows for faster operation. To ensure sufficient lifetime for digital circuitry and to keep power consumption at an acceptable level, the dimension-reduction is accompanied by lowering of nominal supply voltages. Due to the reduction of supply voltage the available signal swing is lowered, fundamentally limiting the achievable dynamic range at reasonable power consumption levels. Additionally, lower supply voltages require biasing at lower operating voltages which results in worse transistor properties, and hence yield circuits with lower performance. To achieve a high linearity, high sampling speed, high dynamic range, with low supply voltages and low power dissipation in ultra-deepsubmicron CMOS technology is a major challenge.

The key limitation of analog circuits is that they operate with electrical variables and not simply with discrete numbers that, in circuit implementations, gives rise of a beneficial noise margin. On the contrary, the accuracy of analog circuits fundamentally relies on matching between components, low noise, offset and low distortions. In this section, the most challenging design issues for low voltage, high-resolution A/D converters in deep submicron technologies such as contrasting the degradation of analog performances caused by requirement for biasing at lower operating voltages, obtaining high dynamic range with low voltage supplies and ensuring good matching for low-offset are reviewed. Additionally, the subsequent remedies to improves the performance of analog circuits and data converters by correcting or calibrating the static and possibly the dynamic limitations through calibration techniques are briefly discussed as well.

With reduction of the supply voltage to ensure suitable overdrive voltage for keeping transistors in saturation, even if the number of transistors stacked-up is kept at the minimum, the swing of signals is low if high resolution is required. Low voltage is also problematic for driving CMOS switches especially for the ones connected to signal nodes as the on-resistance can become very high or at the limit the switch does not close at all in some interval of the input amplitude. One solution is the multi-chip solution, where digital functions are implemented in a single or multiple chips and the analog processing is obtained by a separate chip with suitably high supply voltage and reduced analog digital interference. The use on the same chip of two supply voltages, one for the digital part with lower and one for the analog part with higher supply voltage is another possibility. The multiple threshold technology is another option.

In general, to achieve a high gain operation, high output impedance is necessary, e.g. drain current should vary only slightly with the applied V_{DS}. With the transistor scaling, the drain assert its influence more strongly due to the growing proximity of gate and drain connections and increase the sensitivity of the drain current to the drain voltage. The rapid degradation of the output resistance at gate lengths below 0.1 μm and the saturation of g_m reduce the device intrinsic gain $g_m r_o$ characteristics. As transistor size is reduced, the fields in the channel increase and the dopant impurity levels increase. Both changes reduce the carrier mobility, and hence the transconductance g_m. Typically, desired high transconductance value is obtained at the cost of an increased bias current. However, for very short channel the carrier velocity quickly reaches the saturation limit at which the transconductance also saturates becoming independent of gate length or bias $g_m = W_{eff}C_{ox}\ v_{sat}/2$. As channel lengths are reduced without proportional reduction in drain voltage, raising the electric field in the channel, the result is velocity saturation of the carriers, limiting the current and the transconductance. A limited transconductance is problematic for analog design: for obtaining high gain it is necessary to use wide transistors at the cost of an increased parasitic capacitances and, consequently, limitations in bandwidth and slew rate. Even using longer lengths obtaining gain with deep submicron technologies is not appropriate; it is typically necessary using cascode structures with stack of transistors or circuits with positive feedback. As transistor's dimension reduction continues, the intrinsic gain keeps decreasing due to a lower output resistance as a result of drain-induced barrier lowering (DIBL) and hot carrier impact ionization. To make devices smaller, junction design has become more complex, leading to higher doping levels, shallower junctions, halo doping, etc. all to decrease drain-induced barrier lowering. To keep these complex junctions in place, the annealing steps formerly used to remove damage and electrically active defects must be curtailed, increasing junction leakage. Heavier doping also is associated with thinner depletion layers and more recombination centers that result in increased leakage current, even without lattice damage. In addition, gate leakage currents in very thin-oxide devices will set an upper bound on the attainable effective output resistance via circuit techniques (such as active cascode). Similarly, as scaling continues, the elevated drain-to-source leakage in an off-switch can adversely affect the switch performance. If the switch is driven by an amplifier, the leakage may lower the output resistance of the amplifier, hence limits its low-frequency gain.

Low-distortion at quasi-*dc* frequencies is relevant for many analog circuits. Typically, quasi-*dc* distortion may be due to the variation of the depletion layer width along the channel, mobility reduction, velocity saturation and nonlinearities in the transistors' transconductances and in their output conductances, which is heavily dependent on biasing, size, technology and typically sees large voltage swings. With scaling higher harmonic components may increase in amplitude despite the smaller signal; the distortion increases significantly. At circuit level the degraded quasi-*dc* performance can be compensated by techniques that boost gain, such as (regulated) cascodes. These are, however, harder to fit within decreasing supply voltages. Other solutions include a more aggressive reduction of signal magnitude which requires a higher power consumption to maintain SNR levels.

The theoretically highest gain-bandwidth of an OTA is almost determined by the cutoff frequency of transistor (see Fig. 2.5b for assessment of GBW for two technological nodes). Assuming that the kT/C noise limit establishes the value of the load capacitance, to achieve required SNR large transconductance is required. Accordingly, the aspect ratio necessary for the input differential pair must be fairly large, in the hundred range. Similarly, since with scaling the gate oxide becomes thinner, the specific capacitance C_{ox} increases as the scaling factor. However, since the gate area decreases as the square of the scaling factor, the gate-to-source and gain-to-drain parasitic capacitance lowers as the process is scaled. The coefficients for the parasitic input and output capacitance, C_{gs} and C_{gd} shown in Fig. 2.6a) have been obtained by

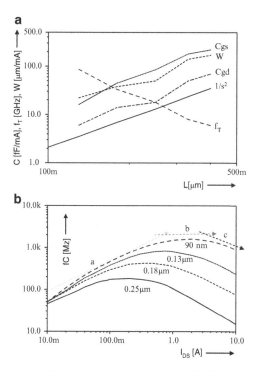

Fig. 2.6 (a) Scaling of gate width and transistor capacitances. (b) Conversion frequency f_c versus drain current for four technological nodes

simulation for conventional foundry processes under the assumption that the overdrive voltage is 0.175 V. Similarly, with technology-scaling the actual junctions become shallower, roughly proportional to the technology feature size. Also, the junction area roughly scales in proportion to the minimum gate-length, while the dope level increase does not significantly increase the capacitance per area. Altogether this leads to a significantly reduced junction capacitance per g_m with newer technologies.

Reducing transistor parasitic capacitance is desired, however, the benefit is contrasted by the increased parasitic capacitance of the interconnection (the capacitance of the wires connecting different parts of the chip). With transistors becoming smaller and more transistors being placed on the chip, interconnect capacitance is becoming a large percentage of total capacitance. The global effect is that scaling does not benefit fully from the scaling in increasing the speed of analog circuit as the position of the non-dominant poles is largely unchanged. Additionally, with the reduced signal swing, to achieve required SNR signal capacitance has to increase proportionally. By examining Fig. 2.6b), it can be seen that the characteristic exhibits convex curve and takes the highest value at the certain sink current (region b). In the region of the current being less than this value (region a), the conversion frequency increases with an increase of the sink current. Similarly, in the region of the current being higher than this value (region c), the conversion frequency decreases with an increase of the sink current.

There are two reasons why this characteristic is exhibited; in the low current region, the g_m is proportional to the sink current, and the parasitic capacitances are smaller than the signal capacitance. At around the peak, at least one of the parasitic capacitances becomes equal to the signal capacitance. In the region of the current being larger than that value, both parasitic capacitances become larger than the signal capacitance and the conversion frequency will decrease with an increase of the sink current.

In mixed signal application the substrate noise and the interference between analog and digital supply voltages caused by the switching of digital sections are problematic. The situation becomes more and more critical as smaller geometries induce higher coupling. Moreover, higher speed and current density augment electro-magnetic issues. The use of submicron technologies with high resistive substrates is advantageous because the coupling from digital sections to regions where the analog circuits are located is partially blocked. However, the issues such as the bounce of the digital supply and ground lines exhibit strong influence on analog circuit behavior. The use of separate analog and digital supplies is a possible remedy but its effectiveness is limited by the internal coupling between close metal interconnections. The substrate and the supply noise cause two main limits: the in-band tones produced by nonlinearities that mix high frequency spurs and the reduction of the analog dynamic range required for accommodating the common-mode part of spurs. Since the substrate coupling is also a problem for pure digital circuit the submicron technologies are evolving toward silicon-on-insulator (SOI) and trench isolation options.

The offset of any analog circuit and the static accuracy of data converters critically depend on the matching between nominally identical devices. With transistors becoming smaller, the number of atoms in the silicon that produce many of the transistor's properties is becoming fewer, with the result that control of dopant numbers and placement is more erratic. During chip manufacturing, random process variations affect all transistor dimensions: length, width, junction depths, oxide thickness etc., and become a greater percentage of overall transistor size as the transistor scales. The stochastic nature of physical and chemical fabrication steps causes a random error in electrical parameters that gives rise to a time independent difference between equally designed elements. The error typically decreases as the area of devices. Transistor matching properties are improved with a thinner oxide [130]. Nevertheless, when the oxide thickness is reduced to a few atomic layers, quantum effects will dominate and matching will degrade. Since many circuit techniques exploit the equality of two components it is important for a given process obtaining the best matching especially for critical devices. Some of the rules that have to be followed to ensure good matching are: firstly, devices to be matched should have the same structure and use the same materials, secondly, the temperature of matched components should be the same, e.g. the devices to be matched should be located on the same isotherm, which is obtained by symmetrical placement with respect to the dissipative devices, thirdly, the distance between matched devices should be minimum for having the maximum spatial correlation of fluctuating physical parameters, common-centroid geometries should be used to cancel the gradient of parameters at the first order. Similarly, the same orientation of devices on chip should be the same to eliminate dissymmetries due to unisotropic fabrication steps, or to the uniostropy of the silicon itself and lastly, the surroundings in the layout, possibly improved by dummy structures should be the same to avoid border mismatches.

Since the use of digital enhancing techniques reduces the need for expensive technologies with special fabrication steps, a side advantage is that the cost of parts is reduced while maintaining good yield, reliability and long-term stability. Indeed, the extra cost of digital processing is normally affordable as the use of submicron mixed signal technologies allows for efficient usage of silicon area even for relatively complex algorithms. The methods can be classified into foreground and background calibration.

The foreground calibration, typical of A/D converters, interrupts the normal operation of the converter for performing the trimming of elements or the mismatch measurement by a dedicated calibration cycle normally performed at power-on or during periods of inactivity of the circuit. Any miscalibration or sudden environmental changes such as power supply or temperature may make the measured errors invalid. Therefore, for devices that operate for long periods it is necessary to have periodic extra calibration cycles. The input switch restores the data converter to normal operational after the mismatch measurement and every conversion period the logic uses the output of the A/D converter to properly address the memory that contains the correction quantity. In order to optimize the memory size the stored data should be the minimum word-length, which depends on technology accuracy and expected A/D linearity. The digital measure of errors, that allows for calibration by digital signal processing, can be at the element, block or entire converter level. The calibration parameters are stored in memories but, in contrast with the trimming case, the content of the memories is frequently used, as they are input of the digital processor.

Methods using background calibration work during the normal operation of the converter by using extra circuitry that functions all the time synchronously with the converter function. Often these circuits use hardware redundancy to perform a background calibration on the fraction of the architecture that is not temporarily used. However, since the use of redundant hardware is effective but costs silicon area and power consumption, other methods aim at obtaining the functionality by borrowing a small fraction of the sampled-data circuit operation for performing the self-calibration.

2.3 A/D Converter Building Blocks

2.3.1 Sample-and-Hold

Inherent to the A/D conversion process is a sample-and-hold (S/H) circuit that resides in the front-end of a converter (and also between stages in a pipeline converter). In addition to suffering from additive circuit noise and signal distortion just as the rest of the converter does, the S/H also requires a precision time base to define the exact acquisition time of the input signal. The dynamic performance degradation of an ADC can often be attributed to the deficiency of the S/H circuit (and the associated buffer amplifier).

The main function of an S/H circuit is to take samples of its input signal and hold its value until the A/D converter can process the information. Typically, the samples are taken at uniform time intervals; thus, the sampling rate (or clock rate) of the circuit can be determined. The operation of an S/H circuit can be divided into sample mode (sometimes also referred as acquisition mode) and hold mode, whose durations need not be equal. In sample mode, the output can either track the input, in which case the circuit is often called a track-and-hold (T/H) circuit or it can be reset to some fixed value. In hold mode an S/H circuit remembers the value of the input signal at the sampling moment and thus it can be considered as an analog memory cell. The basic circuit elements that can be employed as memories are capacitors and inductors, of which the capacitors store the signal as a voltage (or charge) and the inductors as a current. Since capacitors and switches with a high off-resistance needed for a voltage memory are far easier to implement in a practical integrated circuit technology than inductors and switches with a very small on-resistance required for a current memory, all sample-and-hold circuits are based on voltage sampling with switched capacitor (SC) technique.

S/H circuit architectures can roughly be divided into open-loop and closed-loop architectures. The main difference between them is that in closed-loop architectures the capacitor, on which the voltage is sampled, is enclosed in a feedback loop, at least in hold mode. Although open-loop S/H architecture provide high speed solution, its accuracy, however, is limited by the harmonic distortion arising from the nonlinear gain of the buffer amplifiers and the signal-dependent charge injection from the switch. These problems are especially emphasized with a CMOS technology as shown in Section 3.3. Enclosing the sampling capacitor in the feedback loop reduces the effects of nonlinear parasitic capacitances and signal-dependent charge injection from the MOS switches. Unfortunately, an inevitable consequence of the use of feedback is reduced speed.

Figure 2.7 illustrates three common configurations for closed-loop switched-capacitor S/H circuits [56, 57, 59, 91]. For simplicity, single-ended configurations

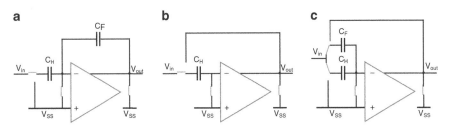

Fig. 2.7 Switched capacitor S/H circuit configurations in sample phase: (**a**) a circuit with separate CH and CF, (**b**) a circuit with one capacitor, and (**c**) a circuit with CF shared as a sampling capacitor

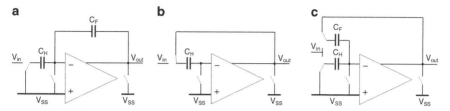

Fig. 2.8 Switched capacitor S/H circuit configurations in hold phase: (**a**) a circuit with separate CH and CF, (**b**) a circuit with one capacitor, and (**c**) a circuit with CF shared as a sampling capacitor

are shown; however in circuit implementation all would be fully differential. In a mixed-signal circuit such as A/D converters, fully differential analog signals are preferred as a means of getting a better power supply rejection and immunity to common mode noise. The operation needs two non-overlapping clock phases – sampling and holding or transferring. Switch configurations shown in Fig. 2.7 are for the sampling phase, while configurations shown in Fig. 2.8 are for hold phase.

In all cases, the basic operations include sampling the signal on the sampling capacitor(s) C_H and transferring the signal charge onto the feedback capacitor C_F by using an op amp in the feedback configuration. In the configuration in Fig. 2.7a, which is often used as an integrator, assuming an ideal op amp and switches, the op amp forces the sampled signal charge on C_H to transfer to C_F. If C_H and C_F are not equal capacitors, the signal charge transferred to C_F will display the voltage at the output of the op amp according to $V_{out} = (C_H/C_F) V_{in}$. In this way, both S/H and gain functions can be implemented within one SC circuit [57,91]. In the configuration shown in Fig. 2.7b, only one capacitor is used as both sampling capacitor and feedback capacitor. This configuration does not implement the gain function, but it can achieve high speed because the feedback factor (the ratio of the feedback capacitor to the total capacitance at the summing node) can be much larger than that of the previous configuration, operating much closer to the unity gain frequency of the amplifier. Furthermore, it does not have the capacitor mismatch limitation as the other two configurations. Here, the sampling is performed passively, i.e. it is done without the op amp, which makes signal acquisition fast. In hold mode the sampling capacitor is disconnected from the input and put in a feedback loop around the op amp. This configuration is often used in the front-end input S/H circuit [56, 59] and will be discussed in more detail in Section 3.3.

Figure 2.7c shows another configuration which is a combined version of the configurations in Fig. 2.7a and Fig. 2.7b. In this configuration, in the sampling phase, the signal is sampled on both C_H and C_F, with the resulting transfer function $V_{out} = (1 + (C_H/C_F)) V_{in}$. In the next phase, the sampled charge in the sampling capacitor is transferred to the feedback capacitor. As a result, the feedback capacitor has the transferred charge from the sampling capacitor as well as the input signal charge. This configuration has a wider bandwidth in comparison to the configuration shown in Fig. 2.7a, although feedback factor is comparable.

Important parameters in determining the bandwidth of the SC circuit are G_m (transconductance of the op amp), feedback factor β, and output load capacitance. In all of these three configurations, the bandwidth is given by $1/\tau = \beta \times G_m/C_L$, where C_L is the total capacitance seen at the op amp output. Since S/H circuit use amplifier as buffer, the acquisition time will be a function of the amplifier own specifications. Similarly, the error tolerance at the output of the S/H is dependent on the amplifier's offset, gain and linearity. Once the hold command is issued, the S/H faces other errors. Pedestal error occurs as a result of charge injection and clock feedthrough. Part of the charge built up in the channel of the switch is distributed onto the capacitor, thus slightly changing its voltage. Also, the clock couples onto the capacitor via overlap capacitance between the gate and the source or drain. Another error that occurs during the hold mode is called droop, which is related to the leakage of current from the capacitor due to parasitic impedances and to the leakage through the reverse-biased diode formed by the drain of the switch. This diode leakage can be minimized by making the drain area as small as can be tolerated. Although the input impedance to the amplifier is very large, the switch has a finite off impedance through which leakage can occur. Current can also leak through the substrate.

A prominent drawback of a simple S/H is the on-resistance variation of the input switch that introduces distortion. Technology scales the supply voltage faster than the threshold voltage, which results in a larger on-resistance variation in a switch. As a result, the bandwidth of the switch becomes increasingly signal dependent. Clock bootstrapping was introduced to keep the switch gate-source voltage constant (Section 3.3.3). Care must be exercised to ensure that the reliability of the circuit is not compromised.

While the scaling of CMOS technology offers a potential for improvement on the operating speed of mixed-signal circuits, the accompanying reduction in the supply voltage and various short-channel effects create both fundamental and practical limitations on the achievable gain, signal swing and noise level of these circuits, particularly under a low power constraint. In the sampling circuit, thermal noise is produced due to finite resistance of a MOS transistor switch and is stored in a sampling capacitor. As the sampling circuit cannot differentiate the noise from the signal, part of this signal acquisition corresponds to the instantaneous value of the noise at the moment the sampling takes place. In this context, when the sample is stored as charge on a capacitor, the root-mean-square (rms) total integrated thermal noise voltage is $\overline{v_{ns}^2} = kT/C_H$, where kT is the thermal energy and C_H is the sampling capacitance. This is often referred to as the kT/C noise. No resistance value at the expression is present, as the increase of thermal noise power caused by increasing the resistance value is cancelled in turn by the decreasing bandwidth.

In the sampling process the kT/C noise usually comprises two major contributions – the channel noise of the switches and the amplifier noise. Since no direct current is conducted by the switch right before a sampling takes place (the bandwidth of the S/H circuit is assumed large and the circuit is assumed settled), the $1/f$ noise is not of concern here; only the thermal noise contributes, which is a function of the channel resistance that is weakly affected by the technology scaling [120]. On the other hand, the amplifier output noise is in most cases dominated by the channel noise of the input transistors, where the thermal noise and the $1/f$ noise both contribute. Because the input transistors of the amplifier are usually biased in saturation region to derive large transconductance (g_m), impact ionization and hot carrier effect tend to enhance their thermal noise level [121, 122]; the $1/f$ noise increases as well due to the reduced gate capacitance resulted from finer lithography and therefore shorter minimum gate length. It follows that, as CMOS technology scaling continues, amplifier increasingly becomes the dominant noise source. Interestingly, the input-referred noise (the total integrated output noise as well) still takes the form of kT/C with some correction factor χ_1, $\overline{v_{ns}^2} = \chi_1 kT/C_H$. Thus a fundamental technique to reduce the noise level, or to increase the signal-to-noise ratio of an S/H circuit, is to increase the size of the sampling capacitors. The penalty associated with this technique is the increased power consumption as larger capacitors demand larger charging/discharging current to keep up the sampling speed.

2.3.2 Operational Amplifier

In front-end S/H amplifiers or multi-stage A/D converters, precision op amps are almost invariably employed to relay the input signal (or the residue signal) to the trailing conversion circuits. Operating on the edge of the performance envelope, op amps exhibit intense trade-offs amongst the dynamic range, linearity, settling speed, stability, and power consumption. As a result, the conversion accuracy and speed are often dictated by the performance of these amplifiers.

Amplifiers with a single gain stage have high output impedance providing an adequate dc gain, which can be further increased with gain boosting techniques. Single-stage architecture offers large bandwidth and a good phase margin with small power consumption. Furthermore, no frequency compensation is needed, since the architecture is self-compensated (the dominant pole is determined by the load capacitance), which makes the footprint on the silicon small. On the other hand, the high output impedance is obtained by sacrificing the output voltage swing, and the noise is rather high as a result of the number of noise-contributing devices and limited voltage head-room for current source biasing.

The simplest approach for the one-stage high-gain operational amplifier is telescopic cascode amplifier [150] of Fig. 2.9a. With this architecture, a high open loop dc gain can be achieved and it is capable of high speed when closed loop gain is low. The number of current legs being only two, the power consumption is small. The biggest disadvantage of a telescopic cascode amplifier is its low maximum output swing, $V_{DD}-5V_{DS,SAT}$, where V_{DD} is the supply voltage and $V_{DS,SAT}$ is the saturation voltage of a transistor. With this maximum possible output swing the input common-mode range is zero. In practice, some input common-mode range, which reduces the output swing, always has to be reserved so as to permit inaccuracy and settling transients in the signal common-mode levels. The high-speed capability of the amplifier is the result of the presence of only n-channel transistors in the signal path and of relatively small capacitance at the source of the cascode transistors.

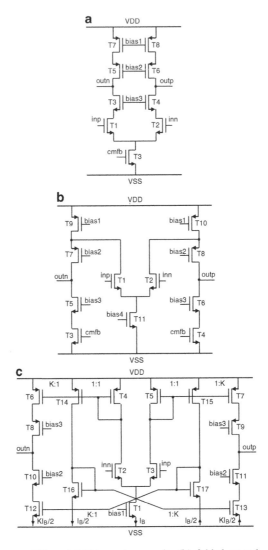

Fig. 2.9 One-stage amplifiers: (**a**) Telescopic cascade, (**b**) folded cascade, and (**c**) push–pull current-mirror amplifier with a cascade output stage

The gain-bandwidth product of the amplifier is given by GBW $= g_{m1}/C_L$, where g_{m1} is the transconductance of transistors T_1 and C_L is the load capacitance. Thus, the GBW is limited by the load capacitance. Due to its the simple topology and dimensioning, the telescopic cascade amplifier is preferred if its output swing is large enough for the specific application. The output signal swing of this architecture has been widened by driving the transistors T_7–T_8 into the linear region [151]. In order to preserve the good common mode rejection ratio and power supply rejection ratio properties of the topology, additional feedback circuits for compensation have been added to these variations. The telescopic cascode amplifier has low current consumption, relatively high gain, low noise and very fast operation. However, as it has five stacked transistors, the topology is not suitable for low supply voltages.

The folded cascode amplifier topology [152] is shown in Fig. 2.9b. The swing of this design is constrained by its cascoded output stage. It provides a larger output swing and input common-mode range than the telescopic amplifier with the same dc gain and without major loss of speed. The output swing is $V_{DD} - 4V_{DS,SAT}$ and is not linked to the input common-mode range, which is $V_{DD} - V_T - 2V_{DS,SAT}$. The second pole of this amplifier is located at g_{m7}/C_{par}, where g_{m7} is the transconductance of T_7 and C_{par} is the sum of the parasitic capacitances from transistors T_1, T_7 and T_9 at the source node of transistor T_7. The frequency response of this amplifier is deteriorated from that of the telescopic cascode amplifier because of a smaller transconductance of the p-channel device and a larger parasitic capacitance. To assure symmetrical slewing, the output stage current is usually made equal to that of the input stage. The GBW of the folded cascode amplifier is also given by g_{m1}/C_L. The open loop dc gain of amplifiers having cascode transistors can be boosted by regulating the gate voltages of the cascode transistors [140]. The regulation is realized by adding an extra gain stage, which reduces the feedback from the output to the drain of the input transistors. In this way, the dc gain of the amplifier can be increased by several orders of magnitude. The increase in power and chip area can be kept very small with appropriate feedback amplifier architecture [140]. The current consumption of the folded cascode is doubled compared to the telescopic cascode amplifier although the output voltage swing is increased since there are only four stacked transistors. The noise of the folded cascode is slightly higher than in the telescopic cascode as a result of the added noise from the current source transistors T_9 and T_{10}. In addition, the folded cascade has a slightly smaller dc gain due to the parallel combination of the output resistance of transistors T_1 and T_9.

A push-pull current-mirror amplifier, shown in Fig. 2.9c, has much better slew-rate properties and potentially larger bandwidth and dc gain than the folded cascode amplifier. The slew rate and dc gain depend on the current-mirror ratio K, which is typically between one and three. However, too large current-mirror ratio increases the parasitic capacitance at the gates of the transistors T_{12} and T_{13}, pushing the non-dominant pole to lower frequencies and limiting the achievable GBW. The non-dominant pole of the current mirror amplifier is much lower than that of the folded cascode amplifier and telescopic amplifiers due to the larger parasitic capacitance at the drains of input transistors. The noise and current consumption of the current-mirror amplifier are larger than in the telescopic cascode amplifier or in the folded cascode amplifier. A current-mirror amplifier with dynamic biasing [153] can be used to make the amplifier biasing be based purely on its small signal behavior, as the slew rate is not limited. In dynamic biasing, the biasing current of the operational amplifier is controlled on the basis of the differential input signal. With large differential input signals, the biasing current is increased to speed up the output settling. Hence, no slew rate limiting occurs, and the GBW requirement is relaxed. As the settling proceeds, the input voltage decreases and the biasing current is reduced. The biasing current needs to be kept only to a level that provides enough GBW for an adequate small-signal performance. In addition to relaxed GBW requirements, the reduced static current consumption makes the design of a high-dc gain amplifier easier. With very low supply voltages, the use of the cascode output stages limits the available output signal swing considerably. Hence, two-stage operational amplifiers are often used, in which the operational amplifier gain is divided into two stages, where the latter stage is typically a common-source output stage. Unfortunately, with the same power dissipation, the speed of the two-stage operational amplifiers is typically lower than that of single-stage operational amplifiers.

Of the several alternative two-stage amplifiers, Fig. 2.10a shows a simple Miller compensated amplifier [154]. With all the transistors in the output stage of this amplifier placed in the saturation region, it has an output swing of $V_{DD} - V_{DS,SAT}$. Since the non-dominant pole, which arises from the output node, is determined dominantly by an explicit load capacitance, the amplifier has a compromised frequency response. The gain bandwidth product of a Miller compensated amplifier is given approximately by GBW $= g_{m1}/C_C$, where g_{m1} is the transconductance of T_1. In general, the open loop dc gain of the basic configuration is not large enough for high-resolution applications. Gain can be enhanced by using cascoding, which has, however, a negative effect on the signal swing and bandwidth. Another drawback of this architecture is a poor power supply rejection at high frequencies because of the connection of V_{DD} through the gate-source capacitance $C_{GS5,6}$ of T_5 and T_6 and C_C. The noise properties of the two-stage Miller-compensated operational amplifier are comparable to those of the telescopic cascode and better than those of the folded cascode amplifier. The speed of a Miller-compensated amplifier is determined by its pole-splitting capacitor C_C. Usually, the position of this non-dominant pole, which is located at the output of the two-stage amplifier, is lower than that of either a folded-cascode or a telescopic amplifier. Thus, in order to push this pole to higher frequencies, the second stage of the amplifier requires higher currents resulting in increased power dissipation. Since the first stage does not need to have a large output voltage swing, it can be a cascode stage, either a telescopic or a folded cascode. However, the current consumption and transistor count are also increased. The advantages of the folded cascode structure are a larger input common-mode range and the avoidance of level shifting between the stages, while the telescopic stage can offer larger bandwidth and lower thermal noise.

Figure 2.10b illustrates a folded cascode amplifier with a common-source output stage and Miller compensation. The noise properties are comparable with those of the folded cascode amplifier. If a cascode input stage is used, the lead-compensation resistor can be merged with the cascode transistors. An example of this is the folded cascode amplifier with a common-source output stage and Ahuja-style compensation [155] shown in Fig. 2.10c. The operation of the Ahuja-style compensated operational amplifier is suitable for larger capacitive loads than the Miller-compensated one and it has a better power supply rejection, since the substrate noise coupling through the gate-source capacitance of the output stage gain transistors is not coupled directly through the pole-splitting capacitors to the operational amplifier output [155].

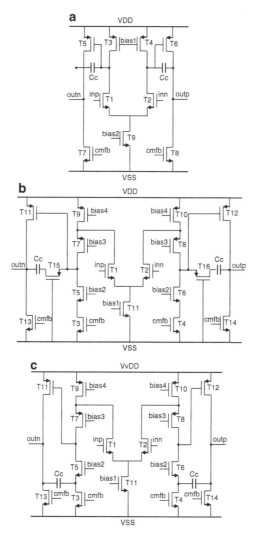

Fig. 2.10 Two-stage amplifiers: (**a**) Miller compensated, (**b**) folded cascode amplifier with a common-source output stage and Miller frequency compensation, and (**c**) folded cascode amplifier with a common-source output stage and Ahuja-style frequency compensation

2.3.3 Latched Comparators

The offset in preamps and comparators constitutes the major source of error in flash-type converters. Simple differential structure with thin oxide devices will keep dominating the preamp architecture in newer technologies. Dynamic performance is crucial at high sample rates with high input frequencies.

Because of its fast response, regenerative latches are used, almost without exception, as comparators for high-speed applications. An ideal latched comparator is composed of a preamplifier with infinite gain and a digital latch circuit. Since the amplifiers used in comparators need not to be either linear or closed-loop, they can incorporate positive feedback to attain virtually infinite gain [171]. Because of its special architecture, working process of a latched comparator could be divided in two stages: tracking and latching stages. In tracking stage the following dynamic latch circuit is disabled, and the input analog differential voltages are amplified by the preamplifier. In the latching stage while the preamplifier is disabled, the latch circuit regenerates the amplified differential signals into a pair of full-scale digital signals with a positive feedback mechanism and latches them at output ends.

Depending on the type of latch employed, the latch comparators can be divided into two groups: static [59, 173, 174], which have a constant current consumption during operation and dynamic [175–177], which does not consume any static power. In general, the type of latch employed is determined by the resolution of the stage. For a low-resolution quantization per stage, a dynamic latch is more customary since it dissipates less power than the static latch (the dynamic latch does not dissipate any power during the resetting period), even though the difference is negligible at high clock rates. While the latch circuits regenerate the difference signals, the large voltage variations on regeneration nodes will introduce the instantaneous large currents. Through parasitic gate-source and gate-drain capacitances of transistors, the instantaneous currents are coupled to the inputs of the comparators, making the disturbances unacceptable. It is so-called kickback noise influence. In flash A/D converters where a large number of comparators are switched on or off at the same time, the summation of variations came from regeneration nodes may become unexpectedly large and directly results in false quantization code output [172]. It is for this reason that the static latch is preferable for higher-resolution implementations.

The static latched comparator from [173] is shown in Fig. 2.11a. When the clock signal is high, T_{10} and T_{11} discharge the latch formed with cross-connected transistors T_{8-9} to the output nodes. When the latch signal goes low, the drain current difference between T_6 and T_7 appears as the output voltage difference. However, some delay in the circuit is present since T_8 and T_{11} have to wait until either side of the output voltage becomes larger than V_T. The other is that there is a static current in the comparators which are close to the threshold after the output is fully developed. Assume the potential of V_{outp} node is higher than that of V_{outn} node. After the short period, T_{11} turns off and the potential of V_{outp} becomes V_{DD}, however, since T_8 is in a linear region, the static current from T_6 will drain during the regeneration period. Since the input transistors are isolated from the regeneration nodes through the current mirror, the kickback noise is reduced. However, the speed of the regeneration circuit is limited by the bias current and is not suitable for low-power high-speed applications.

Figure 2.11b illustrates the schematic of the comparator given in [59]. The circuit consists of a folded-cascode amplifier (T_1–T_7) where the load has been replaced by a current-triggered latch (T_8–T_{10}). When the latch signal is high (resetting period),

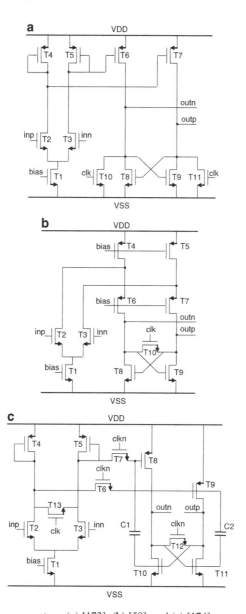

Fig. 2.11 Static latch comparators: (**a**) [173], (**b**) [59], and (**c**) [174]

transistor T_{10} shorts both the latch outputs. In addition, the on-resistance R of T_{10} can give an extra gain A_{reset} at the latch output, $A_{reset} = (g_{m1,2} \times R)/(2 - g_{m8,9} \times R)$, which speed up the regeneration process. However, the on-resistance R, should be chosen such that $g_{m8,9} \times R < 2$ and should be small enough to reset the output at the clock rate. Since all transistors are in active region, the latch can start regenerating right after the latch signal goes low. The one disadvantage of this scheme is the large kickback noise. The folding nodes (drains of T_4 and T_5) have to jump up to V_{DD} in every clock cycle since the latch output does the full swing. Because of this, there are substantial amounts of kickback noise into the inputs through the gate-drain capacitor of input transistors T_1 and T_2 (C_{GD1}, C_{GD2}). To reduce kickback noise, the clamping diode has been inserted at the output nodes [178].

In Fig. 2.11c design shown in [174] is illustrated. Here, when the latch signal is low (resetting period), the amplified input signal is stored at gate of T_8 and T_9 and T_{12} shorts both V_{outp} and V_{outn}. When the latch signal goes high, the cross-coupled transistors T_{10} and T_{11} make a positive feedback latch. In addition, the positive feedback capacitors, C_1 and C_2, boost up the regeneration speed by switching T_8 and T_9 from an input dependant current source during resetting period to a cross-coupled latch during the regeneration period. Because of C_1 and C_2, the $T_8 \sim T_{11}$ work like a cross-coupled inverter so that the latch does not dissipate the static power once it completes the regeneration period. However, there is a large amount of kickback noise through the positive feedback capacitors, C_1 and C_2. The switches (T_6, T_7 and T_{13}) have been added to isolate the preamplifier from the latch. Therefore, the relatively large chip area is required due to the positive feedback capacitors (C_1, C_2), isolation switches (T_6, T_7 and T_{13}) and complementary latch signals.

The concept of a dynamic comparator exhibits potential for low power and small area implementation and, in this context, is restricted to single-stage topologies without static power dissipation. A widely used dynamic comparator is based on a differential sensing amplifier as shown in Fig. 2.12a was introduced in [175]. Transistors T_{1-4}, biased in linear region, adjust the threshold resistively and above them transistors T_{5-12} form a latch. When the latch control signal is low, the transistors T_9 and T_{12} are conducting and T_7 and T_8 are cut off, which forces both differential outputs to V_{DD} and no current path exists between the supply voltages. Simultaneously T_{10} and T_{11} are cut off and the transistors T_5 and T_6 conduct. This implies that T_7 and T_8 have a voltage of V_{DD} over them. When the comparator is latched T_7 and T_8 are turned on. Immediately after the regeneration moment, the gates of the transistors T_5 and T_6 are still at V_{DD} and they enter saturation, amplifying the voltage difference between their sources. If all transistors T_{5-12} are assumed to be perfectly matched, the imbalance of the conductances of the left and right input branches, formed by T_{1-2} and T_{3-4}, determines which of the outputs goes to V_{DD} and which to 0 V. After a static situation is reached (V_{clk} is high), both branches are cut off and the outputs preserve their values until the comparator is reset again by switching V_{clk} to 0 V. The transistors T_{1-4} connected to the input and reference are in the triode region and act like voltage controlled resistors. The transconductance of the transistors T_{1-4} operating in the linear region, is directly proportional to the drain-source voltage of the corresponding transistor V_{DS1-4}, while for the transistors T_{5-6} the transconductance is proportional to $V_{GS5,6}$-V_T. At the beginning of the latching process, $V_{DS1-4} \approx 0$ while $V_{GS5,6} - V_T \approx V_{DD}$. Thus, $g_{m5,6} \gg g_{m1-4}$, which makes the matching of T_5 and T_6 dominant in determining the latching balance. As small transistors are preferred, offset voltages of a few hundred millivolts are easily resulted. Mismatch in transistors T_{7-12} are attenuated by the gain of T_5 and T_6, which makes them less critical. To cope with the mismatch problem, the layout of the critical transistors must be drawn as symmetric as possible. In addition to the

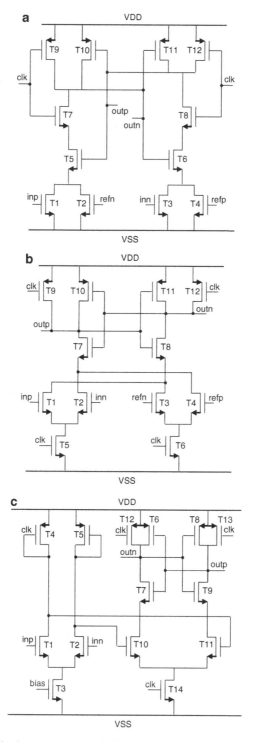

Fig. 2.12 Dynamic latch comparators: (**a**) [175], (**b**) [176], and (**c**) [177]

mismatch sensitivity, the latch is also very sensitive to an asymmetry in the load capacitance. This can be avoided by adding an extra latch or inverters as a buffering stage after the comparator core outputs. The resistive divider dynamic comparator topology has one clear benefit, which is its low kickback noise. This results from the fact that the voltage variation at the drains of the input transistors T_{1-4} is very small. On the other hand, the speed and resolution of the topology are relatively poor because of the small gain of the transistors biased in the linear region.

A fully differential dynamic comparator based on two cross-coupled differential pairs with switched current sources loaded with a CMOS latch is shown in Fig. 2.12b [176]. The trip point of the comparator can be set by introducing imbalance between the source-coupled pairs. Because of the dynamic current sources together with the latch, connected directly between the differential pairs and the supply voltage, the comparator does not dissipate dc-power. When the comparator is inactive the latch signal is low, which means that the current source transistors T_5 and T_6 are switched off and no current path between the supply voltages exists. Simultaneously, the p-channel switch transistors T_9 and T_{12} reset the outputs by shorting them to V_{DD}. The n-channel transistors T_7 and T_8 of the latch conduct and also force the drains of all the input transistors T_{1-4} to V_{DD}, while the drain voltage of T_5 and T_6 are dependent on the comparator input voltages. When clock signal is raised to V_{DD}, the outputs are disconnected from the positive supply, the switching current sources T_5 and T_6 turn on and T_{1-4} compare $V_{inp} - V_{inn}$ with $V_{refp} - V_{refn}$. Since the latch devices T_{7-8} are conducting, the circuit regeneratively amplifies the voltage difference at the drains of the input pairs. The threshold voltage of the comparator is determined by the current division in the differential pairs and between the cross-coupled branches. The threshold level of the comparator can be derived using large signal current equations for the differential pairs. The effect of the mismatches of the other transistors T_{7-12} is in this topology not completely critical, because the input is amplified by T_{1-4} before T_{7-12} latch. The drains of the cross-coupled differential pairs are high impedance nodes and the transconductances of the threshold-voltage-determining transistors T_{1-4} large. A drawback of the differential pair dynamic comparator is its high kickback noise: large transients in the drain nodes of the input transistors are coupled to the input nodes through the parasitic gate-drain capacitances. However, there are techniques to reduce the kickback noise, e.g. by cross-coupling dummy transistors from the differential inputs to the drain nodes [13]. The differential pair topology achieves a high speed and resolution, which results from the built-in dynamic amplification.

Figure 2.12c illustrates the schematic of the dynamic latch given in [177]. The dynamic latch consists of pre-charge transistors T_{12} and T_{13}, cross-coupled inverter T_{6-9}, differential pair T_{10} and T_{11} and switch T_{14} which prevent the static current flow at the resetting period. When the latch signal is low (resetting period), the drain voltages of T_{10-11} are $V_{DD}-V_T$ and their source voltage is V_T below the latch input common mode voltage. Therefore, once the latch signal goes high, the n-channel transistors $T_{7, 9-11}$ immediately go into the active region. Because each transistor in one of the cross-coupled inverters turns off, there is no static power dissipation from the latch once the latch outputs are fully developed.

2.4 A/D Converters: Summary

Practical realizations from Section 2.1.4 (Fig. 2.13) show the trend that to a first order, converter power is directly proportional to sampling rate f_S when f_S is much lower than the device technology transition frequency f_T. However, power dissipation required becomes nonlinear as the speed capabilities of a process technology are pushed to the limit. In the case of *constant current-density* designs, there is an optimum power point when total intrinsic capacitance equals total extrinsic capacitance, beyond which power increases yield diminishing returns in speed improvements.

Power dependence on converter resolution is not as straight-forward as its dependence on f_S because converter architecture also varies with resolution. Flash converters dominate at the high-speed low-resolution end of the spectrum while pipelined converters are usually employed at the low-speed high-resolution end typically from 8–12 bits without calibration and up to 15 bits with calibration. The inter-stage gain makes it possible to scale the components along the pipeline,

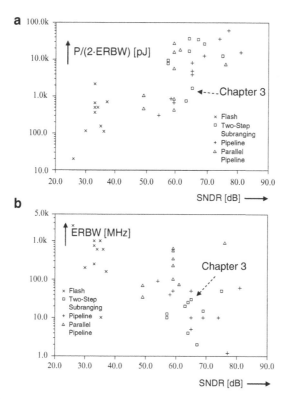

Fig. 2.13 (a) Energy versus SNDR for A/D converters shown in Tables 2.1, 2.2, 2.3, and 2.4. (b) Effective resolution bandwidth versus SNDR for A/D converters shown in Tables 2.1, 2.2, 2.3, and 2.4

which leads to low power consumption. Resolutions up to 15 bits can be covered with two-step/multi-step/subranging converters or with an architecture which is a combination of these. Pipeline and two-step/multi-step converters tend to be the most efficient at achieving a given resolution and sampling rate specification. Its power proficiency for high resolution will be demonstrated with prototype [37] in the next chapter.

Chapter 3
Design of Multi-Step Analog to Digital Converters

In highly integrated telecommunication systems, moving analog-to-digital converters capable of IF sampling towards the high frequency front-end maximize economic-value exploiting system complexity and full integration. The goal of the A/D converter is to minimize analog filtering and to replace it with better controllable digital functions. System-on-chip, *SoC*, realizations require an A/D converter embedded in a large digital IC. To achieve the lowest cost, the system-on-chip has to be implemented in state-of-the-art CMOS technologies and must be area and power efficient and avoid the need for trimming to achieve the required accuracy. The rapidly decreasing feature size and power supply voltage of deep-submicron CMOS technology increases the pressure on converter requirements. As the supply voltage is scaled down, the voltage available to represent the signal is reduced. To maintain the same dynamic range on a lower supply voltage, the thermal noise of the circuit must also be proportionally reduced. In analog circuits, decreasing voltage supply consequently reduces the output swing, which reduces the SNR. This results in an increase of power consumption, which is determined by the SNR at a given frequency. Although several architectures can be embedded in *SoC* realizations, some of the drawbacks such as large amount of silicon and power of flash A/D converters, the matching of the inputs of several tens of folding amplifiers in the folding A/D converters and the larger latency for certain digital feedback loops in pipelined A/D converters, limits the choice to sub-ranging or two-step/multi-step architecture.

3.1 Multi-Step A/D Converter Architecture

Demanding requirements are placed on high-performance analog-to-digital converters and analog components in most digital receivers. In cellular base station digital receivers for example, sufficient dynamic range is needed to handle high-level interferers (or blockers) while properly demodulating the lower level desired signal. A cellular base station consists of many different hardware modules including one that performs the receiver and transmitter functionality. Today,

A. Zjajo and J. Pineda de Gyvez, *Low-Power High-Resolution Analog to Digital Converters*, Analog Circuits and Signal Processing, DOI 10.1007/978-90-481-9725-5_3, © Springer Science+Business Media B.V. 2011

Table 3.1 Requirements for
IF conversion

Technology	Digital CMOS
Resolution	12 bit
Supply voltage	Single supply
Sample rate	>50 Msample/s
Effective bandwidth	25 MHz
SNR	>66 dB
SFDR	>75 dB
THD	>70 dB
Power dissipation	<150 mW
Area	<1 mm^2

analog technology is being replaced by CDMA and WCDMA worldwide and Europe adapted GSM over a decade ago.

To verify the effective resolution bandwidth versus power efficiency of a multi-step architecture, a GSM base-station application (Table 3.1) has been targeted as an appropriate vehicle for a prototype described in this Chapter. Various CDMA and GSM designs exist today and methods to reduce cost and power are continuously being sought by base transceiver station manufacturers. Optimizing single-carrier solutions or developing multi-carrier receivers can accomplish this. For the sub-sampling receiver architecture, which is commonly used in base transceiver station equipment, stringent noise and distortion requirements are placed on the A/D converter. In receiver applications, the lower level desired signal is digitized alone or in the presence of an unwanted signal(s) that can be significantly larger in amplitude. To properly design the receiver, the A/D converter effective noise figure must be determined under these two signal extremes. The converter noise figure is determined by comparing its total noise power to the thermal noise floor. For small analog input signals, the thermal plus quantization noise power dominate the A/D converter noise floor, which is used to approximate the A/D converter effective noise figure. In practice, once the A/D converter effective noise figure is known in the small signal condition and the cascaded noise figure of the analog circuitry (RF and IF) is determined, the minimum power gain ahead of the A/D converter is selected to meet the required receiver noise figure. The amount of power gain places an upper limit on the highest interference level the receiver can tolerate before the A/D converter overloads. The sub-sampling architecture can be used with a single down conversion architecture if sufficient SNR and SFDR performance can be obtained from the converter at higher IF frequencies. Distortion causes inter-modulation of large unwanted signals, the resulting products of which can fall in the wanted channel band. A/D converter with the 12-bit accuracy is sufficient together with the amount of channel filtering and the gain of the automatic gain control, which is included to relax dynamic range requirements of the A/D converter. To be able to deal with the complete GSM band, the sample rate of the A/D converter has to be 50–60 MSample/s with an effective resolution bandwidth of 25 MHz. The SINAD of the GSM signal is only 9 dB but in order to handle the large neighboring channels the SNR of the converter needs to be 66 dB. The spurious tones generated by large interfering unwanted channels can disturb the reception of a small

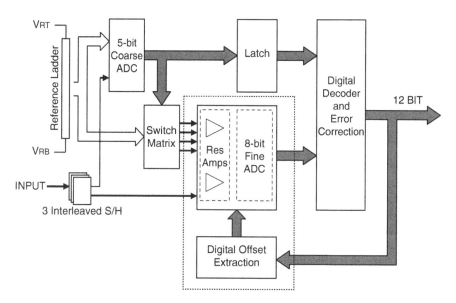

Fig. 3.1 Prototype two-step/multi-step A/D converter

wanted channel. The SFDR must therefore be below 75 dB. The target power dissipation is chosen as 150 mW.

A detailed block diagram of the two-step A/D converter is shown in Fig. 3.1. Because the five to eight partitioning (the A/D converter utilizing 5-bit coarse and 8-bit fine quantizer) offers a lower transistor count for the required accuracy, this topology was selected over the six to seven approach. The differential input signal is sampled with three time-interleaved sample-and-hold circuits. Noise generated and sampled in the S/H deteriorates the analog input signal which has to be quantized by the A/D converter. This generated and sampled noise has to be sufficiently low to meet SNR requirements. The resulting analog signal is processed into the 5-bit coarse A/D converter, which compares the differential input signal with a resistor reference ladder. Because of their simple structure and high-speed capability, flash architectures are natural choices for coarse A/D converter. To perform a proper total conversion the coarse A/D converter has to have accuracy such that the resulting residue signal is always in the range of the fine A/D converter. If offset voltage is the main inaccuracy, then input device area needs to be increased in inverse proportion to the required precision, increasing power of the fine comparators in a square fashion. Alternatively, some offset cancellation technique can be used (with some timing and speed impact), increasing power only by the extra circuitry needed to do the cancellation. The acquired signal from this quantization is stored in a latch and is also applied to a switch unit. The switching-matrix voltage reference will add power similar to that required by the reference ladder for a full flash converter. The power of the resistor ladders in the switching-matrix should also follow the same characteristics as will be shown in the coarse section. This switch unit selects, according to the coarse quantization, four reference signals, from the same resistor reference ladder used for the coarse quantization. These selected reference signals are combined with the held input signals from S/H in two residue amplifiers. Reducing the offsets of these amplifiers is necessary since residue signals have to be accurate at the overall A/D converter accuracy. After amplification, the residue signals are placed to fine resistor ladder using fine buffers. The 8 bit of the fine ADC can be generated with sufficient accuracy without using compensation by using a folding and interpolating A/D converter.

If there is an error in coarse A/D converter, missing codes can occur, which can be caused by insufficient settling in the coarse A/D converter or by mismatch in the coarse comparators. Therefore over-range of 1 bit is applied in the fine A/D converter. The S/H circuit must exhibit linearity consistent with 12-bit operation. For a 12-bit linear reconstruction D/A converter (consisting of resistor reference ladder and switch matrix), the coarse A/D converter needs only to be 5-bits linear since errors in its quantization levels generate over or under-range values in the residue signal that are eventually digitized by the fine A/D converter. However, before inputs of the residue amplifier are subtracted, they must retain 12-bit linearity, a very important constraint when one input must be converted from voltage to current before subtraction from the other. The overall linearity of the residue amplifier and fine A/D converter must exhibit linearity consistent with 8 bits.

The overall A/D converter consists primarily of non-critical low-power components, such as low-resolution quantizers, switches and open-loop amplifiers. Although a multi-step A/D converter makes use of considerable amount of digital logic, most of its signal-processing functions are executed in the analog domain. The conversion process therefore is susceptible to analog circuit and device impairments. Most of these errors will be further enlightened in the following section.

3.2   Design Considerations for Non-Ideal Multi-Step A/D Converter

Besides timing errors, the primary error sources present in a multi-step A/D converter are offset and gain errors in the S/H circuit, coarse A/D converter nonlinearity, D/A converter nonlinearity, residue amplifier offset and gain errors and fine A/D converter nonlinearity. As shown in Sections 2.1 and 2.2 the use of redundancy and digital correction has emerged as an effective means of coping with the some of the errors present in a multi-step A/D converter, namely those originating from high offsets of amplifiers and comparators [59, 111, 112]. The key advantages are that these techniques reduce the sensitivity of the linearity of such A/D converters to offsets in their comparators and offsets in their inter-stage analog processing. Since nonlinearity and gain errors in the coarse A/D converter provoke over-range problems and code level shifting, the approaches to apply digital correction are based on either increasing the input range of the next stage and using extra comparators or using the partial codes in the next stages to correct the code of the present stage.

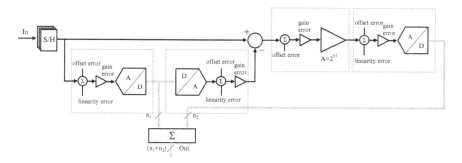

Fig. 3.2 Block diagram of a two-stage A/D converter with offset, gain and linearity errors

Usually, calibration has to be applied only to the first or, at the most, to the first and second converter stages, depending on the required resolution [62]. With digital correction, the effects of offset, gain and coarse A/D converter nonlinearity are reduced or eliminated; therefore, the D/A converter nonlinearity and residue amplifier gain and offset errors limit the performance of multi-step converters.

The block diagram of a two-stage A/D converter with offset, gain and linearity errors is shown in Fig. 3.2. The sample-and-hold errors are not shown since the requirements of this block in the two-step architecture are not different from the requirements in other architectures. The non-ideal coarse A/D converter, D/A converter, residue amplifier and fine A/D converter are replaced by ideal elements in series with gain, offset and linearity errors.

Offset and gain errors are combined result of two physical effects: (i) noise, which includes charge injection noise in analog switches, thermal, shot and flicker noise, and noise coupled from digital circuitry (via crosstalk or substrate) and (ii) on-chip process parameter variation, e.g. device mismatch. The offset errors include offset caused by either component mismatch, self heating effects, comparator hysteresis or noise. The gain error group includes all the errors in the amplifying circuit, including technology variations and finite gain and offset of the operational amplifier. The D/A converter errors are caused by resistor ladder variations and noise, as well as by errors in the switch matrix, which are mainly due to charge injection in the CMOS transmission gates.

Thermal noise is caused by the random motion of electrons. Since electrons carry charge, the thermal motion of electrons results in a random current that increases with temperature. This noise current is present in all circuits and corrupts any signals passing through. In a multi-step analog to digital converter the sample and hold circuit is the most important source of noise. Since thermal noise is random from one sample to the next, it is not easily corrected by calibration, although it can be alleviated by using large components or by over-sampling. However, for a fixed input bandwidth specification, both of these remedies increase the power dissipation. Thus, a fundamental trade-off exists between thermal noise, speed, and power dissipation.

The offset error in the S/H as well as sub-range offset error shifts the entire correct subrange and is equivalent to the combination of two offsets in the preceding stage: one in the input branch and the other of opposite polarity in the coarse A/D converter branch. The offset of the residue amplifier can be moved towards the input of the converter. To obtain identical results, a second offset voltage of opposite polarity has to be inserted into the A/D sub-converter branch [54]. Thus, the offset at the gain stage input is effectively split into two parts. The offset error of the first sample-and-hold circuit at the input of the A/D converter only cause input-referred offset, but does not affect the linearity of the multi-step converter. The accuracy requirement for the coarse A/D converter is equal to their effective stage resolution if the over-range is applied. The performance of a coarse A/D converter is in turn limited primarily by the accuracy of the comparators and secondarily by the accuracy of the reference. Both of these can be modeled as an offset in the comparator threshold level. The threshold level of a comparator is normally determined directly by the reference voltage at its input. The different reference voltage levels needed are usually implemented with a resistor string, which rely on the relative matching of the resistors. When the over-range compensation is applied [55, 59] the matching requirements can easily be achieved without any high-precision components. The relaxed offset voltage specifications allow the adjustment of the threshold voltage level also with a built-in adjusting circuitry in some comparator topologies. If the comparator makes a wrong decision the wrong reference is subtracted from the input. The result is a residue that is out of range of the next stage of the multi-step ADC when amplified by residue amplifier. If the correction range is not exceeded by the combination of this coarse A/D converter offset and the nonlinearity present in the coarse A/D converter, the effect of the coarse A/D converter offset is eliminated by the digital correction, leaving the input-referred offset as the only effect of a sub-range offset.

The effect of coarse A/D converter offset is studied by examining plots of the ideal residue versus the input in Fig. 3.3a, residue versus input with coarse A/D converter offset in Fig. 3.3b and residue versus input with coarse A/D converter offset error when over-range is applied in Fig. 3.3c. In Fig. 3.3a, both, the coarse A/D converter and the D/A converter are assumed to be ideal. When the input is between the decision levels determined by the coarse A/D converter, the coarse ADC and DAC outputs are constant; therefore, the residue rises with the input. When the input crosses a decision level, the coarse A/D converter and D/A converter outputs increase by 1 LSB at a 2-bit level, so the residue decreases by digital value of conversion range of fine ADC. When the coarse A/D converter has some nonlinearity, with D/A converter still ideal, as shown in Fig. 3.3b for a similar example, two of the coarse ADC decision levels are shifted, one by $-1\frac{1}{2}$ LSB ($n + 1$ error) and the other by $+2$ LSB ($n + 2$ error). When the input crosses a shifted decision level the residue decreases by digital value of conversion range of fine A/D converter. If the conversion range of the second stage is increased to handle the larger residues, they can be encoded and the errors corrected (Fig. 3.3c) [55, 59].

The effect of an offset error in a comparator on a stage transfer function is shown in Fig. 3.3d. The dotted line represents an ideal transfer function, and the solid line shows a transfer function with an offset voltage in a comparator.

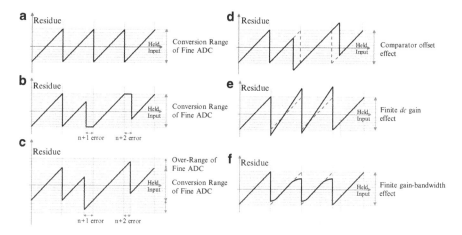

Fig. 3.3 (**a**) Ideal residue versus input. (**b**) Residue versus input with coarse A/D converter offset. (**c**) Residue versus input with coarse A/D converter offset errors when over-range is applied. (**d**) Effect of comparator offset. (**e**) Finite dc gain effect on transfer function, and (**f**) Finite gain-bandwidth effect on transfer function

Matching of the reference ladder resistors in a D/A converter is adequate for 12-bit level [118], while several techniques for reducing comparator offsets, such as auto-zeroing [48], averaging [50] or capacitive interpolation [62] have been developed. Alternatively, the offsets and input capacitance can be reduced by means of distributed pre-amplification combined with averaging [108, 109], and possibly also with interpolation [110]. An offset on the residue amplifier gives a *dc* shift of the fine A/D converter reference with respect to the coarse A/D converter and D/A converter range. The offset error in the residue amplifier does not affect linearity if calibration is used as shown in Section 3.5.2.

The gain requirements are straightforward. Any gain error of the input sample-and-hold stage due to finite gain of the operational amplifier only causes a change in the conversion range of the entire A/D converter, but does not affect linearity. In a multi-step analog to digital converter, errors in per stage gain (Fig. 3.3e) can be caused by a variety of sources such as non-infinite op-amp gains, resistor mismatches, etc. An error in per stage gain causes non-linearity in the transfer characteristic from input to output of the multi-step A/D converter. As in the case of offset, the gain error of the sample-and-hold circuit only alters the conversion range of the multi-step A/D converter but does not affect linearity. A gain error in the residue amplifier scales the total range of residue signal (signal as a result of the subtraction of the input signal and the D/A converter signal) and causes an error in the analog input to the next stage when applied to any nonzero residue, which will result in residue signal not fitting in the fine A/D converter range. If the error in the analog input to the next stage is more than one part in 2^r (where r is the resolution remaining after the residue amplifier), it will result in a conversion error that is not removed by digital correction. Moreover, if the residue amplifier gain is smaller than the ideal value, missing codes can occur [55]. Since all nonzero residues are affected by residue amplifier gain errors, the conversion-range boundary has no special significance from a gain-error standpoint. Gain errors cannot be corrected by digital error correction since shifting of the A/D sub-converter and D/A converter levels increases the absolute value of the residue for small input values [59]. Due to the larger residue, the non-linearity caused by a gain error increases, although, in practice, this disadvantage is small. Every linearity error in the A/D sub-converter increases the residue and thus worsens gain error effects. Therefore, the increase in non-linearity due to the increased residue only applies to ideal A/D converters while for real implementations the performance of the two error correcting algorithms is almost identical. Dual-residue signal processing [61] spreads the errors of the residue amplifiers over the whole fine range, which results in an improved linearity. An error in the range of the fine A/D converter results in an error similar to a residue amplifier gain error. The gain of the residue amplifier should therefore be lined with the fine A/D converter range.

Sample and Hold tracking nonlinearity refers to the distortion created by the sample and hold circuit as it tracks the input signal. This type of distortion tends to increase as the input frequency increases. In a switched capacitor sample and hold circuit the nonlinearity is primarily caused by the nonlinear resistance of the MOS switch and the nonlinear junction capacitance associated with the source and drain diffusions. More in depth analysis of S/H nonlinearity is given in Section 3.3.

Non-ideality in coarse A/D converter, D/A converter and fine A/D converter is modeled as an input-referred linearity error. The coarse A/D converter quantization determines the selection of the sub-ranges and the D/A converter setting. When an error is made in this quantization the wrong sub-range is selected, which will result in missing codes. By adding over-range to the fine A/D converter the accuracy requirements of the coarse A/D converter is reduced significantly. However, since the output of the fine A/D converter is not corrected, coarse A/D converter errors there do cause A/D converter nonlinearity but in amount that is diminished by the combined inter-stage gain before fine A/D converter. To build multi-step ADC's with a large tolerance to component non-idealities, redundancy is introduced by making the sum of the individual stage resolutions greater than the total resolution. When the redundancy is eliminated by a digital-correction algorithm, it can be used to eliminate the effects of coarse A/D converter nonlinearity and inter-stage offset on the overall linearity [118].

The references of the D/A converter and the subtraction of the input signal and the D/A converter output determine the achievable accuracy of the total A/D converter. Non-uniform spacing of the D/A converter reference levels also contributes to the nonlinearity of the analog to digital converter. The residue signal is incorrect exactly by the amount of the D/A converter nonlinearity. The linearity of the fine A/D converter determines the overall achievable linearity of the A/D converter. However, since the residue amplifier provides gain, the linearity requirements are reduced by this gain factor. An error in this quantization increases the differential-nonlinearity of the total A/D converter. Influence of the finite gain-bandwidth product (Fig. 3.3f) of each stage on total A/D converter resolution will be further argued in following sections.

3.3 Time-Interleaved Front-End Sample-and-Hold Circuit

The front-end sample-and-hold plays a crucial role in the performance of multi-step A/D converters. Without an S/H, the maximum allowable frequency of the input signal would be substantially lower than its theoretical Nyquist requirement $(f_{in} < f_S/2)$ and limited to $f_{in} < (3f_S)/(\pi 2^N)$, where N is resolution in bits and f_S is sampling frequency. However, linearity and dynamic range of the S/H directly affect those of the overall system, while the speed-precision trade-offs limit the conversion rate. This trade-off is further discussed in this section.

3.3.1 Time-Interleaved Architecture

The sampling rate of a system can be increased by a using time-interleaved technique [90–106], where a higher sampling rate is obtained by running the system in parallel as illustrated in Fig. 3.4a, although at different clock phases. The total three-time sample-and-hold circuit consists of three sample-and-hold units. Since the clock signals are one high out of three, each sample-and-hold unit is sampling during one clock period and holding during the other two enabling three actions (sampling, coarse and fine decision) to proceed at the same time on different analog samples as shown in Fig. 3.4b. In this design, only one clock cycle of the master clock for sampling is used implying that the total capacitance as seen from the input is reduced since now only one instead of three hold capacitors C_H is connected to the input simultaneously. If sample-and-hold unit one $S/H1$ is tracking the analog input signal n, $S/H3$ performs the coarse quantization of the previous sample $n - 1$

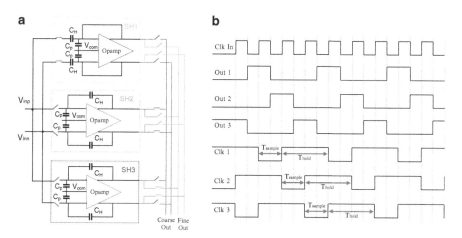

Fig. 3.4 (a) Total sample-and-hold circuit consisting of three time-interleaved sample-and-hold units at the moment when S/H1 is sampling and S/H2 and S/H3 are holding. (b) Timing diagram of interleaved principle, signals out 1–3 are internal signals of clock generator

while $S/H2$ completes the fine quantization of two samples earlier $n - 2$ and so on. In sample mode, the output switches are open and the load capacitance of the opamp is small. Hence, the opamp bandwidth is high and output can now follow the input. During operation, the input signal has to pass different paths from the input to the digital output. However, if offset, gain, bandwidth or time mismatch occur between the individual units, the path for the input signal changes each time it is switched from one unit to another, which gives rise to fixed-pattern noise at the output of the sample-and-hold circuit. Extensive discussion on the effects of these errors and possible remedies are given in the next section.

The basic architecture of each sample-and-hold unit circuit is shown in Fig. 2.7b (Section 2.3.1). During the sample phase the unity-gain feedback around the operational amplifier establishes virtual ground at the input nodes of the amplifier. The input signals charge the sampling capacitors C_H with the bottom plates of these capacitors tied directly to the input signals via the two sampling switches. Special attention has to be paid to these input switches since they are exposed to the full input voltage swing. Therefore, the on-resistance changes substantially for varying input signals. The variation can be reduced by using a transmission gate that is designed for symmetry such that the on-resistance is equal for maximum and minimum input voltage and/or to make the switch gate-source voltage constant as shown in Section 3.3.3.2. At this particular instance in time, the circuit tracks the input signal until the switches turn off and reset the output to a common-mode value. The voltage across the hold capacitor, ΔV_H, after turn-off of the sampling switch is $\Delta V_H = V_{in} + V_{ped}$, where the pedestal voltage can be split into a gain error and an offset, $V_{ped} = \varepsilon V_{in} + V_{off}$: a non-unity gain ε caused by a signal dependent charge injection and a constant dc offset voltage V_{off} caused by the signal independent charge injection and the clock feed-through. The effect of the input voltage and sampling time instant on the amplitude of the pedestal voltage is illustrated in Fig. 3.5 for both sample and hold phases.

At the end of the sample phase, the feedback switches (realized as transmission gates to ensure that the switches are conducting throughout the node voltage swing) open first, while the input sampling switches are driven by a delayed version of the switch-control clock. Such switching sequence reduces the signal-dependent charge injection [125]. The opening of the feedback switches inject charge trapped in the switch on the hold capacitors C_H. Since this effect appears as a common-mode signal to the amplifier, it is reduced by the common-mode rejection ratio of the amplifier. The clock feed-through is corrected as well to a first order by using a differential signal path. As long as the error is present on both signal inputs and is the same magnitude, it can be cancelled by taking the input differentially, although, this technique depends on the absolute matching of transistors. The second order effect due to the switch on-resistance R_{on} of the input switches does not appear as a common mode and is therefore not rejected by the op-amp since the positive input switch sees an equal but opposite input voltage as that seen by negative input switch. As a result, the switch on-resistance R_{on} of positive input switch is different from that of the negative input switch. This error is minimized by using small feedback switches and large bottom plate input switches.

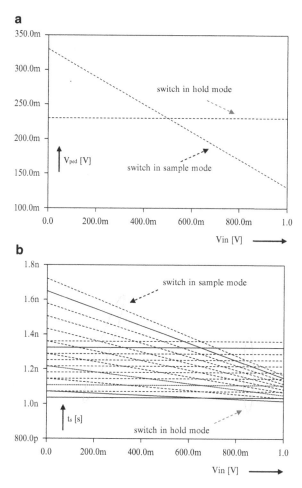

Fig. 3.5 (**a**) Pedestal voltage as a function of input voltage. (**b**) Sampling time t_s as a function of input voltage for different fall times of gate control signals

In hold mode capacitors are connected to the output of the operational amplifier and the sampled and held input voltage is restored at the amplifier output. Due to the fact that the outputs of the amplifier always return to the common-mode voltage during the sample phase (caused by the unity-gain feedback), the maximum output voltage step the operational amplifier has to perform is only half of the full scale voltage. In the sample mode the op-amp is switched as a follower, which means that at one side of the sample capacitor the *dc* value of the ladder is present but so is the offset of the op-amp. Due to conservation of charge, the sampled input appears at the op amp output. Because the summing nodes of the op amp are driven to the same potential, no differential signal charge is stored on the parasitic capacitance at the op amp input. When the sample-and-hold unit switches to hold mode, the sampling capacitor, which now contains the signal value and the offset of the op-amp, is connected across the op-amp. Now the offset is applied with reversed sign to the capacitor, thereby compensating for the sampled offset. At the output, the sampled value is seen without offset. This offset cancellation ability is one of the main advantages of the closed loop sampling architecture, e.g. when the op-amp open loop gain A_O is large, the residual offset error is proportional to $1/A_O\beta$. Although in a modern CMOS technology, the intrinsic gain of a transistor, defined as the product of the transconductance g_m and the output resistance r_o is typically low due to both channel length modulation and drain-induced barrier lowering effects, the open loop gain of the op-amp can be improved by a number of techniques including active cascode [138–141], positive feedback [142–144], and replica-amp gain enhancement [145]. An added advantage of the closed loop offset cancellation ability is that low frequency noise components such as the $1/f$ noise associated with the input CMOS differential pair is also largely cancelled, provided that the sampling frequency is much larger than the corner frequency of the $1/f$ noise.

The type and the dimensions of the switches not only determine the charge injection that can cause non-linearity and offset, but also the bandwidth and the stability of the sample-and-hold circuit are affected. In the hold mode, feedback switches are not involved in the more sensitive sampling process and thus their requirements are somewhat relaxed. The effective capacitance C_{eff}, which can be derived from the hold capacitance C_H, the input capacitance C_G of the operational amplifier and the parasitic capacitance C_{par} together with the switch on-resistance, forms a pole in addition to the dominant pole of the operational amplifier. The optimum speed is achieved if both poles are real and identical, e.g., $\omega_s = 4\omega_{GBW}$ [127], where ω_s denotes the pole caused by the switch and ω_{GBW} is the gain-bandwidth product of the amplifier. The necessary on-resistance of the feedback switches can thus be found as $r_{on} = 1/(4\omega_{GBW} C_{eff})$. In a differential implementation, during the sample mode, four switches are closed: two switches to provide the virtual ground nodes at the input of the operational amplifier and two switches at the input to connect the input signal to the hold capacitors. Due to the second resistance in the loop, the transfer function and thus the location of the poles differs from the results obtained for hold mode. In contrast to the situation in the hold mode, there is one more degree of freedom in this case: the on-resistances of the input switches and switches, which provide the virtual ground, can be set independently. A realistic choice from an implementation point of view is to choose both on-resistances to be equal. To obtain two identical poles, if it is assumed that both on-resistances are equal, will lead to $\omega_s \approx 3\omega_{GBW}$. Although this seems to be a relaxed requirement compared to the hold mode, the on-resistance of the switches in the sample mode must be much smaller due to the larger effective capacitance.

During the hold phase, the switched capacitor sample and hold amplifier behaves as a feedback circuit with a unit step applied to the input. The output response of this circuit is a step response that requires a finite amount of time to settle to a given accuracy. Failure to settle accurately, which is caused by an inadequate hold time, results in errors that degrade the performance of the A/D converter. If the settling is linear, the error is proportional to the input, and the result is a fixed gain error. If the settling is nonlinear, the effect is a signal dependent error [126]. Choosing the amplifier bandwidth too high increases the amplifier's wide-band noise and additionally demands unnecessarily low on-resistance of the switches and thus large transistor dimensions which in turn increases charge injection. The output voltage has to fulfill two settling requirements. By the time the converter makes its decision, the output signal of the sample-and-hold circuit must have settled to 5 and 8 bit accuracy, respectively. The settling to the full 12 bit accuracy can take place simultaneously with the settling of the D/A converter and is only necessary at the input of the residue amplifier. In the sample mode, the speed is dictated by the required acquisition time. For the sample-and-hold circuit this is the time the operational amplifier needs to re-establish virtual ground at its input nodes after switching from hold mode to sample mode. When the settling time is dictated by the op amp gain-bandwidth ω_1, the speed of the circuit decreases with the square of the supply voltage if the dynamic range is kept constant. Increasing the op amp transconductance g_m compensates for the bandwidth loss.

At the sampling phase, the sampling behavior depends on the unit gainbandwidth and phase margin of the circuit. But at the amplify phase, the settling behavior mostly depends on the closed-loop time constant of the amplifier only. Because the feedback factor is smaller than one, the phase margin is not an issue at the amplifying phase. If $(V_{GS}-V_T)$ is fixed to meet an output swing requirement and g_m (proportional to $C_{ox}(W/L)$ is fixed to meet a speed requirement, C_G will be proportional to L. Then, scaling of L with advanced CMOS technologies will reduce C_G. However, the noise mostly originate from the sampling capacitor given its weak dependence on C_G and further reduction on the capacitor value is not expected even with scaled technologies as a result. As shown in Fig. 3.6a, the optimum time constant remains constant regardless of the S/H circuit size (or I_D) because C_L scales together with C_H and parasitic capacitance C_p. So, if speed is the only constraint, the power dissipation can be reduced by scaling down the capacitor size until the speed is limited by other practical considerations, such as layout, matching, etc. The time constant is normalized to the τ_t ($=1/f_{t,\,intrinsic}$) of the device which is approximately (C_G/g_m). The parasitic-loading effects worsen as the conversion rate increases. In a high-speed converter, the parasitic capacitance can be comparable to the total sampling capacitance. At each point on the curve, there are two possible bias conditions, and only the curve with a positive slope will be considered here on since it is the low power solution.

Noise is an important constraint in high resolution front-end S/H circuits, thus, an appropriate hold capacitor size must be first chosen in order to reduce its kT/C noise level down below a given noise requirement. The choice of the hold capacitor value is a trade-off between noise requirements on the one hand and speed and power consumption on the other hand. The sampling action adds kT/C noise to the system which can only be reduced by increasing the hold capacitance C_H. A large capacitance, on the other hand, increases the load of the operational amplifier and thus decreases the unity-gain bandwidth for a given power consumption. The selected value of the hold capacitor therefore is always a compromise. Next, the op amp size and its bias current can be determined for a given speed requirement and minimum power dissipation using τ-versus-C_H curves as in Fig. 3.6a. Notice

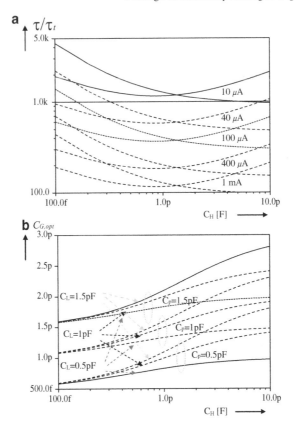

Fig. 3.6 (**a**) Closed loop normalized time constant versus hold capacitance C_H for different biasing conditions; case for $L = 0.18$ μm, $C_H = 3C_L$, $C_L = C_p$. (**b**) Optimum gate capacitance $C_{G,opt}$ versus hold capacitance C_H for different loading and parasitic conditions

that for low frequency operation (where τ/τ_t is large) the C_G that achieves the minimum power dissipation for given settling time and noise requirements usually does not correspond to the minimum time constant point. This is because fixing the C_H/C_G ratio of the circuit to the minimum time constant point requires larger C_G resulting in power increase and excessive bandwidth.

Near the speed limit of the given technology (where the ratio τ/τ_t is small), however, the difference in power between the minimum power point and the minimum time constant point becomes smaller as the stringent settling time requirement forces the C_H/C_G ratio to be at its optimum value to achieve the maximum bandwidth. The C_H/C_G (Fig. 3.6b) ratio can be expressed as $C_{G,opt} = \chi_2 C_H$, where χ_2 is a circuit-dependent proportionality factor. For a given speed requirement and signal swing, a two times reduction in noise voltage (in σ) requires a four times increase in the sampling capacitance value and the op amp size. Conversely, a two times increase in the supply voltage and the signal swing results in a four times smaller S/H circuit, and therefore, a two times smaller op amp power dissipation.

Fig. 3.7 (a) Maximum achievable SNR for different sampling capacitor values and resolutions. (**b**) SNR versus power dissipation

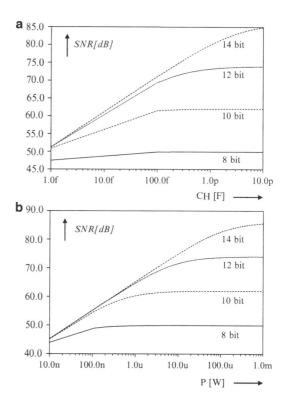

This means that the S/H circuit power *quadruples* for every additional bit resolved for a given speed requirement and supply voltage as illustrated in Fig. 3.7. Notice that for a small sampling capacitor values, thermal noise limits the SNR, while for a large sampling capacitor, the SNR is limited by the quantization noise and the curve flattens out.

3.3.2 Matching of Sample-and-Hold Units

The uniformity of the sample-and-hold units is affected by two principal sources of error (Fig. 3.8): inaccuracies in the fabrication process [130–132] and control signal feed-through. Although the sample-and-hold units are designed to be nominally identical, the degree of component matching is limited by imperfections in the fabrication process. Capacitor matching is determined by variations in the area of the capacitor plates and the thickness of the dielectric. The matching of transistor characteristics on a chip is determined by the matching of threshold voltages, mobilities, oxide and gate-overlap capacitances, and the widths and lengths of the transistor gates.

Fig. 3.8 Sources of errors in
the three times interleaved
sample-and-hold

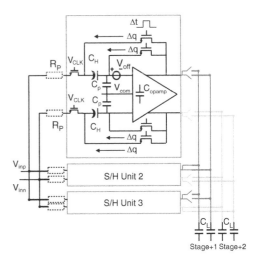

In general, effect of the non-uniformity of the sampling units coherent to a time-interleaved system can be divided into static (offset) and dynamic (time, gain, and bandwidth) mismatch errors [133–136]. Static mismatch causes fixed pattern noise in the sample-and-hold. It arises from op-amp offset mismatches and charge injection mismatches across the sample and hold units. For a *dc* input, each sample and hold unit may produce a different output and the period of this error signal is N/f_s as illustrated in Figs. 3.9a and 3.11a. Since the offset voltage under consideration is that of a CMOS source-coupled differential amplifier, it is a function of the threshold voltage V_T, aspect ratio W/L and g_m/I_{bias}. Of these three terms, the process determines the first two, while the third is proportional to the square root of the bias current. Under fast turn-off conditions the variation in channel charge dominates the charge injection mismatch q [131] and can be modeled as part of the mismatch of two geometric parameters, the channel width and length. The difference in channel area, $\delta(WL)$, of two mismatched transistors is $\delta(WL) = (W + \delta W)(L + \delta L)-WL$, where δW and δL are the variations in channel width and length. The voltage error due to non-uniform charge injection, δV_{ped}, is thus proportional to the channel width and length of the sampling transistor, which implies that the size of the error voltage and the input time constant are related. The switches with smaller W/L ratio yield smaller pedestal mismatches but limit the signal bandwidth of the sampling unit. Charge injection is deposited on the hold capacitor C_H, introducing an error in the voltage stored on the capacitor, which appears as a distortion at the output with period of N/f_s. Another potential source of non-uniformity of the pedestal voltages resulting from charge injection mismatch is the variation in sampling capacitance. However, typically, this error is small compared to the error from the mismatch of the sampling switches and can therefore be neglected.

Offset mismatch will distort the signal every time a sample is taken with different offset than the one used to take the previous sample. Offset mismatch can be measured

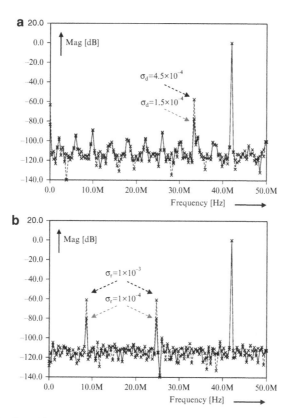

Fig. 3.9 $N = 3$, $f_s = 100$ MS/s: (**a**) Simulated offset mismatch at $\sigma_d = 1.5 \times 10^{-4}$ and $\sigma_r = 4.5 \times 10^{-4}$. (**b**) Simulated time mismatch at $\sigma_r = 1 \times 10^{-4}$ and $\sigma_r = 1 \times 10^{-3}$

by sending signals with the same number of periods to all interleaved channels and studying the difference in average values. Once the offset is measured, it can be compensated for and thus the offset errors corrected. As shown in *Appendix A*, assuming that d_n, $n = 0,1,2,...,N - 1$ are independent and identically distributed random variables with zero mean and standard deviation σ_d, power density of a spurious due to the offset, can be expressed as

$$P_d^{spur}(k) = \frac{1}{N^2} \left| \sum_{n=0}^{N-1} d_n \cdot e^{-jkn(2\pi/N)} \right|^2 = \frac{1}{N^2} \sigma_d^2 \tag{3.1}$$

The offset mismatch in time-interleaved systems can be cancelled with analog techniques or digitally. One limitation of the digital method [91, 97] from the system point of view is the fact that the static offset has to be measured before the calibration.

Alternatively, chopping of the channel input signals, controlled with a pseudo-random signal, together with de-chopping of the channel A/D converter outputs, can

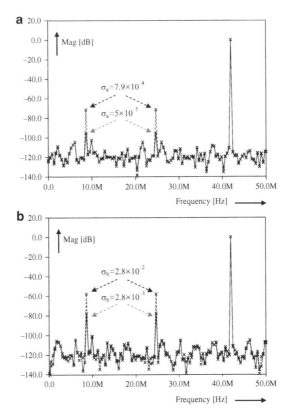

Fig. 3.10 $N = 3$, $f_s = 100$ MS/s: (**a**) Simulated gain mismatch at $\sigma_a = 5 \times 10^{-5}$ and $\sigma_a = 7.9 \times 10^{-4}$. (**b**) Simulated bandwidth mismatch at $\sigma_d = 2.8 \times 10^{-2}$ and $\sigma_r = 2.8 \times 10^{-3}$

be employed [98]. One analog offset cancellation techniques employ an auxiliary input stage to store a representation of the operational amplifier offset and the charge injection offset on capacitors [137]. In this implementation with design percussions, such as differential signal path, bottom plate sampling, small feedback switches, opamp high common-mode rejection ratio and by using the closed loop sampling architecture the resulting *dc* offset is mainly cancelled, such that consequent offset mismatch is sufficiently low for the required resolution.

The uniformity of the sampling unit transfer characteristics is also affected by feed-through of control signals via the substrate and parasitic inter-layer capacitances. In addition, perturbations (e.g. ringing) on common power, ground, signal, or signal-return lines, induced by control signals, can be the cause of S/H unit performance mismatch. Therefore, the layout of the circuit is carefully designed and extensively shielded so as to minimize coupling through parasitic inter-layer capacitances and through the substrate.

In the front-end sample and hold, where the clock is used to sample a continuous time signal, any deviation of the sampling moment from its ideal value results in an

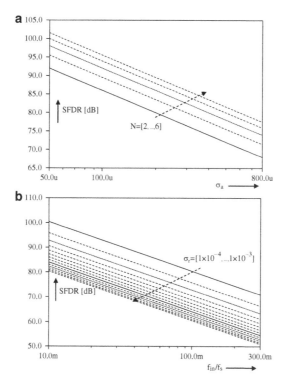

Fig. 3.11 SFDR versus (**a**) offset mismatch ranging from $\sigma_d = 1.5 \times 10^{-4}$ and $\sigma_d = 4.5 \times 10^{-4}$ for different N, (**b**) f_{in}/f_s ratio for different time mismatch ranging from $\sigma_r = 1 \times 10^{-3}$ and $\sigma_r = 1 \times 10^{-4}$

error voltage in the sampled signal equal to the signal change between these two moments. Thus, the sampling clock has to be regarded as a sensitive analog signal and treated accordingly. Beside the random variations in the phase (mainly due to device noise and random noise coupled from the power supply and substrate and the signals present on the chip or the circuit board that can couple to the clock), time-interleaved front-end S/H suffer also from frequency-dependent sample time deviations (systematic error, clock skew), which originates mainly due to device mismatch and asymmetric layout of the clock generator and clock lines. The error signal caused by clock skew is the largest at the zero-crossings with a period of N/f_s and is modulated by the input frequency f_{in} as illustrated in Figs. 3.9b and 3.11b. As shown in *Appendix A*, assuming that r_n, $n = 0,1,2,\ldots,N-1$ are independent and identically distributed random variables with zero mean and standard deviation σ_r, power density of a spurious due to the time mismatch can be expressed as

$$P_r^{spur}(k) = \frac{A^2}{2N^2}\left|\sum_{n=0}^{N-1}(1 - jr_n\omega_{in}T)\cdot e^{-jkn(2\pi/N)}\right|^2 = \frac{A^2\omega_{in}^2T^2}{2N^2}\sigma_r^2 \qquad (3.2)$$

where A is the amplitude of the input sinusoid. The clock skew between sampling clocks of distributed S/H circuits can be calibrated by measuring its value and controlling tunable delays of a DLL [147]. However, in general, calibration of the skew between S/H circuits has two significant drawbacks. First, measurement of the skew is complex and second, tuning of the delays requires high accuracy from the calibration hardware and algorithm. Alternatively, timing alignment within required accuracy can be obtained by using a master clock [126] to synchronize the different sampling instants and careful design by matching the channels clock and input signals lines [148]. In this design, similar approach is followed: besides extensive shielding of the clock lines, differences in line lengths, in passive interconnect parameters (such as line resistivity and capacitance and line dimensions and via/contact resistance) and in delays of any active buffers within the clock distribution network are kept to the minimum.

Similarly to clock skew, if the gains of each S/H unit are different, the basic error occurs with a period of N/fs but the magnitude of the error is modulated by the input frequency f_{in}. In both cases (clock skew and gain mismatch), noise spectrum peaks at $fs/N \pm f_{in}$ (Figs. 3.10c and 3.12c). Assuming that a_n, $n = 0,1,2,\ldots,N-1$ are independent and identically distributed random variables with zero mean and standard deviation σ_a, power density of a spurious due to the gain mismatch can be expressed as (Appendix A)

$$P_a^{spur}(k) = \frac{A^2}{2N^2} \left| \sum_{n=0}^{N-1} a_n \cdot e^{-jkn(2\pi/N)} \right|^2 = \frac{A^2}{2N^2} \sigma_a^2 \tag{3.3}$$

Gain and timing offset mismatch will have a similar effect on the signal. By pure observation, it is not possible to distinguish whether the signal is distorted by gain or timing errors. One possible method to analyze the underlying behavior is to vary the input frequency; e.g. gain errors will not vary with input frequency, however, timing errors will increase linearly as the sampling frequency increases. Gain and timing offset errors also introduce errors at every sampling point.

As opposed to the offset error, however, the magnitudes of these errors depend on the phase of the sampled signal. Gain errors, for example, are small if the signal is sampled near its zero-crossing and large if it is sampled near its peaks. The gain mismatch can be calibrated digitally by measuring the reference levels and storing them in a memory. The ideal output code can be recovered using these measured reference levels [63]. However, by dimensioning the open loop dc-gain of the operational amplifiers large enough, the effect of their mismatch is suppressed below the quantization noise level. With careful sizing and layout, capacitor matching sufficient for 12-bit level, depending on the process, is achieved.

The random variation in S/H unit's internal capacitance C_G, transconductance g_m, switch on-resistance R_{on}, hold capacitance C_H as well as input capacitance and kickback noise of the subsequent stages, as seen from one S/H unit, degrade output

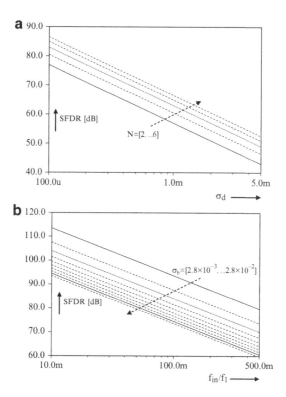

Fig. 3.12 SFDR versus (**a**) gain mismatch ranging from $\sigma_a = 5 \times 10^{-5}$ and $\sigma_a = 7.9 \times 10^{-4}$ for different N, (**b**) different bandwidth mismatch ranging from $\sigma_b = 2.8 \times 10^{-2}$ and $\sigma_b = 2.8 \times 10^{-3}$

settling behavior and circuit gain-bandwidth product differently. This error occurs with a period of N/f_s, but the magnitude of the error, which is frequency and amplitude dependant, is modulated by the input frequency f_{in} with noise spectrum spurious at $f_s/N \pm f_{in}$ (Figs. 3-10d and 3-12d). Assuming that b_n, $n = 0,1,2,\ldots,N-1$ are independent and identically distributed random variables with zero mean and standard deviation σ_b, power density of a spurious due to bandwidth mismatch in one-pole system can be expressed as (Appendix A)

$$P_b^{spur}(k) = \frac{A^2}{2N^2} \left| \sum_{n=0}^{N-1} b_n e^{-jkn(2\pi/N)} \right| = \frac{A^2 f_{in}^2}{2N^2 f_1^2} \sigma_b^2 \tag{3.4}$$

By severely increasing the bandwidth, the impact of the bandwidth mismatch at the signal frequency becomes lower. For this reason, the bandwidth of each sample-and-hold unit has been chosen larger than what is required when just looking at signal attenuation.

3.3.3 Circuit Design

3.3.3.1 Folded-Cascode with Gain-Boosting Auxiliary Amplifier

The maximum speed and, to a large extent, the power consumption of S/H is determined by the operational amplifier. In general, the amplifiers open loop dc gain limits the settling accuracy of the amplifier output, while the bandwidth and slew rate of the amplifier determine the maximal clock frequency. The operational amplifiers in S/H circuit have some unique requirements, the most important of which is the input impedance, which must be purely capacitive so as to guarantee the conservation of charge. Consequently, the operational amplifier input has to be either in the common source or the source follower configuration. Another characteristic feature of S/H circuit is the load at the amplifier output, which is typically purely capacitive and as a result, the amplifier output impedance can be high. The benefit of driving solely capacitive loads is that no output voltage buffers are required. In addition, if all the amplifier internal nodes have low impedance, and only the output node has high impedance, the speed of the amplifier can be maximized. Unfortunately, an output stage with very high output impedance cannot usually provide high signal swing.

The ultimate settling accuracy is limited by the finite amplifier dc gain. What the exact settling error is depends not only on the gain but also on the feedback factor in the circuit utilizing the amplifier. A very widely-used method to improve the dc gain is based on local negative feedback [56, 138, 165, 166]. In addition to this cascode regulation other techniques for increasing the dc gain have been proposed as well. Gain boosting with positive feedback has been investigated [142, 143]. In [167] dynamic biasing, where the op amp current is decreased toward the end of the settling phase, is used to increase the dc gain. It exploits the fact that current reduction lowers the transistor $g_{DS,}$ which increases the dc gain. By regulating the gate voltages of the cascode transistors [140] by adding an extra gain stage, the dc gain of the amplifier can be increased by several orders of magnitude.

Besides the amplifier bandwidth, the settling time is limited by the fact that the amplifier can supply only a finite current to the load capacitor. Consequently, the output cannot change faster than the slew rate. When designing an amplifier, the load capacitor is known and the required slew rate $SR = k \cdot V_{max}/T_S$ can be calculated from the largest voltage step V_{max} and the clock period T_S. A commonly used rule of thumb suggests that one third of the settling time should be reserved for slewing, resulting in k of 6. The required slewing current is $I_{SR} = (k \cdot V_{max} C_L)/T_S$. It is linearly dependent on the clock frequency, while the current needed to obtain the amplifier bandwidth has a quadratic dependence.

In order to use the amplifier in a closed loop configuration, its frequency response should be close to the single pole response. Thus, the phase margin at the unity gain frequency has an effect on the settling time as well and the choice between an n-channel device and p-channel device input pair is made on the basis of this requirement. The p-channel device input architecture (Fig. 3.13) offers lower gain-bandwidth product (g_{m1}/C_L) due to the low p-channel device

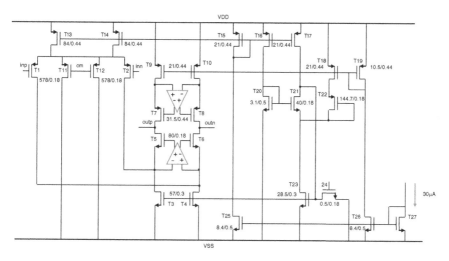

Fig. 3.13 Folded-cascode amplifier with gain-boosting auxiliary amplifier

transconductance, but the highest non-dominant pole (g_{m5}/C_1) associated with the n-channel device cascode devices at the folding node (source of transistor T_5). On the other hand, utilizing a n-channel device input pair gives higher gain-bandwidth product, but the non-dominant pole is lower. Feed-forward capacitors can be used to bypass the cascode transistors at high frequencies to improve the phase margin [156–159]. In principle, the technique produces a zero, which is used to cancel the pole associated with the cascode node. It is, however, not possible to place this zero exactly on top of the pole. Thus, there is a sufficiently closely spaced pole-zero pair, a doublet, which is known to introduce a slowly settling component in the step response [160]. It is possible to employ an n-channel device and a p-channel device input pair in parallel, which increases the slew rate by 1/3 (with the same total current consumption). However, at the same time that will increase the input capacitance and thermal noise and lower the non-dominant pole.

In fully differential amplifiers the common mode voltage level is not automatically determined. To set it to the wanted level, the input stage has been provided with common-mode circuit, consisting of two extra transistors T_{11} and T_{12} in a common-source connection, whose gates are connected to a desired common-mode reference V_{cm} at the input, and their drains connected to the ground [140]. If a positive dc feedback between the output and input exists, the V_{cm} level at the output is regulated so that the V_{cm} level at the input is equal to V_{cm} at the gates of T_{11} and T_{12}. If the V_{cm} level at the input rises, the common-mode current in T_1 and T_2 is lowered and taken away by T_{11} and T_{12} to ground. The result is that the common-mode current in T_5 and T_6 increases pulling the V_{cm} level at the output back down. The advantage of this solution is that the common-mode range at the output is not restricted by a regulation circuit and can approach a rail-to-rail behavior very closely.

The transistors of the output stage have three constrains: the sum of the saturation voltage for the transistors in one of the output branches must fit into the voltage headroom, resulting as the difference between the voltage supply and the desired output voltage swing. Second, the transconductance of the cascading transistors $T_{5,6}$ must be high enough, in order to boost the output resistance of the cascode, allowing a high enough dc gain. Finally the saturation voltage of the active loads $T_{3,4}$ and $T_{9,10}$ must be maximized, in order to reduce the extra noise contribution of the output stage. These considerations underline a tradeoff between fitting the saturation voltage into the voltage headroom and minimizing the noise contribution. A good compromise is to make the cascading transistors larger than the active loads: in such a way the transconductance of the cascading transistors is maximized, boosting the dc gain, while their saturation voltage is reduced, allowing for a larger saturation voltage for the active loads, without exceeding the voltage headroom.

The op amp unity gain frequency ω_1 (Fig. 3.14a) can be made larger increasing g_{m1} by means of making the transistors bigger; however this does not necessarily imply a faster op amp. The parasitic capacitance C_G is also increased, therefore feedback factor $\beta = C_H/(C_H + C_p + C_G)$ becomes smaller and dominant pole $\omega_p = \beta\omega_1$ is pushed towards lower frequencies. Therefore, a tradeoff between the increase of g_{m1}

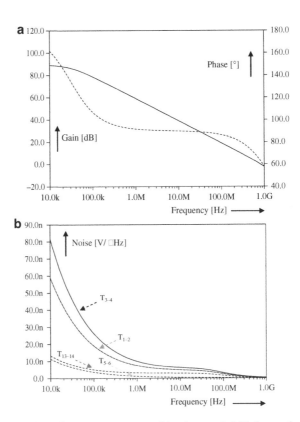

Fig. 3.14 (a) Open-loop frequency response of implemented folded-cascode amplifier with auxiliary gain-boosting amplifier. (b) Noise contribution of the individual transistors

and C_G exists. This suggests that an optimum size for the input pair exist, which maximizes the transconductance of the op amp by avoiding to make the input capacitance dominant on the feedback factor (Section 3.3.1).

The total noise contribution of all the devices in the amplifier is usually combined as a single voltage source at the amplifier input. Assuming the noise sources to be uncorrelated, the total noise is obtained as a root of the sum of the squares of the individual input-referred noise sources. The noise contribution of the devices in the amplifier's first stage is the most significant (Fig. 3.14b), and usually the noise of the other stages can be neglected, since it is attenuated by the preceding voltage gain. The input-referred noise of the amplifier input pair is reduced by increasing the transconductance, increasing the current, or increasing the aspect ratio of the devices. The effect of the last method, however, is partially canceled by the increase in the noise excess factor γ. When referred to the amplifier input, the noise voltages of the transistors used as current sources (or mirrors) in the first stages are multiplied by the transconductance of the device itself and divided by the transconductance of the input transistor, which again suggests that maximizing input pair transconductance minimizes noise. It can be further reduced by decreasing the transconductances of the current sources. Since the current is usually set by other requirements, the only possibility is to decrease the aspect ratio of the device. This leads to an increase in the gate overdrive voltage, which, as a positive side effect, also decreases γ. It should be noticed that the overdrive voltage is equal to $V_{DS,SAT}$. Consequently, obtaining low noise with low supply voltage is difficult, especially with single stage amplifiers, where the output signal swing does not permit large $V_{DS,SAT}$. Increasing L to avoid short channel effects is also possible, but with a constant aspect ratio it increases the parasitic capacitances, reducing the amplifier bandwidth. Cascode transistors do not make a significant contribution to noise, because their noise voltage is transformed into current through the high output impedance of the underlying current source.

3.3.3.2 Bootstrap Circuit

In standard CMOS technologies, the threshold voltage of MOS transistors does not scale with the supply voltage and it becomes a significant problem when MOS transistors are used as switches at low voltages. When the signal amplitudes are large, accuracy and signal bandwidth are limited by distortion, which originates from the fact that switch on-resistance are not constant but vary as functions of drain and source voltages. The on-resistance is expressed as $R_{on} = L/(\mu C_{ox}W(V_{GS}-V_T))$, if V_{DS} is small. In the equation two different signal-dependent terms can be identified. The first and dominant one is the gate-source voltage V_{GS}. The second is the threshold voltage V_T dependency on the source-bulk. Although large transistor switches can be used for the worst case V_T design, the switch parasitic capacitance can significantly overload the output of the circuit. Therefore, increasing $V_{GS}-V_T$ is desirable to implement low on-resistance switch without adding too much parasitic capacitance.

Several methods allow increase of this gate voltage drive. One method is to reduce V_T by including an extra low-threshold transistor in the process, although it

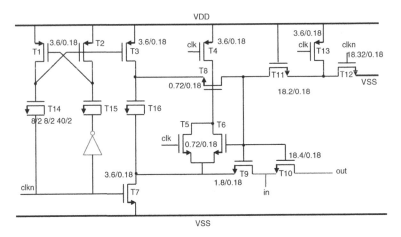

Fig. 3.15 Bootstrap circuit to boost the clock voltage

will add to process complexity. Another method is to increase V_{GS} by using one large supply created from chip supply to drive all switches on the chip, but potential problems including possible cross-talk to some sensitive nodes through the shared supply and difficulty in estimating the total charge drain to drive all switches renders this method absolvent.

Another viable solution to avoid major source of non-linearity is to make the switch gate-source voltage constant, by making the gate voltage track the source voltage with an offset ΔV_{off_in}, which is, at its maximum, equal to the supply voltage. This technique, which is implemented in this design, is called bootstrapping [67]. In this case, bootstrap circuit shown in Fig. 3.15 drives each switch that use the same clock to avoid the problem of crosstalk through the clock line. A ΔV_{off_in} can be generated with a switched capacitor, which is pre-charged in every clock cycle. During the clock phase when the transistor is non-conductive the switched capacitor is pre-charged to ΔV_{off_in}. To turn the switch on, the capacitor is switched between the input voltage and the transistor gate.

The capacitor values are chosen as small as possible for area considerations but large enough to sufficiently charge the load to the desired voltage levels. The device sizes are chosen to create sufficiently fast rise and fall times at the load. The load consists of the gate capacitance of the switching device T_{10} and any parasitic capacitance due to inter-connect between the bootstrap circuit and the switching device. Therefore, it is desirable in the layout to minimize the distance between the bootstrap circuit and the switch or to insert shielding protection. The output waveforms of the bootstrap circuit are shown in Fig. 3.16a.

When the switch T_{10} is on, its gate voltage V_G is greater than the analog input signal V_{in} by a fixed difference of $\Delta V_{off\_in} = V_{DD}$. Although the absolute voltage applied to the gate may exceed for a positive input signal, none of the terminal-to-terminal device voltages exceeds V_{DD}. A single-phase clock *clk* turns the switch T_{10} on and off. During the off phase, *clk* is low discharging the gate of the

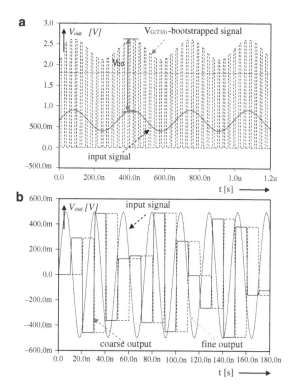

Fig. 3.16 (**a**) Bootstrap circuit output. (**b**) Sample-and-Hold output

switch to ground through devices T_{11} and T_{12}. At the same time, V_{DD} is applied by T_3 and T_7 across as capacitor connected transistor T_{16}, which act as the battery across the gate and source during the on phase. T_8 and T_9 isolate the switch from the capacitance while it is charging. When *clkn* goes high, T_6 pulls down the gate of T_8, allowing charge from the battery capacitor to flow onto gate of T_{10}. This turns on both T_9 and T_{10}. T_9 enables gate of T_{10} to track the input voltage applied at the source of T_{10} shifted by V_{DD}, keeping the gate-source voltage constant regardless of the input signal. For completeness of this section, the output waveforms of the total sample-and-hold circuit are shown in Fig. 3.16b.

3.4 Multi-Step A/D Converter Stage Design

3.4.1 Coarse Quantization

To maximize the settling time of the sub-D/A converter output, i.e. to achieve a high conversion speed, the coarse A/D converter should be able to provide its output to the sub-D/A converter as soon as possible after the S/H circuit samples the input and enters the hold mode. Therefore, almost without exception, the coarse A/D converter of multi-step A/D converter is of parallel, flash type [1–21] as it provides the highest throughput rate. As mentioned in Section 2.1, in the flash architecture the analog signal is simultaneously compared to every threshold voltage of the A/D converter by a bank of comparator circuits. The threshold levels are usually generated by resistively dividing one or more references into a series of equally spaced voltages, which are applied to one input of each comparator. All of the drawbacks of flash converters stem from the exponential dependence of comparator count on resolution. The large number of comparators required, $2^N - 1$, where N is the resolution of the A/D converter causes various detrimental effects: large die size which implies high cost, large device count leading to low yield, complicated clock and single distribution with significant capacitive loading, large input capacitance requiring high power dissipation in the S/H driving the coarse A/D converter and degrading dynamic linearity, high power supply noise due to large digital switching current and significant errors in threshold voltages caused by comparator input bias current flowing through the resistive reference ladder. These factors make implementation of flash converters above 8 bits very difficult, especially if low power dissipation is required.

The accuracy requirement for the coarse A/D converter is equal to their effective stage resolution if the over-range is applied. The performance of a low-resolution flash A/D converter is in turn limited primarily by the accuracy of the comparators and secondarily by the accuracy of the reference. To ease the problem of the large input capacitance, the difference between the analog input and each reference voltage can be quantized at the output of each preamplifier, which is possible because of preamplifier finite gain (non-zero linear input range). This indicates that interpolating between the outputs of preamplifiers can increase the equivalent resolution of a flash stage [168]. The interpolation technique of Fig. 3.17 substantially reduces the input capacitance (from $2^N - 1$ to $2^{N-Ninterpolation}$), power dissipation and area of flash converters, while preserving the one-step nature of the architecture, since all of the signals arrive at the input of the latches simultaneously and hence can be captured on one clock edge.

Interpolation can generally be viewed as analog to digital conversion in terms of zero-crossing points rather than direct amplitude quantization; in essence, interpolation adds zero crossings to the set of input/output characteristics of a flash stage. The interpolation principle makes use of the fact that the preamplifiers are non-ideal. Instead of switching from low to high instantaneously when the input voltage exceeds the reference voltage, they follow the input signal over a limited range in a more or less linear way. Since the information contained in the preamplifiers output signals is not affected as long as the position of the zero crossing of the output signal remains unchanged, the accuracy of the converter is not affected. By scaling the amplifiers in the analog preprocessing chain from front to back also the overall power consumption can be optimized under the given gain/bandwidth constraints.

The gain in the pre-amplifiers reduces power consumption of preamplifiers, which drives the comparators, as well as the required accuracy and thereby, the power

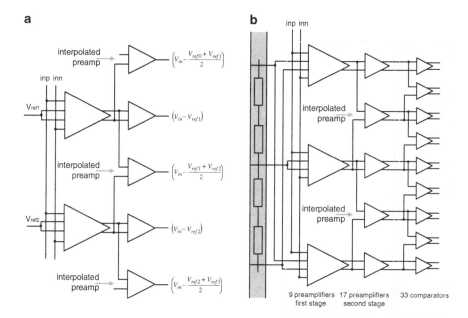

Fig. 3.17 (a) Interpolation principle. (b) Interpolation in coarse A/D converter implementation

consumption of the comparators. By their parallel nature, the power dissipated by a flash converter increases by the number of quantization levels desired. So to a first order, power increases exponentially as a power of two for every additional bit of resolution (Fig. 3.18). The relationship between power dissipation and sampling rate depends upon the process used and the method used to vary the circuit speed. The power of a circuit is a function the fixed extrinsic capacitance (any capacitance that does not scale with transistor width, internally or off-chip) C_{fixed} and maximum sample frequency $f_{S(max)}$, which is some factor (4–50) lower than the device f_T (Fig. 3.19). When f_S is much lower than $f_{S(max)}$, the power is directly proportional to f_S. This occurs when the intrinsic capacitance C_{scaled}, which scales with transistor W, is much smaller than C_{fixed}. When C_{scaled} is much larger than C_{fixed}, the asymptotic behavior takes over as f_S approaches $f_{S(max)}$. In this regime, power increase results in a diminishing increase in f_S and is inefficient from a power utilization standpoint. The transition or break-even point between these two regimes occurs when C_{scaled} equals C_{fixed}. In a power-speed efficiency sense, this is the optimum power point, P_{opt}. The minimal power consumption is limited by the matching quality of the technology for a given speed and accuracy.

The accuracy requirements of the 5 bits coarse A/D converter is limited to only 6 bits because the fine A/D converter is able to correct errors up to half a sub-range. In the implementation of the overall A/D converter the differential outputs from S/H are compared with a static reference ladder to obtain the coarse quantization. To be able to compare two signals with a static reference a comparator is necessary, which does not have the trip point at the zero crossing but at a certain reference

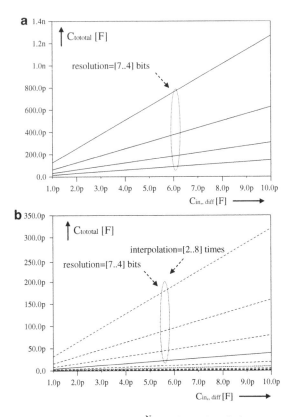

Fig. 3.18 (**a**) Left, total capacitance $C_{tot} = \chi_3(2^N - 1)*C_{in,diff}$ in a flash converter as a function of intrinsic preamplifier capacitance for different converter resolutions. (**b**) Right, total capacitance $C_{tot} = \chi_3(2^{N-Ninterpolating})*C_{in,diff}$ in a interpolating converter as a function of intrinsic preamplifier capacitance for different converter resolutions

voltage. Therefore pre-amplifiers are used with four inputs: two inputs are connected to the S/H analog outputs and two inputs are connected to the reference ladder. By using different references for each pre-amp all zero-crossings are generated. The maximum non-linearity errors occur at the differential amplifiers whose reference voltage is furthest from the input voltage and the non-linearity errors of the preamplifiers, which have the reference voltage closest to the input voltage and are responsible for A/D conversion, are minimized.

Behind this first pre-amplifier stage, interpolation is applied and by combination of the output signals from adjacent pre-amplifiers the additional zero-crossings are generated. Interpolation lends itself to implementation in sub-micron technologies since the preamplifiers do not need to have an accurate gain, high linearity or large output swings and can be made simple to maximize the speed (Fig. 3.20). Interpolation is applied again (Fig. 3.22a) and these amplifiers drive 33 comparators. Each of the comparators compares the difference of the S/H outputs voltages with the

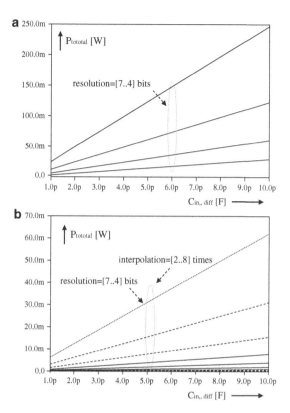

Fig. 3.19 (a) Left, total power consumed in a flash converter according $P_{tot} = C_{tot} \, V_{DD}^2 \, fs$ at $fs = 60MS/s$ as a function of intrinsic preamplifier capacitance for different converter resolutions. (b) Right, total power consumed in a interpolating converter according $P_{tot} = C_{tot} \, V_{DD}^2 \, fs$ at $fs = 60MS/s$ as a function of intrinsic preamplifier capacitance for different converter resolutions

difference of the reference voltages prior to digital code encoding. Since this conversion technique needs only static *dc* reference, it is therefore naturally free from the *RC* delay. The most important comparator specifications are offset, gain, speed, power consumption and immunity to noise and mismatch just as for op amps. As very small transistors are preferred to minimize power and area, comparators are inevitably sensitive for larger offsets. To lessen the impact of the offset voltage of comparators on linearity of A/D converter, several schemes, such as inserting a preamplifier [1], auto-zero scheme [2] or digital background calibration [18], have been developed. In the fully differential comparator shown in Fig. 3.21, the consequent low offset is achieved and therefore no offset compensation is required as result of four measures. The input signal is relatively large due to signal amplification in the preamplifier circuits. The large transresistance of the current-to-voltage conversion causes a large LSB voltage at the input of the voltage

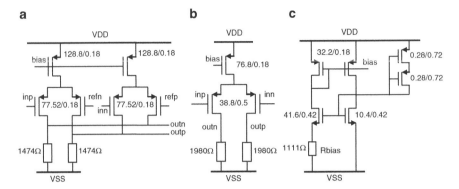

Fig. 3.20 Transistor level implementation of coarse A/D converter preamplifiers: (**a**) first stage, (**b**) second stage, and (**c**) bias circuit

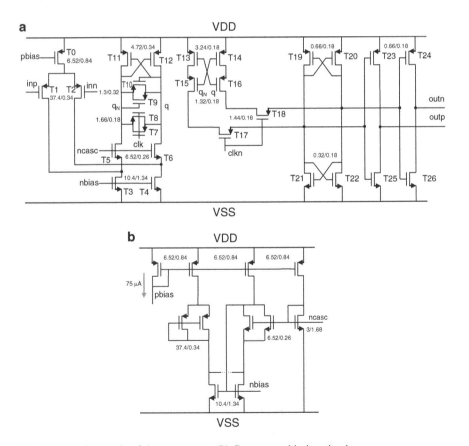

Fig. 3.21 (**a**) Schematic of the comparator. (**b**) Comparator biasing circuit

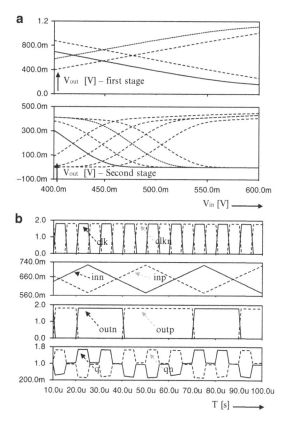

Fig. 3.22 (a) Left, some interpolated signals of the preamplifiers first and second stage. (b) Right, the comparator switching

decision circuit. The two-phase clocking scheme [179] reduces the number of devices that contribute to the offset, and finally the choice of appropriate g_m-ratios further reduces the input-referred offset. As a result of the absence of offset compensation, the clock frequency is high.

The current-input latch which results from combining the input current-to-voltage conversion and the subsequent voltage decision circuit is characterized by high speed and low offset. The speed is achieved by clamping the input voltage swing in the current-to-voltage conversion circuit and by optimizing the design of the regenerative circuit. When *clk* is high, the output of the decision circuit is dependent on the input signals. The regenerative action of the clock combined with the preamplifier causes an imbalance in the decision circuit, forcing the outputs into a state determined by $V_{(q)}$ and $V_{(qN)}$.

When *clk* is low, cross-coupled inverters are isolated from decision circuit; the comparator stops comparing and remembers the status of the inputs at the instant *clk* is switched low. Current flows through the closed resetting switches $T_{7,8}$, which forces the previous two logic state voltages to be equalized. After the input stage settles on its decision, a voltage proportional to the input voltage difference is established between nodes q and q_N in the end (Fig. 3.22b). This voltage will act as the initial imbalance for the following decision interval. The operation speed of the latch is determined by the regeneration time constant $\tau = C_{tot}/(g_{mn} + g_{mp})$ where g_{mn} is transconductance of $T_{5,6}$, g_{mp} of $T_{11,12}$ and C_{tot} total capacitance at q and q_N. The lengths of switches $T_{7,8}$ should be small as possible because it adds parasitic junction capacitances and can introduce undesirable gain (q/q_N). However, its width should be large enough to reset node q and q_N at the end of the reset phase to prevent hysteresis. As diode connected transistors $T_{9,10}$ limit the decision circuit result. During the decision-making they are switched off. At the output of the comparator, two inverters $T_{23,25}$ and $T_{24,26}$ are placed to buffer this information and produce a digital signal.

3.4.2 Fine Quantization

The coarse quantizer digitizes the input signal with low resolution, and applies the resultant digital value to a reconstruction D/A converter. The analog output of the D/A converter is then subtracted from the held output of S/H to form a residue signals. The four residue signals are connected to four fine buffers, which apply the signals to both the top and bottom of two moving ladders in the fine A/D converter. Although the full-parallel system (flash) implementation for the 8-bit fine A/D converter would provide a one-step operation, large number of comparators required, renders this architecture obsolete. On the other hand, folding and interpolation technique have been shown to be an effective mean of digitization of high bandwidth signals at intermediate resolution [169, 180].

3.4.2.1 Concept of Folding

A two-step A/D converter gains efficiency by partitioning an N-bit quantization into two lower-resolution quantizations (Fig. 3.23a). The object of a folding A/D converter is to form the residue signal with simple analog circuits thereby obviating the need for the coarse quantizer, D/A converter, and sub-tractor components. In such an implementation as shown in Fig. 3.23b), the low dynamic range residue signal generated by the analog folding circuit directly drives the fine quantizer. The folded signal is similar to the residue signal in a sub-ranging A/D converter, except for the fact that the residue signal is not generated from the output results of the coarse quantizer. Because of the periodic nature the residue signal, however, the digitized output from the fine quantizer is ambiguous and a coarse quantizer is still necessary to ascertain in which period of the folding circuits transfer characteristic the quantizer input signal lies. A high conversion rate is achieved due to the parallelism. The open-loop design of the folding amplifiers also speeds up the converter.

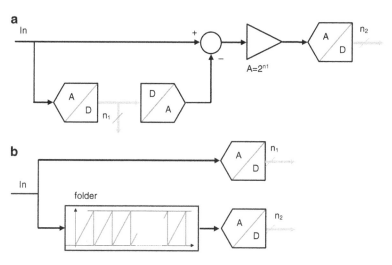

Fig. 3.23 (a) Top, two-step partitioning. (b) Bottom, folding topology

The ideal folding operation maps the input signal into successive linear segments, in a periodic fashion. The input-output characteristic of the analog folding circuit can be parameterized by the number of piece-wise linear segments, or folds, which it contains. The coarse quantizer is used essentially as a pointer to the segment of the fold where the input value lies, thus determining the most significant bits, while the fine part is used to resolve the voltage range at the output of the folder. The number of folds (F_F), e.g. the number of linear segments the input is folded into, determines the resolution of both the coarse and fine pointer thresholds. Its resolution is $n_1 = log_2 F_F$ for coarse pointer, while the fine pointer requires $2^N/F_F$ thresholds so its resolution is $n_2 = log_2(2^N/F_F)$.

Folding A/D converters based on the architecture of Fig. 3.23b would be possible if simple analog circuits could easily realize the piece-wise linear input-output characteristics indicated. The saw-tooth shaped transfer characteristic is not easy to implement due to its discontinuity. At these discontinued points the slew rate should be infinite, thus a triangular characteristic is preferred. Nevertheless, the perfectly linear triangular curve is difficult to generate and its corners tend to become rounded. In order to compensate for this problem, two folding circuits are used in parallel to generate two folding signals that have a carefully calculated mutual offset. This will guarantee that at least one signal is operating in a reasonably linear region for all inputs, as illustrated in Fig. 3.24a. Each folding signal only has to stay linear within the region which it is in use. It is also important to notice that when two signals are used, the range of V_{out} to be detected is further reduced by a factor of two. This demonstrates that the number of voltage levels that need to be distinguished per folding signal can be interchanged to the number of folding signals used. Yet, in practice, with a differential folding design, the folding signal is linear for a small section around the zero-crossing. As a result, the idea of using parallel folding signals is further expanded into a zero-crossing detection scheme.

As more folding signals are used in parallel as in Fig. 3.24b, the area each folding signal needs to stay linear is reduced. Eventually, if the folding signals have a mutual offset of one LSB with respect to the input voltage, instead of detecting voltage levels in the folding signals, the locations of the zero-crossings are used to determine the code transitions.

Implementation of such a zero-crossing detection scheme is more robust than a voltage level detection since it does not require extremely linear signals. As long as the comparator can determine the sign of a folding signal, the shape of the signal is of less importance. Figure 3.25 explains the difference between zero-crossing detection and voltage level detection with an example for three-bit resolution. As more folding signals are used, it starts to impose limitations because each folding signal requires a different folding amplifier. Thus, the associated hardware will increase, and the complexity and power consumption will be comparable to flash converter.

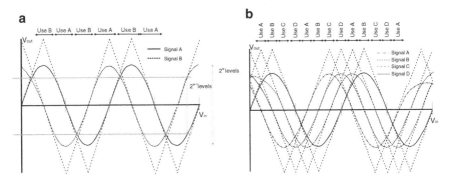

Fig. 3.24 (a) Use of a second folding signal for overcoming nonlinearity in corner region. (b) The use of four folding signals for further shrinking the linear region required for each folding signal

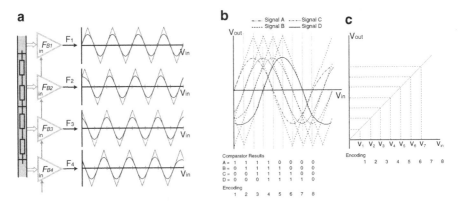

Fig. 3.25 (a) Left, parallel folding. (b) Right, a zero-crossing detection scheme where the locations of zero-crossings are used to determine the code transitions, which alleviate the problem of requiring a perfectly linear signals. (c) A voltage level detection scheme where the linear signal is compared to reference voltages (v_1 to v_7) to determine code transitions

3.4.2.2 Design of Folding and Interpolating A/D Converter

From a power perspective, the preamplifier stages in the folding block consume more power than a comparator at a given speed for two reasons; the output node capacitance of the folding block is about $n/2$ times larger due to the n preamp stages in a n times folder and the bandwidth of the folding block needs to be about $n/2$ times higher than a comparator from the $n/2$ times input signal frequency multiplication by the folding action. From the prior description of the folding converter, an $m+l$ bit converter with m most significant bits and l least significant bits, requires 2^m-1 MSB comparators, 2^m times folding per block and 2^l-1 folding blocks and LSB comparators. It is clear from Fig. 3.26 that the folding converter consumes more power and is slower than a regular flash converter. To improve the performance of the folding converter, interpolation is used to generate half or more

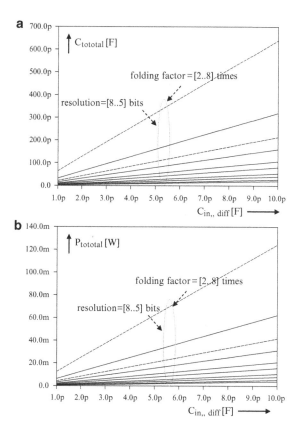

Fig. 3.26 (a) Left, total capacitance C_{tot} in a folding and interpolating converter as a function of intrinsic preamplifier capacitance for different converter resolutions. (b) Right, total power consumed in a folding and interpolating converter according $P_{tot} = C_{tot} V_{DD}^2$ fs at fs $= 60$MS/s as a function of intrinsic preamplifier capacitance for different converter resolutions and folding factors

of the folded waveforms. Since only the zero-crossing point is important, the interpolated waveform could replace the folding block waveform, allowing its removal. Although in this case only one block is eliminated, in the case of eight-times folding, three of seven blocks could be removed using interpolation, thus approaching a saving of half the power and area as the number of folds increases, yielding a more favorable power performance than for a flash converter. For this reason, most folding converter use the modified architecture for low-distortion coupled with interpolation to reduce the power and area.

As discussed in the previous section, a folding and interpolating A/D converter determines the digital output code transitions based on the zero-crossing locations. Therefore, in order to increase the A/D converters resolution, more zero-crossings have to be created across the input range. One way to achieve this is by increasing the number of folding amplifiers at the input. Yet, this approach will increase more parallelism, with associated power and speed penalty. The folding amplifiers also have high sensitivity to transistors matching which in turn can affect the A/D converter's performance. An alternative method to increase the number of zero-crossings is by increasing the interpolation factor. However, this approach has several drawbacks. First, more fine comparators are required at the output, which results in increased power consumption, larger area and possibly degraded speed performance. Second, as presented in [184], the signals generated by interpolation have an amplitude mismatch from the original folding signals. The folding signal obtained by interpolation only provides good approximation of the ideal folding signal around the zero-crossing. The interpolated signal itself has a different amplitude and slope. In particular, the mismatch in amplitude leads to displacement in zero-crossings when the interpolation factor is more than 2. Another possible solution to circumvent the problem is to increase the number of foldings in each folding signal before interpolation.

However, inserting too many differential pairs in one folding amplifier reduces the gain of the folding amplifier. This is because an increase in the number of foldings per folding signal reduces the voltage difference between two consecutive differential pairs in the folding amplifier. The g_m curves of the differential pairs starts overlapping; thus, deteriorating the gain of the folding amplifier. Given the limitations described in this section, the focus on improving the folding and interpolating architecture is placed on increasing the number of foldings per signal. Although, cascaded folding and interpolating architecture [109, 184], alleviates the problem of overlapping g_m curves, since folding is conducted at a lower frequency in each stage, its inherited speed-precision-area-time to market trade-off did not warrant the choice, bearing in mind fine A/D converter design specifications.

Architecture overview of the fine A/D converter is shown in Fig. 3.27. Since error correction is employed in the fine A/D converter as explained in Section 3.2, additional required ranges have to be created from the residues, which are the input signals for the fine converter. By making use of dual-residue technique as will be shown in next section, fine A/D converter will not have fixed reference voltages; zero-crossings for the fine conversion are generated by interpolation with two resistor ladders.

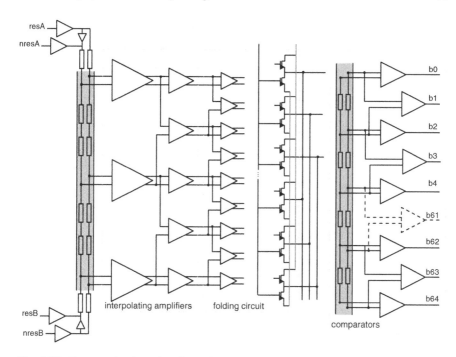

Fig. 3.27 Conceptual schematic of the folding and interpolating fine A/D converter, MSB generation not shown

To reduce power dissipation and capacitive loading, two stages of the interpolation preamplifiers (consisting of nine preamplifiers in the first and 17 preamplifiers in the second stage, similar to coarse quantizer interpolation preamplifiers described in previous section) are placed in front of folding preamplifiers. At the output of the first stage preamplifiers, additional, interpolated, zero-crossings are generated between the differential output voltages of two adjacent preamplifiers as illustrated in Fig. 3.28a. This second interpolation stage is controlled by *dir* signal, which change from high to low depending of the direction of the ladder current. The first folding stage consists of 33 folding amplifiers. These amplifiers generate a bell shaped signal (Fig. 3.28b), which are interpolated to produce additional folds as shown in Fig. 3.29.

Transistor level implementation of first and second preamplifier stage, folding amplifier differential pair and bias circuit is shown in Fig. 3.30. To illustrate the implemented folding principle into more detail, consider typical and implemented folder principle shown in Fig. 3.31. In a typical folder implementation, the inputs of the differential pairs are connected to the converter input voltage and reference voltages generated by a resistor ladder. Cross-connecting the output of every other differential pair produces the periodic transfer characteristic for this folder, depicted in Fig. 3.24. The current output of the folder is converted into a voltage through the load resistors. The transfer characteristic realized in this way resembles the ideal one shown in Fig. 3.23, but its peaks are flattened, which would result in

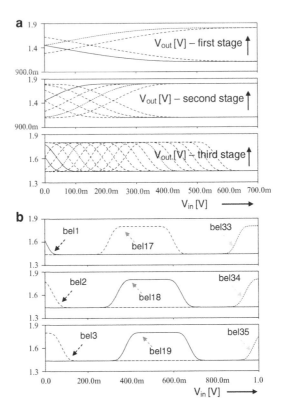

Fig. 3.28 (a) Left, interpolation of the first, second and third preamplifiers stages. (b) Right, forming of bell-shaped folding signals in folding encoder

excessive distortion in the response of the overall A/D converter if the full output range of the folder were to be resolved. This rounding problem is solved by offset parallel folding, where zero-crossing detection replaces precise level quantization. Parallel use of folding blocks increases the resolution of the A/D converter without increasing the folding rate of the system. All folding blocks have exactly the same analog behavior, although they use slightly shifted reference voltages. As the number of folds available increases, the number of levels that need to be resolved at each fold decreases proportionately for a given overall resolution. Although, employing straightforward parallel folding would lead to large number of folders to obtain sufficient resolution, which in itself would lead to large input capacitance, the interpolation technique renders this limitation to acceptable levels.

Ideally the folding amplifier should have piecewise linear transfer characteristics, but due to the mismatch between different differential pairs in the folding amplifier, the slopes of each linear segment may be different. However, mismatch between different circuit elements cause an unwanted drift in output zero-crossing and appears as if it is the offset of the differential pair. Errors such as the difference between current of the constant current sources, the deviation of the tail current

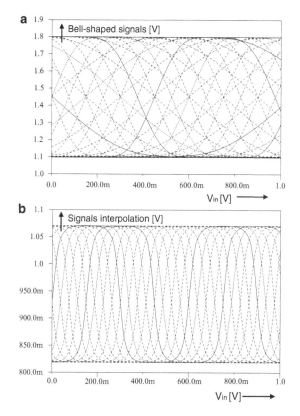

Fig. 3.29 (a) *Left*: Bell-shaped signals at the input of the folding encoder. (b) *Right*: signal interpolation

from their ideal values, offsets of differential pairs, the nonlinear distribution of the reference voltages, and the mismatch between output loads contribute to the equivalent input offset of each folding cell. Nevertheless, due to the gain in the preamplifiers, offset in the differential pairs of the folding amplifier have only limited impact. Transistors with smaller dimensions can be employed limiting total capacitance at the output nodes of the folding block, and hence ensuring a large bandwidth in the folding preprocessing.

Additionally, the gain in the preamplifiers enables large V_{DSAT} for folders, which in turn enables folding under low supply voltages. As a consequence, folder requires only differential pairs, which offers simple, fast and low supply voltage compatible solution (Fig. 3.30). The folder input–output characteristic implies that the bandwidth of the signal at the output of the folder will be larger than that at the input. Second, the slew rate of the folding amplifier and interpolators also should be large enough to prevent the signal skew. Both large bandwidth and slew rate demand large bias current. Band-limiting affects the output of the folder in three ways [191]: (i) it attenuates the waveform; (ii) it introduces group delay; and (iii) it alters the relative position of the zero-crossings. In an actual

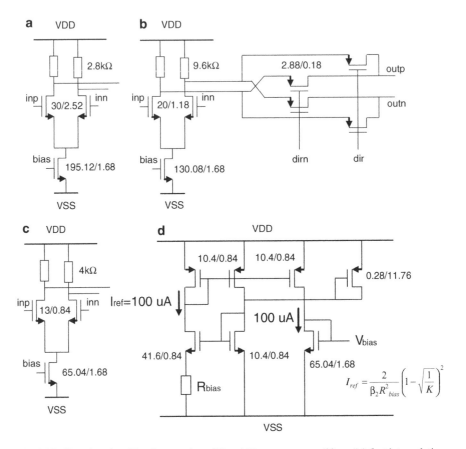

Fig. 3.30 Transistor level implementation of fine A/D converter amplifiers: (**a**) first interpolation stage, (**b**) second interpolation stage preamplifier, (**c**) folding amplifier differential pair, and (**d**) bias circuit

implementation, simple amplification is used to compensate for the attenuation without affecting the position of the zero-crossings, while the group delay is simply an overall delay in the folder output that does not influence the linearity of the converter. However, the remaining displacements of the zero-crossings, correspond to variations in the thresholds used to sample the input signal and, thus, introduce nonlinearity into the conversion process. The narrower the bandwidth of the filter at the output of the folder, the more severe the displacements, and thus the distortion of the signal, become. If the bandwidth of folding amplifier (or other analog preprocess blocks) is not large enough, the high frequency internal signal will cause the degradation of the dynamic performance. The capacitance loading the output nodes of the folder consists of the input capacitance of the following stage and the parasitic capacitances at the drains of the differential pair transistors. The latter can be quite significant [178–182]. The terminating resistors and the capacitances at the output of a folder form a bandwidth-limiting network that

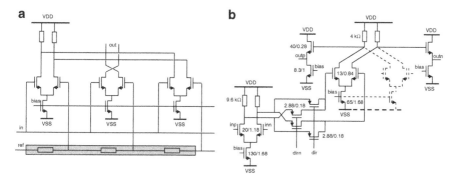

Fig. 3.31 (a) Typical and (b) Implemented folder with second stage preamplifier

filters the output waveform. To combat analog bandwidth limitations a transresistance amplifier is employed [178]. The input and output impedance of the transresistance stage are both $1/g_m$ and are made low, and thus, the analog bandwidth is increased by a factor $g_m R$. An additional advantage is its low output impedance, which facilitates driving the next stage.

The low-ohmic outputs *outp* and *outn* of the folding amplifier are connected to the resistive interpolation ladder. Resistive [169], current [182] or active [183] interpolation can be used to produce additional folds. In current mode interpolation the interpolating currents are split with cascode current mirrors into various fractions proportional to the current mirror size and are summed to form the fine current divisions [194]. However, the current offsets from the interpolating devices (i.e., the current mirrors) cause error in the interpolated zero crossing points. A large channel length is favorable because it yields a larger effective gate voltage, which makes the threshold offset less significant referred to the signal input. In comparison with voltage mode interpolation, current mode interpolation circuits' delay variation is much smaller. In active interpolation [183] differential pairs of folding amplifiers from Fig. 3.31 are replaced by four transistor structure implementing interpolating differential pair. The drain and source connection of two extra transistors are connected to drain and sources of the original differential pair. In the output current, which is a function of both input signals, a zero crossing is realized in between the zero crossings of the two input signals.

Resistive interpolation, on the other hand, offers simplicity of realization and is more power efficient solution in comparison with current interpolation. In the resistive interpolation, the linear portion of two interpolating folding waveforms must extend to the zero crossing point of each other to avoid error in the interpolated folding waveforms. The interpolatable region is half of the linear region of folding waveforms. To improve linearity in the resistive interpolation, several techniques based on resistive mesh network [184, 193] and averaging [108, 109] are available. However, a common problem in such architectures is the need for over-range comparators to maintain linearity at the edges of the conversion range. A circuit technique in [170] allows reduction of the number of over-

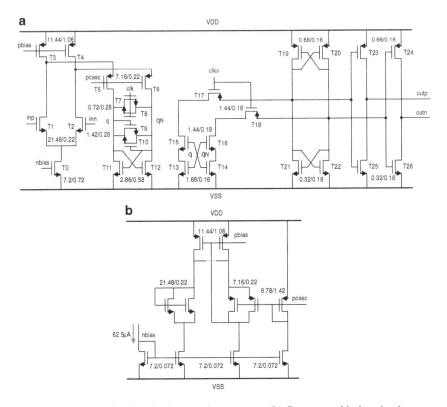

Fig. 3.32 (a) Schematic of the implemented comparator. (b) Comparator biasing circuit

range comparators, although the technique relies on matching the termination resistor with the output resistance of the over-range blocks. Nevertheless, a special case of interpolation is two-times interpolation, where nonlinearity does not affect the accuracy of the interpolated zero crossing point, so long the interpolating folding waveforms possess symmetry and are identical in shape and is therefore utilized in this design. Additionally, in comparison to higher interpolation factor, two-times interpolation reduce the delay difference, which is caused by different impedances looking back into the interpolator from the input terminals of each comparator and similarly, relaxes the bandwidth limitation. With two-times interpolation 32 complementary signals from source follower are converted to 64 zero crossings that could drive 64 differential comparators corresponding to the six least significant bits. To distinguish eight possible input voltages that correspond to the same folding signal output, over-range signals are applied to additional four comparators to generate two most significant bits. Comparators of both stages are similar to the one described in Section 3.2 (Fig. 3.32).

The low output impedance of the source follower circuit can drive the resistors directly. The *dc* voltage drop produced by the source follower bias current across each ladder defines half of the differential full-scale range; therefore, the quantizer input range is controlled by varying the ladder bias current. Each tap on both ladders must follow the full excursion of the input signal; however, the lower taps settle prohibitively slowly due to the distributed *RC* delay of the comparators loading the ladders. However, signal interpolation reduces this capacitive loading on the ladder.

3.5 Inter-Stage Design and Calibration

3.5.1 Sub-D/A Converter Design

After the coarse decision is completed and a thermometer code is generated, a combination of *exor* gates compares every two adjacent bits of the thermometer code. The results then turn on corresponding switch in the switching matrix, which selects certain sub-range *sub* from the static resistor reference ladder. Suppose *sub(n)* is selected as shown in Fig. 3.33a. By switching the proper switches four references closest to the differential input signal are selected. These references are combined together with the differential input signal to generate two differential pairs of residue signals according to: *resA = inp-refA; nresA = inn-nrefA; nresB = inp-refB; resB = inn-nrefB*. Both pairs of residue signals are connected to both the top and bottom of two resistor ladders as shown in previous section. The signal changes from *sub(n)* into *sub(n + 1)*. One pair of reference signals (*refB* en *nrefB*) remains connected by the switches, while the other pair (*refA* en *nrefA*) changes taps. As seen in Fig. 3.33b, insufficient settling in the coarse A/D converter or

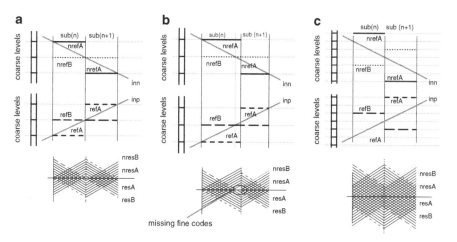

Fig. 3.33 (a) Switching of reference signals and generation of residue signals. (b) Signal switching with coarse ADC error, without over-range. (c) Signal switching with over-range and residue signals with over-range

mismatch in the coarse comparators is translated into a quantization error and appears as a shift in the location of the quantization step causing missing codes. As mentioned in Section 3.2, use of over-range and digital correction has emerged as an effective means of coping with these errors [59, 111, 112]. To generate this over-range, the fine A/D converter does not use the same references as employed by the coarse decision; it connects to reference taps shifted half a sub-range as shown in Fig. 3.33c.

The pairs of residue signals are generated in a similar way as described above. The residue signal has an in-range part as well as an under- and over-range part. When the coarse A/D converter makes an error, there are still fine comparators outside the sub-range to quantize this level. Reconstruction of the out-of-range levels is performed by adding up the properly delayed stage outputs with 1-bit overlap: the *MSB* of fine stage is added to the *LSB* of the coarse stage. The *LSB* of the coarse stage is not corrected, which suggests that the coarse stage must be a full flash without over-range. The excess in hardware caused by the over-range correction is very small. In a fine stage, the number of comparators is increased, but in the sub-D/A converter, only a few extra switches are required. However, as the comparator specifications are simultaneously relaxed significantly, the effects on area and power minimization are positive. For the reconstruction in the digital domain, only a small adder is required. It is noticeable, that the S/H operation, sub D/A conversion and subtraction of the residue have to fulfill an accuracy requirement equal to the total resolution of the multi-step A/D converter. Thus, the resolution of a multi-step A/D converter is limited by the accuracy of the sub-D/A converter, i.e. settling and component matching.

In a standard CMOS process a medium resolution sub-D/A conversion can be performed on several ways such as based on binary weighted current sources [196] or by using current division in R-2R ladder with MOS switches [197]. The resistor-ladder architecture is by far the simplest D/A converter implementation. Additionally, they are inherently monotonic as long as the switching elements are designed correctly and the DNL is relatively low compared to other architectures. The resistor-ladder D/A converter is essentially a string of identical resistors in series, with the top resistor tied to power supply and the bottom resistor tied to ground (Fig. 3.34). The nodes in between each resistor have different voltages depending on their proximity to power supply and by using thermometer or binary decoding on the digital signal one specific node can be selected as the correct analog voltage. The number of resistor elements determines the resolution of the resistor-ladder D/A converter; an n-bit D/A converter requires a ladder with 2^n resistors.

In high-speed operation, parasitic capacitors at a tap point create the voltage glitch. This transient has to settle out to the given accuracy within a given period of time, and the worst case settling occurs at the middle tap where the equivalent R-value is one half of the total resistance plus the switch on-resistance. This transient causes the signal dependent settling of the D/A converter and can translate into harmonic distortion. Therefore, the R-value is designed small enough so that the worst case transient settles within 12 bit accuracy. A limit on R-value is set, however, by mismatches of individual resistors, which determine the overall accuracy of the generated reference voltages. It can be shown [217] assuming that the resistor values

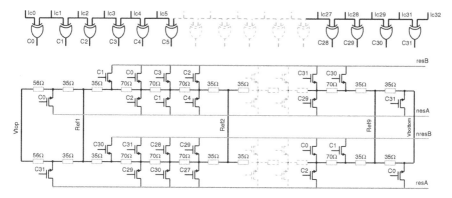

Fig. 3.34 Resistor string D/A converter

are normally distributed with mean R and standard deviation σ_R that the maximum mismatch σ_R/R allowed for 5-bit resolution and 12-bit precision is $\leq 0.1\%$.

3.5.2 Residue Amplifier

3.5.2.1 Offset Calibration

To build multi-step A/D converter with a large tolerance to component non-idealities (comparator offsets, etc.) redundancy is introduced by making the sum of the individual stage resolutions greater than the total resolution. The conversion accuracy thus solely relies on the precision of the residue signals; the conversion speed, on the other hand, is largely determined by the settling speed of the residue amplifier. When the redundancy is eliminated by a digital-correction algorithm, it can be used to eliminate the effects of inter-stage offset on the overall linearity [118]. However, a gain error in the residue amplifier is still critical [200]. The accumulative inter-stage gain relaxes the impact of circuit non-idealities, such as noise, nonlinearity, and offset, of later stages on the overall conversion accuracy. Consider a classical single-residue processing in multi-step A/D converter illustrated in Fig. 3.35a. A gain error in the residue amplifier scales the total range of residue signal and causes an error in the analog input to the next stage when applied to any nonzero residue, resulting in residue signal not fitting in the fine A/D converter range. If the error in the analog input to the fine ADC stage is more than one part in 2^r (where r is the resolution remaining after the residue amplifier gain error), it will result in a conversion error, which can lead to non-monotonicity or missing codes [200], that is not removed by digital correction. If the references for fine converter experience the same gain as the residue signal this conversion error can be reduced [64, 201, 202], although it will still limit the achievable accuracy to around 10 bits.

To overcome this limitation dual-residue processing [61] as illustrated in Fig. 3.35b have been employed. According to coarse quantization decision, a first

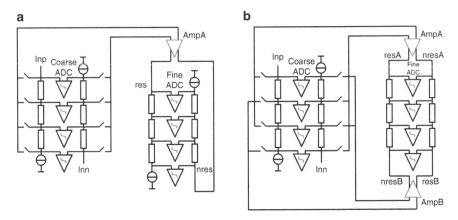

Fig. 3.35 (a) Single-residue and (b) dual-residue signal processing

and a second residue amplifier pass the difference between the analog signal and the closest and the second closest quantization level, respectively. By passing both residues to subsequent stages, information is propagated about the exact size of the quantization step, because the sum of the two residues is equal to the difference between the two quantization levels. The absolute gain of the two residue amplifiers is therefore not important, providing that both residue amplifiers match and have sufficient signal amplitude to overcome finite comparator resolution. By making use of dual-residue technique, fine A/D converter does not have fixed reference voltages. Conceptually, dual-residue system can be considered like two amplifiers generating zero-crossing at the edges of the sub-range. Baring that in mind, the additional required zero-crossing for the comparators can be generated by resistive interpolation. Thus, to summarize, the dominant error contributing components in the signal path before gain is applied are the S/H, the reference ladder, the switches in the switch unit and the offset on the residue amplifiers. Sufficient power is spent to meet the noise and linearity requirements of the S/H and matching of the reference ladder resistors is adequate for 12-bit level. Since switches in switch matrix are simple CMOS switches designed to have low enough on-resistance to provide sufficient bandwidth for 12 bit settling of the reference signals on the residue amplifiers, the offset on the residue amplifiers remains as the only accuracy-limiting component. Therefore, to maintain speed in the residue amplifiers while accomplishing 12 bit linearity requirement, offset calibration is applied.

A wide variety of calibration techniques to minimize or correct the steps causing discontinuities in the A/D converter's stage transfer functions has been proposed [29, 95, 96, 117, 203–211]. The mismatch and error attached to each step can either be averaged out, or their magnitude can be measured and corrected. Analog calibration methods include in this context the techniques in which adjusting or compensation of component values is performed with analog circuitry, while the calculation and storing of the correction coefficient can be digital. When no idle time exists in the system to update the coefficients, the calibration measurements

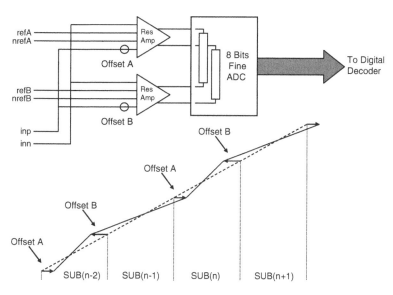

Fig. 3.36 Transfer curve of ADC with residue amplifier offset

must run in the background without interrupting the normal operation. Typically, the background calibration techniques [34, 95, 96, 208–211] are developed from the same algorithms as the foreground methods by adding hardware or software to perform the calibration coefficient measurements transparent to the normal operation. Mixed-signal chopping and calibration technique [34] applied to the residue amplifiers is one such technique, where the digital processing capability of the CMOS technology is used to extract the offset from the A/D converter output.

Because of the way of switching in dual-residue signal processing, offset on both amplifiers will have the effect on the transfer curve as shown in Fig. 3.36, which gives a deterministic repeated pattern in the INL curve of the A/D converter. Thus, the offsets can be measured in the digital domain by observing at the regular digital signal at the output of the ADC. Since offset is determined randomly, it can have any value. However, contributions to overall offset can be partitioned into two components: a common ($V_{common} = (V_{offsetA} + V_{offsetB})/2$), which is an equal value for both amplifiers, and a differential component ($V_{diff} = (V_{offsetA} - V_{offsetB})/2$), which is also equal but has the opposite sign. A common offset component reduce the over-range capability in the fine A/D converter, which results in a smaller allowable offset in the coarse A/D converter. The differential offset component, however, directly reduces the non-linearity of the two-step A/D converter. The total compensation loop is shown in Fig. 3.37. To extract both common and differential offset components and to distinguish the offset of the residue amplifiers from the *dc* value of the input signal of the A/D converter the chopping method [98] is applied. In general, dynamic offset-cancellation techniques can be subdivided into auto-zeroing and chopping techniques [212]. The fundamental difference between them

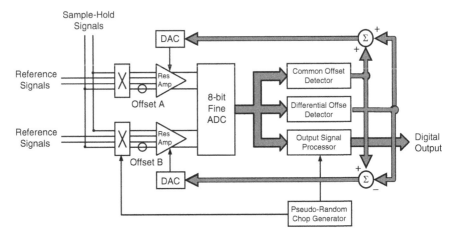

Fig. 3.37 Compensation loop

is the offset handling. While the autozero principle first measures the offset and subtracts it in a next phase, chopping modulates the offset to higher frequencies. Many derivatives of these two basic offset-cancellation techniques can be found, like correlated double-sampling [212], chopper-stabilization, self-calibrating opamps or the two or three-signal approach [213] as examples of autozeroing techniques and synchronous detection, chopper amplifier, chopper-stabilization and dynamic element matching (DEM) as illustration of chopping techniques. The main characteristic of the autozero technique and its derivatives is that the offset cancellation is done in two phases. A sampling phase when the offset is measured and sampled on and an amplification phase when the sampled offset is subtracted from the input signal and amplified. However, as a consequence of the sampling action, high-frequency components are folded back to the lower-frequencies and as a result the thermal noise floor is increased by the ratio of the unity-gain bandwidth of the amplifier and the autozeroing frequency [212]. In the chopper technique the input signal is modulated to the chopping frequency, amplified and modulated back to the lower frequency. The offset is modulated only once and appears at the chopping frequency and its odd harmonics. These frequency components are then removed by a low-pass filter. In contrast to the increased white noise component of autozero amplifiers, the low-frequency noise of chopper amplifiers is almost equal to the wideband thermal noise, assuming that the chopping frequency is higher than the $1/f$ noise corner frequency. The lower noise of the chopper technique is the main reason to use this technique for calibration of the residue amplifier.

After amplification by the residue amplifiers, the signal is quantized in the fine A/D converter. In the digital domain the data is chopped back to retrieve the original input signal, which is applied to a common and a differential offset extractor. Although, the common and differential offset component require different processing in the digital domain, both can be detected by integrating the signal at the output of the A/D converter, where the sign of both common and

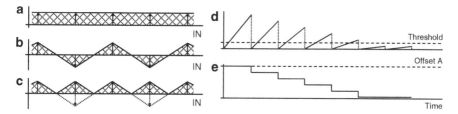

Fig. 3.38 (**a**) Common offset, (**b**) differential offset, (**c**) differential offset with sub-range dependent processing, (**d**) integrator signalm, and (**e**) residue amplifier error

differential offset are extracted [34]. After a certain integration period the content of the integrator is positive if the common offset component shown in Fig. 3.38a is positive and vice versa. A differential offset component detection require additional processing, since the areas of the error curve illustrated in Fig. 3.38b will compensate each other after integration. The necessary additional processing firstly rectifies the error areas, Fig. 3.38c, before integrating the signal. The resulting sign is used to change the values of the up and down counter. The output of the counter is fed to the offset compensation D/A converter located inside the residue amplifiers, which compensates in the analog domain for the offset present in the residue amplifier. Figure 3.38d and e illustrate how the offset is removed after a few integration cycles. The calibration does not change the values on the compensation D/A converter when the content of the integrator is smaller than a threshold value.

The different effects of the common and the differential offset extractors can be seen at the adders. A change caused by the common offset extractor gives both D/A converter values a step of the same sign, while a change caused by the differential offset extractor gives a step of the opposite sign. These D/A converters close the compensation loop, thereby removing the offset in the residue amplifiers.

3.5.2.2 Circuit Design

The residue amplifier two differential pairs act as sub-tractors (Fig. 3.39) [211], since the circuit performs the subtraction of the input signal with the selected references from the reference ladder. The differences between the analog inputs and their respective references are subtracted in the current domain. The circuit gain of the residue amplifier reduces the accuracy (noise and matching) requirements of all circuits after this amplifier with the gain factor. The residue amplifiers provide a gain of eight before the residue signals are applied to the 8 bit fine A/D converter. To reduce the dependence of the circuit gain upon the input level, which usually translates into making the gain relatively independent of the biasing currents, source degeneration of the differential input pair is applied through linear poly-resistor.

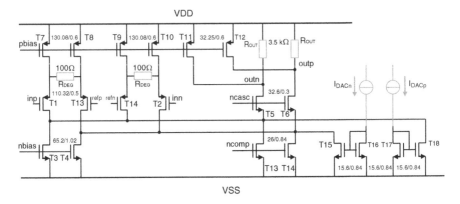

Fig. 3.39 Residue amplifier and part of offset compensation circuit

In this case, the circuit gain is determined by the ratio of the degeneration resistors R_{DEG} and the output resistors R_{OUT}, and thus, the gain matching between both residue amplifiers will be predominantly determined by poly-resistor matching. The effect of the finite output impedances will lead to modulation of the total bias current and therefore affect the gain accuracy and linearity. However, since the matching of poly-resistors is sufficient, it will not limit the performance of the complete A/D converter. Additionally, as shown in previous section, the absolute gain requirement of the residue amplifiers is not essential as long as the gain values of both amplifiers have the same gain value. A current source (transistors $T_{11,12}$) is added to the two output resistors to generate a convenient common-mode signal. The accuracy of this common-mode output signal level will be determined by the process spread of the output resistors and the accuracy of the biasing current.

The biasing and common-mode feedback circuit is shown in Fig. 3.40. Note that R_1 and R_2 have to be sufficiently large to ensure proper conditions for source followers T_{25} and T_{29} when a large differential swing appears at the output. The input-referred offset of the residue amplifier, which determines the accuracy of the total A/D converter, arises from mismatch between the input transistors, mismatch between current sources and poly-resistor matching. The digital offset extraction block determines the digital code, which is a measure for the offset of the amplifier, applied to the current-steering compensation D/A converter (Fig. 3.41). The reference source is simply replicated in each branch of the D/A converter and each branch current is switched on or off based on the input code. For the binary version, the reference current is multiplied by a power of two, creating larger currents to represent higher magnitude digital signals. The compensation D/A converter drive a current via folding nodes and further via a low ohmic cascode node to the output resistors of the amplifier to remove the offset. A nine-bit resolution has been found to be sufficient resolution of the offset compensation. Since this D/A converter only has to provide a current to compensate for the offset error its linearity is not an issue.

From the residue signals, which are the input signals for the fine converter, zero-crossings for the fine conversion are generated by interpolation with two resistor

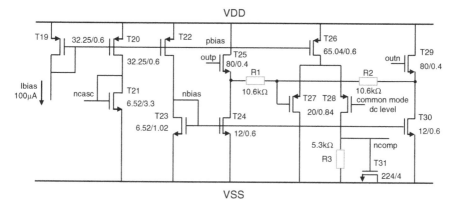

Fig. 3.40 Residue amplifier biasing and common-mode feedback circuit using source followers

ladders as shown in Section 3.4.2.3. The residue signals are set on these resistor ladders with buffers as illustrated in Fig. 3.27.

Transistor level implementation of fine buffers is shown in Fig. 3.42. It consists of first folded cascode stage (T_0–T_7) and source follower output stage (T_9–T_{10}). The source follower is implemented in p-channel MOS in order to eliminate the non-linearity due to the body effect. Sufficient phase margin can be obtained by making the non-dominant pole $p_2 \approx g_{m5}/C_{GS5}$ as large as possible. This can be achieved by increasing the transconductance of the cascode device T_5 by either increasing the width of the device or the current flowing through it. Increasing the width also increases C_{gs5}. Increasing the biasing current increases the power consumption. Furthermore, care have to be taken as to large increase in the bias current can cause the transistor acting as current source T_3 to move out of the strong inversion region of operation. The dominant pole is decreased by inserting the C_{T8} to improve stability of the circuit.

The direction of the current flowing in the fine ladders is determined by the selected sub-range and by the chop state. Two buffers therefore have to sink the ladder current, and two buffers have to source the ladder current, depending on the sub-range and chop state. The current is reversed in a sub-range transition or a chop state change, which can cause a jump in the output signal of the buffers due to the large change in current. To restrict this effect, two current sources (Figs. 3.27 and 3.43) have been added at the top and bottom of both ladders to sink or source the ladder current. Now the buffers only have to deliver the error current, which is much less than the *dc* current through the ladders.

3.6   Experimental Results

The prototype of the two-step/multi-step A/D converter illustrated in Fig. 3.44 and described in this chapter was fabricated in a five-metal layers 0.18-μm CMOS process. The chip has three independent power supplies and grounds: two for

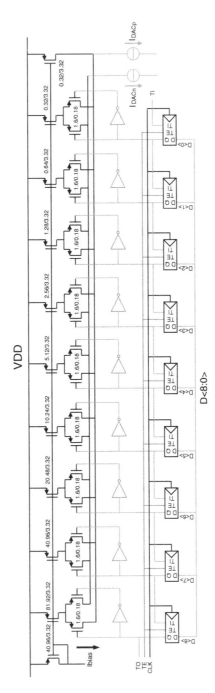

Fig. 3.41 Offset compensation current-steering D/A converter

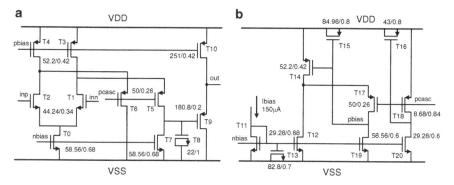

Fig. 3.42 (a) Fine buffer schematic. (b) Fine buffer biasing circuit

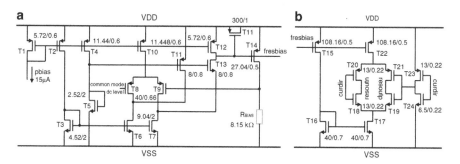

Fig. 3.43 (a) Biasing circuit of the additional current source. (b) Additional current source

analog and digital blocks and one for the output drivers. The supply voltages are provided by *HP3631A* and the reference current sources are generated by *Keithley 224*. Potentiometers are used to adjust the reference voltages and the common mode voltage. The clock reference for signal generation, A/D converter clock generation and data capturing are generated by the *Agilent 81134A*. A single frequency, sinusoidal input signal is generated by an arbitrary waveform generator (*Tektronix AWG2021*) and applied at the first to a narrow, band-pass filter to remove any harmonic distortion and extraneous noise, and then to the test board. The signal is connected via 50 Ω coaxial cables to minimize external interference. On the test circuit board, the single-ended signal is converted to a balanced, differential signal using a transformer. The common-mode voltage of the test signal going into the A/D converter is set through matching resistors connected to a voltage reference. The digital output of the A/D converter is buffered with an output buffer to the drive large parasitic capacitance of the lines on the board and probes from the logic analyzer. The digital outputs are captured by the logic analyzer (*Agilent 1682AD*). A clock signal is also provided to the logic analyzer to synchronize with the A/D converter. All the equipment is set by a LabView program, which also does the signal analysis.

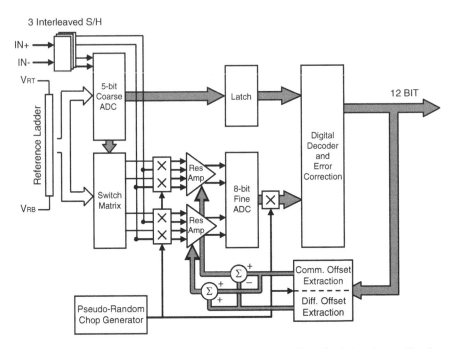

Fig. 3.44 Prototype two-step/multi-step A/D converter with mixed-chopping calibration algorithm

The A/D converter itself must deal with issues such: routing of critical paths, power and ground isolation, noise coupling from the digital sections to the analog sections, shielding of clock lines, etc. In order to reduce wiring capacitance in the input path, front-end sample and hold input is routed as symmetrical and short as possible. Power and ground lines with maximal allowed width are overall routed to reduce the voltage drop. Since the analog sections are susceptible to noise coupling from the digital sections, they are constantly kept separated. Critical clock lines have been provided with shielding. In mixed-signal chips, it is often unclear what the best strategy is for minimizing the impact of noise coupling from the digital circuitry to the sensitive analog circuitry via the common substrate. The most effective way to reduce substrate noise is to create a low-impedance path from the $p+$ substrate to ground (or the lowest potential in the chip). Typically, however, the backside of the die is oxidized by exposure to air, which increases its resistance. Thus even if the die is conductively attached to a package with a grounded cavity, the resistance to the substrate is high. If cost permits, the backside of the die can be back-lapped (ground down), gold-coated, and conductively attached to package. In this prototype, the following approach was taken: separate supply rails were used for the digital and analog signals, which were named V_{DDD}, V_{SSD}, V_{DDA}, and V_{SSA} respectively. Since an n-well process was used, the digital and analog $p-$ channel transistors were naturally isolated by separate wells. The n-channel transistors, however, interact via the common, high-resistance $p+$ substrate. The $p+$ substrate has the advantage that it makes it difficult to create latch-up, which is critical for digital circuits and creating a high resistance path for the coupling of undesired signals. Because noise travels almost exclusively in the $p+$ region, traditional isolation using grounded n-well guard rings to collect noise is not effective [215]. For the analog n-channel transistors, it is important that the source-to-body voltage is constant. Otherwise, if these voltages move relative to each other, the drain current is modulated through the body effect. Therefore, it is important to locally have a low-resistance path from body to source. In the layout, a $p+$ substrate ring was placed around each n-channel analog transistor.

This ring was then contacted to V_{SSA}, which is the same potential as the source for common-source devices. This helps keep the potential of the source and the body the same. For cascode n-channel devices, this arrangement helps reduce fluctuations on the body terminal, but it cannot guarantee that the source and body will move together (since the source is not at ground potential). For cascode devices, however, the relationship between drain current and V_{SB} is much weaker due to source degeneration. Enough of these V_{SSA} substrate contacts are used to make the source-to-body path low resistance. V_{DDA} and V_{DDD} were generated by separate but equal voltage regulators to allow the supply currents to be measured independently. All analog paths were differential to increase the rejection of common mode noise, such as substrate noise and supply voltage fluctuations. To reduce switching noise on power supply lines on-chip decoupling capacitors are employed. Decoupling capacitors act like local power supplies during the switching instant. Thus, most of the current can be drawn from decoupling capacitors instead of directly from the power supplies, which would signal a reduction in the current from an off-chip supply flowing through the parasitic impedance of a package and reduction of the switching noise associated with the current.

A chip microphotograph of the A/D converter is shown in Fig. 3.45. The sample-and-hold circuit with hold capacitance is clearly visible in left part, coarse part with switching matrix is situated right of S/H circuit, while fine part, residue amplifiers, choppers, biasing, calibration and the rest of the circuits are placed further up right. Digital parts are situated on the top right side of layout. Routing of the digital signals such as clocks and data outputs is placed between digital parts. Digital outputs leave the chip from the lower and the right sides of the chip. Using this total arrangement there is a minimum of analog and digital signal lines cross-ings. The sample and hold input is the most critical node of the realized integrated circuit. Therefore, a great deal of care was taken to shield this node from sources of interference. The total sample-and-hold consists of three identical interleaved sample-and-hold units. S/H units, input signals, critical clock lines and output signal lines have been all provided with shielding and routed as short and symmet-rical as possible. In the coarse A/D converter the nine preamplifiers of the first stage must align with 17 preamplifiers of the second stage and 33 coarse comparators implying that a high aspect ratio is necessary for the preamplifier and comparator layout. The switch unit is placed near the reference ladder to reduce the resistor-ladder D/A converter settling time. Although the resistor ladder is placed at some distance from the comparators, this is bearable, since the comparators should have only 6 bits precision. Selected reference signals from the switch unit are routed as

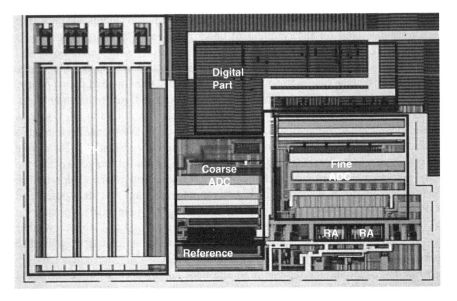

Fig. 3.45 Chip micrograph

short as possible, since the delay due to the wiring capacitance increases the residue amplifier settling time. The delay due to the wiring capacitance causes residue amplifier to momentarily develop its output in the wrong direction until the correct selection switch closes. After the correct switch is selected, the output starts to converge in the correct direction.

Therefore, the wiring capacitance increases the residue amplifier settling time due to the wires. If reference ladder is placed nearer, the reference signals for the comparators could be easily corrupted due to the coupling of the large digital signals traveling nearby. The three stages of the preamplifiers and folding encoder are laid out in a linear array, similar to the coarse A/D converter preamplifiers, and connected to the comparator array by abutment. The comparator array must align with preamplifiers, implying that high aspect ratio is necessary for the comparator layout. Locating these arrays close to each other greatly reduces wiring capacitance providing maximum speed. To keep the comparator array small and the wires short, data is driven out of the array immediately after amplification to full swing. Clocks are distributed from right to left, to partially cancel the sample time variation with reference level that increases from left to right. The comparators with complementary clocks are interleaved, sharing the same input, reference and supply wires so that charge kickback and supply noise are cancelled to first order. The die area is 0.9 by 0.75 mm excluding the bond pads. The complete A/D converter core draws 53 mA from a 1.8 V voltage supply, excluding output buffers, resulting in a less than 100 mW power consumption, of which 6.6 mW is drawn by the digital core for sample frequencies up to 80 MS/s. Measurements across 25 samples show ±0.2 ENOB variations. A code density test [35] was conducted to obtain static linearity of the A/D converter.

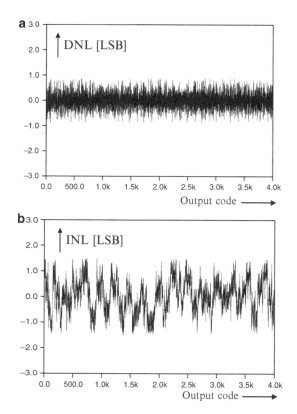

Fig. 3.46 (a) Measured DNL, (b) measured INL

The measured DNL and INL are shown in Fig. 3.46. From the figures, the maximum value of DNL and INL are 0.9 LSB and 1.5 LSB at 12 bit level, respectively.

The dynamic performance of the A/D converter is measured by analyzing a Fast Fourier Transform (FFT) of the digital output codes for a single input signal. Figure 3.47a illustrates the spectrum of the output codes of the A/D converter with an input frequency at 21 MHz sampled at 60 MHz. The largest spike, other than the fundamental input signal, is the spurious harmonic which appears at $fs/3 \pm f_{in}$ and is about 78 dB below the fundamental signal. The SNR, SFDR and THD as a function of input frequency are shown in Fig. 3.47b. All measurements were performed with a 1.8 V supply at room temperature (25°C). The measured results are summarized in Table 3.2. The degradation with a higher input signal is mainly due to the parasitic capacitance, clock non-idealities and substrate switching noise. Parasitic capacitance decreases the feedback factor resulting in an increased settling time constant. Clock skew, which is the difference between the real arrival time of a clock edge and its ideal arrival time of a clock edge, can also be caused by parasitic capacitance of a clock interconnection wire. The non-idealities of clock such as clock jitter, non-overlapping period time, finite rising and fall time, unsymmetrical duty cycle are another reason for this degradation.

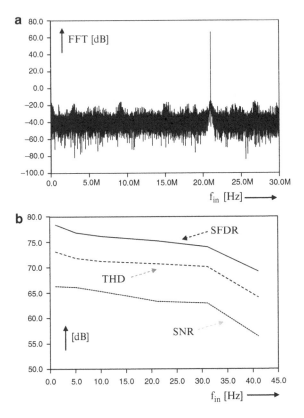

Fig. 3.47 (**a**) Measured frequency spectrum at 60 MS/s with an input frequency at 21 MHz. (**b**) Measured SNR, THD, SFDR as a function of input frequency

Table 3.2 Measured performance of a 12-bit prototype

Technology	Digital CMOS 0.18 µm
Resolution	12 bit
Supply voltage	1.8 Volt
Sample rate	>60 MSample/s
Effective bandwidth	30 MHz
DNL	±0.9 LSB
INL	±1.5 LSB
SNR	66.3 dB
SFDR	78.4 dB
THD	73.1 dB
SNDR	65.1 dB
Power dissipation	100 mW
Area	0.67 mm^2

The three latter errors reduce the time allocated for the setting time. These errors either increase the noise floor or cause distortion in the digital output spectrum resulting in decreased SNR and SNDR. As an input frequency and resolution increase, the requirement for clock jitter [216] is getting more stringent. In other words, a clock jitter error will degrade the SNR even more as an input frequency approaches Nyquist input frequency. A locked histogram test revealed a 3.2-ps rms jitter in the system including the clock generator, the synthesizer, the A/D comparator chip and the board, which translates to a 64-dB SNR at 30 MHz approximately. This confirms the observation that the performance of this converter is limited by the clock jitter at high input frequencies.

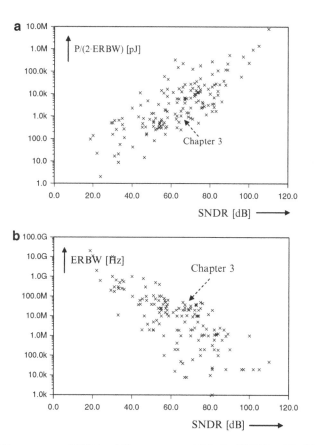

Fig. 3.48 (a) Energy versus SNDR for A/D converters published at ISSCC and ESSIRC in the last 10 years. (b) Effective resolution bandwidth versus SNDR for A/D converters published at ISSCC and ESSIRC in the last 10 years

3.7 Conclusion

The explosive growth of portable multimedia devices has generated great demand for low power A/D converters. With an increasing trend to a system-on-chip, an A/D converter has to be implemented in a low-voltage submicron CMOS technology in order to achieve low manufacturing cost while being able to integrate with other digital circuits. Fundamental limitations to the power dissipation of key functions for high speed A/D converters, such as sampling, quantization and reference generation in each case are dictated by accuracy consideration. The power proficiency for high resolution of multi-step converter by combining parallelism and calibration and exploiting low-voltage circuit techniques is demonstrated with a 1.8 V, 12-bit, 80 MS/s, 100 mW analog to-digital converter fabricated in five-metal layers 0.18-μm CMOS process. As shown in Fig. 3.48, comparison of the figure of merit, as in Section 2.1.4, of all A/D converters (flash, folding and interpolating, multi-step, pipelined, parallel pipelined, successive approximation and sigma-delta) published at *ISSCC* and *ESSIRC* in the last 10 years shows that this prototype ranks among the best reported.

Chapter 4
Multi-Step Analog to Digital Converter Testing

4.1 Analog ATPG for Quasi-Static Structural Test

Complex System-on-Chip (SoC) products include analog and mixed-signal IPs which needs to be testable. Since these IPs are embedded in the SoC, it is difficult to access all of their ports and as such existing test practices are not always applicable, or need to be revised. This implies also that test times need to be reduced to acceptable limits within the digital-testing time domain; it also implies the incorporation of Design-for-Testability (DfT), Built-in-Self-Test (BIST) and silicon debug techniques. For these SoCs, many of the tests exercised at final test are being migrated to wafer test, partly because of the need to deliver known good dies before packaging, and partly because of the need to lower analog test costs.

A typical test flow allocates test times to wafer test and final test. More traditionally, a wafer test consists primarily of *dc* tests with current/voltage checks per pin under most operating conditions and with the test limits properly adjusted and in some cases some low-frequency tests to ensure functionality. A wafer test is geared to check open/short circuits, *dc* biases, charge-pump currents, and logic leakage among other parameters. A final test consists of checking device functionality by exercising tests to cover important circuit parameters. However, with the advent of new packaging techniques and pressure on test costs, tra0ditional functional tests at package level are being pushed backwards to wafer level. Under this new scenario, wafer testing is performed to determine the true performance of the die independent of the packaging.

Structural, fault-orientated testing [218–223] is a convenient mean to avoid functional testing at wafer-level test. These low-frequency test techniques depart from the traditional role of current/voltage *dc* checks and are rather applicable for substituting or complementing functional tests. Fault-oriented test involves the use of a fault model to describe the behavior of a real defect, or set of defects. The fault model is then used to establish the faulty behavior of the circuit so that new tests can be derived or the effectiveness of existing tests investigated. The fault model cannot only describe the fault effect clearly and offer clues to derive the test stimuli, but

A. Zjajo and J. Pineda de Gyvez, *Low-Power High-Resolution Analog to Digital Converters*, Analog Circuits and Signal Processing, DOI 10.1007/978-90-481-9725-5_4, © Springer Science+Business Media B.V. 2011

also makes it feasible to modify input stimuli and estimate fault coverage. This approach is well established for digital circuits for which there are numerous test pattern generators and fault simulators based on a zero/one decision, as is the case of the single-stuck-at fault model. However, due to the continuous nature of analog circuits, the distinction between the fault-free circuit and the faulty circuit is not as clear as in the digital case. The statistical variations of the process parameters and component tolerances make that tolerance windows on the measurements have to be used to make a decision during fault detection. In addition, derivation of an acceptable tolerance window is aggravated with the presence of the overlap regions in the measurement values of the fault-free and faulty circuits, resulting in ambiguity regions for fault detection.

Several studies [224–229] have revealed that faults which shift the operating point of a transistor-level analog circuit can be detected by inexpensive dc testing or power supply current monitoring. A model in [224] tests analog circuits by measuring the dc voltage at different nodes. The relationship between the parameters, such as voltage and current, at component interconnections represents their behavior. This model deduces the values of parameters within the circuit by propagating the effect of the measurement through this model. In [225] a dc test selection procedure was presented where the detection criteria included the effect of parameter tolerance with a linear approximation around the nominal values. In [226], to include the effect of parameter tolerance during testing, the test generation problem is formulated as a min-max optimization problem and solved iteratively as successive linear programming problems. An approach for the fault detection based on Bayes decision rule for dc testing is presented in [227] by combining the a priori information and the information from testing. Principle component analysis is applied for the calculation of the discrimination function in the case of the measurements being dependent. However, the tests are obtained using simulation of a number of circuits and recording the percentage of good and faulty devices misclassified. The parametric fault simulation and test vector generation in [228] utilizes the process information and the sensitivity of the circuit principal components in order to generate statistical models of the fault-free and faulty circuit. The Bayes risk is computed for all stimuli and for each fault in the fault list. The stimuli for which the Bayes risk is minimal, is taken as the test vector for the fault under consideration. In [229], the measurement events are classified according to the regions that data fall into and the statistical profiles of the measurable parameters for each parametric fault are obtained. By iteratively conducting the tests and applying the Bayesian analysis, the occurrence probability of each fault is found.

The Neyman-Person statistical detector [230], which is a special case of the Bayes test, provides a workable solution when the a priori probabilities may be unknown, or the Bayes risk may be difficult to evaluate or set objectively. The study in this section [231] utilizes those findings. In this approach altering the circuit's biasing conditions generate various functional faults encountered in analog circuits. As seen from experimental results [232] acting on the circuit in such way a satisfactory level of correlation between structural and functional testing can be accomplished.

After parameter extraction, sets of bounds of signal values that can occur for fault-free circuits and circuits with specific faults, respectively, are generated and the deviations of the circuit's quasi-static node voltages are calculated through the accessed variations of the extracted process parameters. With Karhunen-Loève expansion method [233] the parameters of the devices are modeled as stochastic processes over the spatial domain of a die, thus making parameters of any two devices on the die, two different correlated random variables. Additionally, a fault model is verified according to performance specifications if one accounts for possible process parameter spread in a computation of a tolerance range. The variation in absolute value of process parameters is considered, as well as the differences between related elements, i.e. matching.

4.1.1   Test Strategy Definition

In this approach the circuit under test is excited with a quasi-static stimulus to sample the response at specified times to detect the presence of a fault. The waveform is systematically formed from piecewise-linear ramp segments that excite the circuit's power supply, biasing, reference and inputs, which forces the majority of the transistors in the circuit to operate in all the regions of operation, and hence, provide bias currents rich in information. To apply the power-supply-current observation concept to analog fault diagnosis, major modifications should be made to the existing current testing techniques, since the method requires more than a simple coarse observation of abnormal currents at the power supply network. Analog faulty behaviours are not so pronounced as those in the digital case, and due to the resolution limitations of the power-supply-current observation technique, the device under test has to be subjected to a design for testability methodology which consists of partitioning the circuit to reach better current observability. The method measures current signatures, not single values. Many faults have unique or near-unique signatures, easing the diagnosis process. Indeed it is independent of the linearity or nonlinearity of the systems, circuit or component.

The top-level test generation flow diagram is shown in Fig. 4.1. Firstly a tolerance window is derived according to test stimuli and test program. The circuit is simulated without any faults and the results of this test are saved in a database. The next step is to sequentially inject the selected faults into the circuit and simulate according to the same test stimuli and test program as used to derive the tolerance window. All simulation results are saved in the database, from where the fault coverage can be calculated in conformance with the tolerance window and discrimination analysis. To derive necessary stimuli, the test stimuli optimization is performed on the results available in the database.

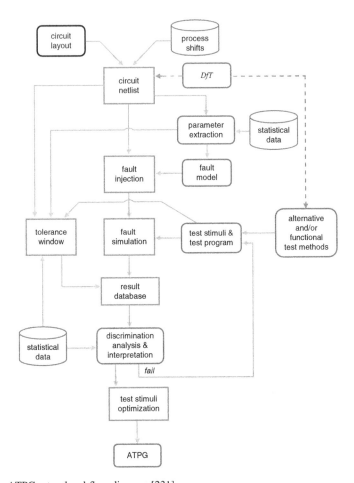

Fig. 4.1 ATPG – top-level flow diagram [231]

4.1.2   Linear Fault Model Based on Quasi-Static Nodal Voltage Approach

4.1.2.1   General Network Analysis

Modern integrated circuits are often distinguished by a very high complexity and a very high packing density. The numerical simulation of such circuits requires modeling techniques that allow an automatic generation of network equations. Furthermore, the number of independent network variables describing the network should be as small as possible. Circuit models have to meet two contradicting demands: they have to describe the physical behavior of a circuit as correct as possible while being simple enough to keep computing time reasonably small. The level of the models ranges from simple algebraic equations, over ordinary and partial differential equations to Boltzmann and Schrodinger equations depending on the effects to be described. Due to the high number of network elements (up to millions of elements) belonging to one circuit one is restricted to relatively simple models. In order to describe the physics as good as possible, so called compact models represent the first choice in network simulation. Complex elements such as transistors are modeled by small circuits containing basic network elements described by algebraic and ordinary differential equations only. The development of such replacement circuits forms its own research field and leads nowadays to transistor models with more than 500 parameters.

   A well established approach to meet both demands to a certain extent is the description of the network by a graph with branches and nodes. Branch currents, branch voltages and node potentials are introduced as variables. The node potentials are defined as voltages with respect to one reference node, usually the ground node. The physical behavior of each network element is modeled by a relation between its branch currents j and its branch voltages v. In order to complete the network model, the topology of the elements has to be taken into account. Assuming the electrical connections between the circuit elements to be ideally conducting and the nodes to be ideal and concentrated, the topology can be described by Kirchhoff's laws (the sum of all branch currents entering a node equals zero and the sum of all branch voltages in a loop equals zero). One elegant way to describe the network topology is by the (reduced) incidence matrix $\mathbf{A} = (a_{ij})$ that express the relation between all nodes (except the ground node) and all branches of the network. It is defined as: $a_{ij} = 1$ if the branch j leaves the node i, $a_{ij} = -1$ if the branch j enters the node i, and $a_{ij} = 0$ otherwise. Let a connected network with n nodes and b branches be given. If $\mathbf{j} = (j_1, j_2, \ldots, j_b)^T$ is the vector of all branch currents of the circuit, then Kirchhoff's current law (KCL) implies

$$\mathbf{A} \cdot \mathbf{j} = 0 \tag{4.1}$$

   The incidence matrix allows, additionally, a simple description of the relation between node potentials and branch voltages of the network. If $\mathbf{v} = (v_1, v_2, \ldots, v_b)^T$ is the vector of all branch voltages and $\mathbf{e} = (e_1, e_2, \ldots, e_{n-1})^T$ denotes the vector of all node potentials, then the relation

$$\mathbf{v} = \mathbf{A}^T \mathbf{e} \tag{4.2}$$

is satisfied. Each individual equation of (3.2) corresponds to one branch voltage. Writing characteristic equations of all network elements as

$$f\left(\frac{dq_C(v,t)}{dt}, \frac{d\phi_L(j,t)}{dt}, \mathbf{v}, \mathbf{j}, t\right) = 0 \tag{4.3}$$

This notation assumes that the terminal equations for capacitors and inductors are defined in terms of charges and fluxes, where $q(t)$ is the charge stored in the capacitors and $\phi(t)$ the flux stored in the inductors. The system (3.1)–(3.3) is a differential algebraic system, e.g. a coupled system of differential and algebraic equations in the network variables j, v and e. The dimension of this system equals $2b + n - 1$. The approach leading to this system is called sparse tableau analysis. The modified nodal analysis (MNA) requires a much smaller number of unknowns. In this case, one replaces the branch currents of all current defining elements in (3.1) by their characteristic equations, and all branch voltages by node voltages using (3.2). The resulting systems represent differential-algebraic equations (DAEs). General differential-algebraic equations have been widely investigated [234–237]. The results cover, among other things, unique solvability, feasibility of numerical methods as well as stability properties. However, most of the results suppose a certain structure (e.g. Hessenberg form), high smoothness and depend mainly on the index of the differential-algebraic equation. Recently, the special structure and the index of the network equations have been investigated [238] and a more general study of different circuit configurations was presented [239]. It clarifies that the index may become arbitrarily high and may also depend on parameters. In [240], electrical networks result in a differential algebraic system with an index not higher than two. In essence, the network branches are a numbered in such a way that the incidence matrix forms a block matrix with blocks describing the different types of network elements. The blocks are then given as $\mathbf{A} = [\mathbf{A}_R, \mathbf{A}_C, \mathbf{A}_L, \mathbf{A}_V, \mathbf{A}_I]$ where the index stands for resistive, capacitive, inductive, voltage source and current source branches, respectively. Replacing the branch currents of all current defining elements in (4.1) by their characteristic equations, and all branch voltages by node voltages using (4.2), the following system can be obtained

$$\mathbf{A}_C \frac{dq_C(\mathbf{A}_C^T \mathbf{e}, t)}{dt} + \mathbf{A}_R g(\mathbf{A}_R^T \mathbf{e}, t) + \mathbf{A}_L \mathbf{j}_L + \mathbf{A}_V \mathbf{j}_V = -\mathbf{A}_I \mathbf{i}_s(t)$$
$$\frac{d\phi_L(\mathbf{j}_L, t)}{dt} - \mathbf{A}_L^T \mathbf{e} = 0 \qquad (4.4)$$
$$\mathbf{A}_V^T \mathbf{e} = \mathbf{v}_s(t)$$

with the unknowns $\mathbf{e}(t)$, $\mathbf{j}_L(t)$ and $\mathbf{j}_V(t)$. The network equations in (3.4) describe linear electric network including capacitors, inductors, resistors, independent voltage and current sources, representing the KCL for each node and the element characteristics of inductors and voltage sources. Denoting the number of nodes by n, the number of inductive branches by n_L and the number of voltage source branches by n_V, the dimension of the system is now $n - 1 + n_L + n_V$. In the charge oriented MNA approach, additionally charges q and fluxes ϕ as unknown variables are introduced. This implies the equivalent system

$$\mathbf{A}_C \frac{dq}{dt} + \mathbf{A}_R g(\mathbf{A}_R^T \mathbf{e}, t) + \mathbf{A}_L \mathbf{j}_L + \mathbf{A}_V \mathbf{j}_V = -\mathbf{A}_I \mathbf{i}_s(t)$$

$$\frac{d\phi}{dt} - \mathbf{A}_L^T \mathbf{e} = 0$$

$$\mathbf{A}_V^T \mathbf{e} = \mathbf{v}_s(t) \tag{4.5}$$

$$q = q_C(\mathbf{A}_C^T \mathbf{e}, t)$$

$$\phi = \phi_L(j_L, t)$$

In general, the charge oriented system (3.5) is, for several reasons, the main approach used in circuit simulators [239] as replacement circuit models for semiconductor elements. The simple form of the equations $q = q_C(\mathbf{A}_C^T \mathbf{e}, t)$ and $\phi = \phi_L(j_L, t)$ involves only function evaluations for the determination of q and ϕ. Consequently, from the computational point of view, the dimension of the charge oriented system equals the dimension of the classical system. To expand the network equation in (4.4) to include the semiconductor devices, the currents of the semiconductor device have to be added to the KCL equation

$$\mathbf{A}_C \frac{dq_C(\mathbf{A}_C^T \mathbf{e}, t)}{dt} + \mathbf{A}_R g(\mathbf{A}_R^T \mathbf{e}, t) + \mathbf{A}_L \mathbf{j}_L + \mathbf{A}_V \mathbf{j}_V + \mathbf{A}_S \mathbf{j}_S + \mathbf{A}_I \mathbf{j}_s = 0$$

$$\frac{d\phi_L(\mathbf{j}_L, t)}{dt} - \mathbf{A}_L^T \mathbf{e} = 0 \tag{4.6}$$

$$\mathbf{A}_V^T \mathbf{e} - \mathbf{v}_s = 0$$

$\mathbf{A}_S$ has the same form as the other incidence matrices $[\mathbf{A}_C, \mathbf{A}_R, \mathbf{A}_L, \mathbf{A}_V, \mathbf{A}_I]$. The entries of the matrix $\mathbf{A}_S$ are defined as $a_{ik} = 1$ if the current j_{Sk} enters node i, $a_{ik} = -1$ if the reference terminal is connected to node i, or $a_{ik} = 0$ otherwise all for $i = 1, \ldots, n - 1$ and $k = 1, \ldots, b_S - 1$. n is the number of nodes of the network and b_S is the number of terminals of the semiconductor. A procedure such as [240], allows decomposing the circuit's unknowns (node voltages, currents through branches) into a differential component $\mathbf{y}$ for time dependent solutions and an algebraic component $\mathbf{z}$ for quasi-static analysis. The nominal voltages and currents $\mathbf{z}_0$ are obtained by [241]

$$\mathbf{z}_0 = -\mathbf{B}^{-1}(\mathbf{C}_Q \mathbf{y}_0 - \mathbf{F}_Q(\mathbf{i}_{(0)}, \mathbf{v}_{(0)})) \tag{4.7}$$

where $\mathbf{B}$, $\mathbf{C}_Q$ and $\mathbf{F}_Q$ are functions of the deterministic initial solution related to linear and non-linear couplings among the circuit's devices, $\mathbf{y}_0$ is an arbitrary initial state of the circuit and $\mathbf{i} \in \mathfrak{R}^{n(I)}$ and $\mathbf{v} \in \mathfrak{R}^{n(V)}$ are the independent current and voltage sources, respectively. It is assumed that for each process parameter p, e.g. threshold voltage, transconductance etc., there is only one solution of $\mathbf{z}_0$.

Table 4.1 MOST key parameters in 0.18 CMOS technology at $V_{BS} = 0V$(a) $I_{DS,lin}$ at $V_{GS} = 1.8V$ and $V_{DS} = 0.1V$ c. $I_{DS,lin}$ at $V_{GS} = -1.8V$ and $V_{DS} = -0.1V$(b) $I_{DS,sat}$ at $V_{GS} = 1.8V$ and $V_{DS} = 1.8V$ d. $I_{DS,sat}$ at $V_{GS} = -1.8V$ and $V_{DS} = -1.8V$

	$W/L = 10/0.18$			$W/L = 10/0.18$		
p	μ	σ	p	μ	σ	Unit
$V_{T0,N}$	516.92	10.44	$V_{T0,P}$	481.148	10.103	mV
$K_{0,N}$	422.53	10.34	$K_{0,P}$	518.538	13.109	mV$^{1/2}$
K_N	446.967	8.461	K_P	451.971	17.434	mV$^{1/2}$
β_N	26.334	1.290	β_P	6.775	0.261	mA/V^2
$W_{eff,N}$	10.034	0.010	$W_{eff,P}$	10.034	0.010	μm
$L_{eff,N}$	0.108	0.005	$L_{eff,P}$	0.143	0.005	μm
$I_{DS,lin}{}^a$	1.354	0.018	$I_{DS,lin}{}^c$	0.402	0.018	mA
$I_{DS,sat}{}^b$	6.035	0.226	$I_{DS,sat}{}^d$	2.914	0.226	mA

However, due to process variations, the manufactured values of process parameters will differ; hence, the manufactured values of the parameters $p_i \in \{p_1, \ldots, p_m\}$ for transistor i are modeled as a random variable

$$p_i = \mu_{p,i} + \sigma_p(d_i) \cdot p(d_i, \theta) \tag{4.8}$$

where $\mu_{p,i}$ and $\sigma_p(d_i)$ are the mean value and standard deviation of the parameter p_i, respectively, $p(d_i,\theta)$ is the stochastic process corresponding to parameter p, d_i denotes the location of transistor i on the die with respect to a point origin and θ is the die on which the transistor lies. This reference point can be located, say in the lower left corner of the die, or in the center, etc. As way of example, Table 4.1 shows some typical transistor parameters p with their mean and spread values.

4.1.2.2 Spatial Correlation Model

The availability of large data sets of process parameters obtained through parameter extraction allows the study and modeling of the variation and correlation between process parameters, which is of crucial importance to obtain realistic values of the modeled circuit unknowns. As an illustration Fig. 4.2a shows the parameter statistics of a batch with three different threshold-adjust implantations (identical for both n- and p-channels). Typical procedures determine parameters sequentially and neglect the interactions between them and, as a result, the fit of the model to measured data may be less than optimum. In addition, the parameters are obtained as they relate to a specific device and, consequently, they correspond to different device sizes. The extraction procedures are also generally specialized to a particular model, and considerable work is required to change or improve these models. For complicated IC models, parameter extraction can be formulated as an optimization problem. The use of direct parameter extraction techniques [242] instead of optimization allows end-of-line compact model parameter determination.

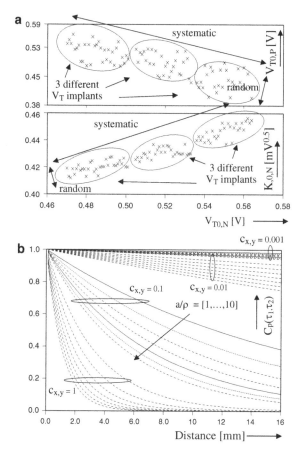

Fig. 4.2 (a) p-Channel threshold voltage, $V_{TO,P}$, versus n-channel threshold voltage, $V_{TO,N}$; measured on 200 transistor pairs from the same batch. (b) Body-effect factor, $K_{0,N}$, versus threshold voltage, $V_{TO,N}$; for 200 n-channel transistors from a batch with three different implantations to adjust the threshold voltage; b) Behavior of modeled covariance functions using $M = 5$ for $a/\rho = [1, \ldots, 10]$ (© IEEE 2009)

The model equations are split up into functionally independent parts, and all parameters are solved using straightforward algebra without iterative procedures or least squares fitting. With the constant downscaling of supply voltage the moderate inversion region becomes more and more important, and an accurate description of this region is thus essential.

The threshold-voltage-based models, such as BSIM and MOS 9, make use of approximate expressions of the drain-source channel current I_{DS} in the weak inversion region (i.e., subthreshold) and in the strong-inversion region (i.e., well above threshold). These approximate equations are tied together using a mathematical smoothing function, resulting in neither a physical nor an accurate description of I_{DS} in the moderate inversion region (i.e., around threshold). The major advantages of surface potential [243] (defined as the electrostatic potential at the gate oxide/substrate interface with respect to the neutral bulk) over threshold voltage based models is that surface potential model does not rely on the regional approach and *I-V* and *C-V* characteristics in all operation regions are expressed/evaluated using a set of unified formulas. In the surface-potential-based model, the channel current I_{DS} is split up in a drift (I_{drift}) and a diffusion (I_{diff}) component, which are a function of the gate bias V_{GB} and the surface potential at the source (v_{s0}) and the drain (v_{sL}) side. In this way I_{DS} can be accurately described using one equation for all operating regions (i.e., weak, moderate and strong-inversion). The numerical progress has also removed a major concern in surface potential modeling: the solution of surface potential either in a closed form (with limited accuracy) exists or as with our use of the second-order Newton iterative method to improve the computational efficiency in MOS model 11 [244].

A random process can be represented as a series expansion of some uncorrelated random variables involving a complete set of deterministic functions with corresponding random coefficients. This method provides a second-moment characterization in terms of random variables and deterministic functions. There are several such series that are widely in use. A commonly used series involves spectral expansion [245], in which the random coefficients are uncorrelated only if the random process is assumed stationary and the length of the random process is infinite or periodic [246]. The use of Karhunen-Loève expansion [233] has generated interest because of its bi-orthogonal property, that is, both the deterministic basis functions and the corresponding random coefficients are orthogonal [247], e.g. the orthogonal deterministic basis function and its magnitude are, respectively, the eigenfunction and eigenvalue of the covariance function. Simulation using Karhunen-Loève expansion can be made efficient if an analytical pre-processing step of the eigen-solution is available, whereby the computational effort is drastically reduced while safeguarding accuracy [246, 247]. Assuming that p_i is a zero-mean Gaussian process and using the Karhunen-Loève expansion, p_i can be written in truncated form (for practical implementation) by a finite number of terms M as

$$p_i = \mu_{p,i} + \sigma_p(d_i) \cdot \sum_{n=1}^{M} \sqrt{\vartheta_{p,n}} \xi_{p,n}(\theta) f_{p,n}(d_i) \qquad (4.9)$$

where $\{\xi_n(\theta)\}$ is a vector of zero-mean uncorrelated Gaussian random variables and $f_{p,n}(d_i)$ and $\vartheta_{p,n}$ are the eigenfunctions and the eigenvalues of the covariance matrix $C_p(d_1, d_2)$ of $p(d_i, \theta)$. Without loss of generality, consider for instance two transistors with given threshold voltages. In this approach, their threshold voltages are modeled as stochastic processes over the spatial domain of a die, thus making the parameters of any two transistors on the die, two differently correlated random variables. The variables can be generated by available established subroutines and then multiplied by the eigenfunctions and eigenvalues derived from eigen-decomposition of the target covariance model. The value of M is governed by the accuracy of the eigen-pairs in representing the covariance function rather than the number of random variables. Unlike previous approaches, which model the covariance of process parameters due to the random effect as a piecewise linear model [248] or through modified Bessel

functions of the second kind [249], here the covariance is represented as (Fig. 4.2b) as
a linearly decreasing exponential function

$$C_p(d_1, d_2) = \left(1 + \varsigma_{d_{x,y}}\right) \cdot \gamma \cdot \left(e^{-c_x|d_{x1} - d_{x2}| \cdot c_y|d_{y1} - d_{y2}|/\rho}\right) \qquad (4.10)$$

where ς is a distance based weight term, γ is the measurement correction factor for
the two transistors located at Euclidian coordinates (x_1, y_1) and (x_2, y_2), respectively,
c_x and c_y are process correction factors depending upon the process maturity. For
instance, in Fig. 4.2b $c_{x,y} = 0.001$ relates to a very mature process, while $c_{x,y} = 1$
indicates that this is a process in a ramp up phase. In (3.10) ρ is the correlation
parameter reflecting the spatial scale of clustering defined in $[-a, a]$, which
regulates the decaying rate of the correlation function with respect to distance
(d_1, d_2). Physically, lower a/ρ implies a highly correlated process and hence, a
smaller number of random variables are needed to represent the random process
and correspondingly, a smaller number of terms in the Karhunen-Loève expansion.
This means that for $c_{x,y} = 0.001$ and $a/\rho = 1$ the number of, transistors that need to
be sampled to assess, say a process parameter such as threshold voltage is much less
than the number that would be required for $c_{x,y} = 1$ and $a/\rho = 10$ because of the
high nonlinearity shown in the correlation function.

To maintain a fixed difference between the theoretical value and the truncated
form, M has to be increased when a increases at constant b. In other words, for a
given M, the accuracy decreases as a/b increases. Eigenvalues $\vartheta_{p,n}$ and eigenfunctions $f_{p,n}(\tau)$ are the solution of the homogeneous Fredholm integral equation of the
second kind indexed on a bounded domain D. To find the numerical solution of
Fredholm integral, each eigenfunction is approximated by a linear combination of a
linearly decreasing exponential function. Resulting approximation error is than
minimized by the Galerkin method.

One example of spatial correlation dependence and model fitting on the available
measurement data of Fig. 4.2a through Karhunen-Loève expansion is given in Fig. 4.3.
The sampling window radius versus on-chip variability as a result of the spatial
filtering analysis is shown in Fig. 4.3a, where a single variability function, which is
tuned for a specific yield expectation, is derived through combination of the distributions of each wafer. The sampling window radius corresponds to the worst-case
distance between the reference point and any other cell within the window area. To
analyze random process variation, dominant systematic effects from the measurement
data are removed by variability decomposition method [250]. For comparison purposes, a grid-based spatial-correlation model (Fig. 4.3b) is intuitively simple and easy
to use, yet, its limitations due to the inherent accuracy-versus-efficiency necessitate a
more flexible approach, especially at short to mid range distances [249].

4.1.2.3 Fault Model Definition

From a statistical modeling perspective, global variations affect all transistors in a
given circuit equally. Thus, systematic parametric variations can be represented by

a deviation in the parameter mean of every transistor in the circuit, which can be

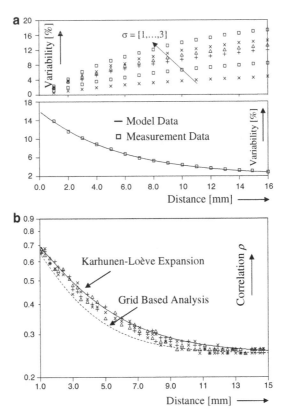

Fig. 4.3 (a) *Top*: systematic effect analysis for 68.3% (σ), 95.4% (2σ), 99%, 99.3%, 99.5% and 99.7% (3σ) yield expectations; *bottom*: measured and modeled random effect. (b) The spatial correlation dependence of Fig. 4.2 (© IEEE 2009)

seen as a defect. A defect model is now introduced, $\eta_p = f(.)$, accounting for voltage and current shifts due to random manufacturing variations in transistor dimensions and process parameters defined as,

$$\eta_p = f(v, W^*, L^*, p^*) \tag{4.11}$$

where $\eta_p = f(.)$ is the function of changes in node voltages and branch currents, v defines a fitting parameter estimated from the extracted data, W^* and L^* represent the geometrical deformation due to manufacturing variations, and p^* models electrical parameter deviations from their corresponding nominal values, as defined in (3.9), e.g. altered transconductance, threshold voltage, etc. This defect model is used to generate a corresponding circuit fault model by including the term η_p of (3.11) into (3.7), written in matrix form as

$$\Xi = \mathbf{z}_0 \times \mathbf{\eta}_p \tag{4.12}$$

where $\mathbf{z}_0$ is a matrix of the nominal data and $\mathbf{\eta}_p$ a random vector accounting for device tolerances. Basically, the fault model of (3.12) shifts the dc nodal voltages (dc branch currents) out of their ideal state based on the random and systematic variations of the process technology. While the functional behavior of a circuit in the frequency domain may not be linear, or even in the dc domain as a result of a nonlinear function between output and input signals, as long as the biasing and input conditions of the circuit under test remain quasi-static, the faulty nodal voltage (branch current) of (3.12) follows a Gaussian distribution as posed in (3.9). An obvious limitation of the fault model of (3.12) is that it cannot capture a faulty transient behavior of the circuit under test.

In general, a circuit design is optimized for parametric yield so that the majority of manufactured circuits meet the performance specifications. The computational cost and complexity of yield estimation, coupled with the iterative nature of the design process, make yield maximization computationally prohibitive. As a result, circuit designs are verified using models corresponding to a set of worst-case conditions of the process parameters. Worst-case analysis refers to the process of determining the values of the process parameters in these worst-case conditions and the corresponding worst-case circuit performance values. Worst-case analysis is very efficient in terms of designer effort, and thus has become the most widely practiced technique for statistical analysis and verification. Algorithms previously proposed for worst-case tolerance analysis fall into four major categories: corner technique, interval analysis, sensitivity-based vertex analysis and Monte Carlo simulation. The most common approach is the corners technique. In this approach, each process parameter value that leads to the worst performance is chosen independently. This method ignores the correlations among the processes parameters, and the simultaneous setting of each process parameter to its extreme value result in simulation at the tails of the joint probability density of the process parameters. Thus, the worst-case performance values obtained are extremely pessimistic. Interval analysis is computationally efficient but leads to overestimated results, i.e., the calculated response space enclose the actual response space, due to the intractable interval expansion caused by dependency among interval operands. Interval splitting techniques have been adopted to reduce the interval expansion, but at the expense of computational complexity. Traditional vertex analysis assumes that the worst case parameter sets are located at the vertices of parameter space, thus the response space can be calculated by taking the union of circuit simulation results at all possible vertices of parameter space. Given a circuit with M uncertain parameters, this will result in a 2^M simulation problem. To further reduce the simulation complexity, sensitivity information computed at the nominal parameter condition is used to find the vertices that correspond to the worst cases of circuit response. The Monte Carlo algorithm takes random combinations of values chosen from within the range of each process parameter and repeatedly performs circuit simulations. The result is an ensemble of responses from which the statistical characteristics are estimated.

Unfortunately, if the number of iterations for the simulation is not very large, the Monte Carlo simulation always underestimates the tolerance window. Accurately determining the bounds on the response requires a large number of simulations, so Monte Carlo method becomes very CPU-time consuming if the chip becomes large. From the test point of view, generally, the fault-free circuit and faulty circuit have different tolerance windows meaning that for the fault-free circuit and each faulty circuit, the corresponding tolerance window should be generated. The total number of tolerance-windows is therefore $(n + 1)$ if the number of faults in the fault list is n.

In our case, based on the central limit theorem, to completely characterize Gaussian data in (3.12) probabilistically, firstly the means and correlations have to be found by calculating the first and second-order moments through expectation. Even if the random variable is not strictly Gaussian, a second-order probabilistic characterization yields sufficient information for most practical problems. To make the problem manageable, the system in (3.12) is linearized by a truncated Taylor approximation assuming that the magnitude of the random defect η_p is sufficiently small to consider the equation as linear in the range of variability of η_p, or that the non-linearity of the electrical fault Ξ in the case of quasi-static dc biasing are so smooth that they might be considered as linear even for a wide range of η_p.

Next, the autocorrelation function of each nodal voltage (branch current) for each of the process parameters has to be calculated. This is necessary to estimate a tolerance window required to make a decision whether the circuit is faulty or not. The autocorrelation of Ξ for a quasi-static time period is then calculated as

$$\mathbf{C}_{\Xi\Xi} = \mathbf{J}_0 \mathbf{C}_{\eta\eta} \mathbf{J}_0^T \tag{4.13}$$

where $\mathbf{J}_0$ is the Jacobian of the initial data $\mathbf{z}_0$ evaluated at p_i and $\mathbf{C}_{\eta\eta}$ is the symmetrical covariance matrix whose diagonal and off-diagonal elements contain the parameter variances and covariances as defined in (3.10), respectively. Following (3.13), the boundaries of quasi-static node voltage Ξ_r with mean value μ_{Ξ_r} are expressed with

$$[\Xi_{r,\min}, \Xi_{r,\max}] = \boldsymbol{\mu}_{\Xi_r} \pm \sum_k \sum_m \left\{ |\mathbf{C}_{\Xi_r \Xi_r}|^{\max} \right\} \tag{4.14}$$

for any $p_i \in \{p_1, \ldots, p_m\}$ of $i \in \{i_1, \ldots, i_k\}$ transistors connected to node $r \in \{r_1, \ldots, r_q\}$. Per definition, setting quasi-static node voltage Ξ_r outside the allowed boundaries in (3.14) designates the faulty behavior. To obtain a closed form of moment equations, Gaussian closure approximations are introduced to truncate the infinite hierarchy. In this scheme higher order moments are expressed in terms of the first and second order moments as if the components of Ξ are Gaussian processes. The method is fast, and comparable to regular nominal circuit simulation. Suppose that there are m-trial Monte Carlo simulation for n faults, the method (using statistical data of the process parameters variations) gains a theoretical speed-up of $m \times n$ over the Monte Carlo method. The precision of the method is illustrated on the quiescent

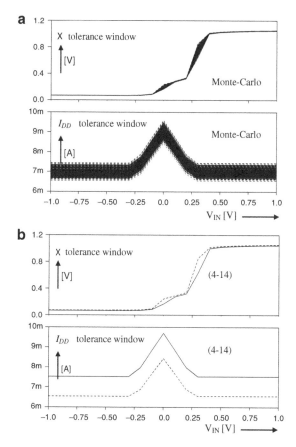

Fig. 4.4 Monte Carlo versus equation (4.14) for the folded node voltage and supply current I_{DD} as function of the input voltage

current and node voltage of folded-cascode amplifier with gain boosting auxiliary amplifier shown in Fig. 3.13. Equation (3.14) versus Monte Carlo analysis (it can be shown that 1,500 iterations are necessary to accurately represent performance function) as a function of the input and supply voltage is illustrated in Figs. 4.4 and 4.5. The difference between the two methods is shown in Figs. 4.6 and 4.7.

4.1.3 *Decision Criteria and Test-Stimuli Optimization*

4.1.3.1 Neyman-Pearson Decision Criteria

As each branch current is a Gaussian random variable in linear combination of parameter variations, the power supply current due to the voltage deviation at the

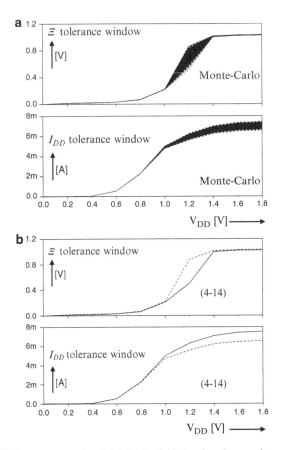

Fig. 4.5 Monte Carlo versus equation (4.14) for the folded node voltage and supply current I_{DD} as function of the supply voltage

node r, denoted as I_{DDn}^r (n samples) is, therefore, also a Gaussian distributed random variable, and its derivatives to all process parameters $\partial I_{DDn}^r / \partial p_i$ can easily be found from its linear expression of parameters. To avoid notation clustering, I_{DDn} denotation is further used. Derivation of an acceptable tolerance window for I_{DDn} is aggravated due to the overlapped regions in the measured values of the error-free and faulty circuits, resulting in ambiguity regions for fault detection. To counter this uncertainty, the Neyman-Pearson test [230, 251], which is based on the critical region $C^* \subseteq \Omega$, where Ω is the sample space of the test statistics, offer the largest power of all tests with significance level α

$$C^* = \{(I_{DD1}, ..., I_{DDn}) : l(I_{DD1}, ..., I_{DDn}|G, F) \leqslant \lambda\} \qquad (4.15)$$

where I_{DD} is an observation sample, $l()$ is a likelihood function, and G and F denote the error-free and faulty responses, respectively. Since both α (the probability

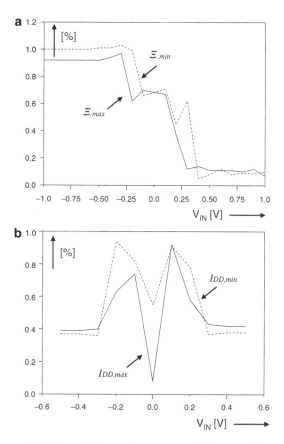

Fig. 4.6 Maximum and minimum difference in percentage (%) between Monte Carlo and (4.14) for the node voltage and supply current I_{DD} as function of the input voltage. $\Xi_{,max}$, $\Xi_{,min}$, I_{DDmax} and I_{DDmin} denote the tails of the probability function

that the fault-free circuit is rejected when it is fault-free) and β (the probability that faulty circuit is accepted when it is faulty) represent probabilities of events from the same decision problem, they are not independent of each other or of the sample size. Of course, it would be desirable to have a decision process such that both α and β are small. However, in general, a decrease in one type of error leads to an increase in the other type for a fixed sample size. The only way to simultaneously reduce both types of errors is to increase the sample size, which proves to be time-consuming process. The Neyman-Pearson test is a special case of the Bayes test, which provides a workable solution when the a priori probabilities may be unknown or the Bayes average costs of making a decision may be difficult to evaluate or set objectively. As the density functions of the I_{DDn} under fault-free and faulty condition $f(I_{DDn}/G)$ and $f(I_{DDn}/F)$, respectively, are often termed likelihood functions, the likelihood ratio is defined as

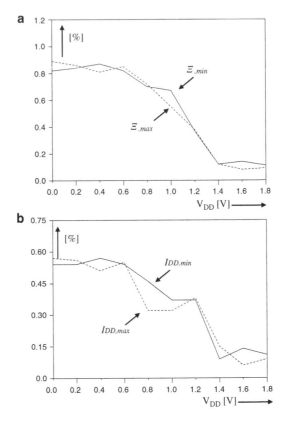

Fig. 4.7 Maximum and minimum difference in percentage (%) between Monte Carlo and this approach for the node voltage and supply current I_{DD} as function of the supply voltage. $\Xi_{,max}$, $\Xi_{,min}$, I_{DDmax} and I_{DDmin} denote the tails of the probability function

$$l((I_{DD1}, ..., I_{DDn})) = \frac{f_{I_{DDn}}((I_{DD1}, ..., I_{DDn})|G)}{f_{I_{DDn}}((I_{DD1}, ..., I_{DDn})|F)} \qquad (4.16)$$

For the threshold λ to be of significance level α we need

$$P = \{(I_{DD1}, ..., I_{DDn}) \in C^*|P(G)\} = \alpha$$
$$P(\bar{I}_{DD} \geqslant \lambda|I_{DD} \sim N(\mu, \sigma^2/n)) = P\left(Z \geqslant \frac{\lambda - \mu_G}{\sigma_G/\sqrt{n}}\right) = \alpha \qquad (4.17)$$

where μ_G and σ_G are mean and variance of the error-free response. $P(Z < z_{(1-\alpha)}) = 1-\alpha$, and $z_{(1-\alpha)}$ is the $(1-\alpha)$-quantile of Z, the standard normal distribution. Recall that if $I_{DD} \sim N(\mu, \sigma^2)$, then $Z = (I_{DD}-\mu/\sigma) \sim N(0,1)$. In the present case, the sample mean of $I_{DD} \sim N(\mu, \sigma^2/n)$, since the variable I_{DD} is assumed to have a normal distribution. From previous equation follows that the test T rejects for

$$T = \frac{\bar{I}_{DD} - \mu_G}{\sigma_G/\sqrt{n}} \geq z_{(1-\alpha)} \tag{4.18}$$

To incorporate the Neyman-Pearson lemma, the first step is to choose and fix the significance level of the test α and establish the critical region of the test corresponding to α. This region depends both on the distribution of the test statistic T and on whether the alternative hypobook is one- or two- sided. After data collection/observation from the measurements, the following step is to calculate the value of T (called t^*) from the data sample. This result is compared with the distribution of T in order to see whether or not it falls in the critical (rejection) region. At the end, decision is made to accept or reject the data sample.

Algorithm

Initialization
– Define probabilities P(G) and P(F) of each observation sample
– Assign significance level α

Data collection
– Collect I_{DDn} sampling points instants for each calculation

Main body
1. Calculate the critical region C^* according to (4.15)
2. Define decision threshold $\lambda^* = \mu_G + z_{(1-\alpha)}\sigma/\sqrt{n}$
3. Calculate the $(1-\alpha)$-quantile of the standard normal distribution $z_{(1-\alpha)}$
4. Calculate the value of T from the data sample according to (4.18)
5. If $T \geq z_{(1-\alpha)}$ reject otherwise accept the observation sample

---

4.1.3.2   Test Stimuli Optimization

All simulation results are stored within a block in the database. This makes it possible to fill-in the database in an incremental way by first investigating results of a certain simulation, and later on adding the results of a different simulation, possibly containing a different set of faults and/or tests. The database-block is viewed as a matrix where a column contains the entire test results for a single fault and a row contains the results for a single test. In this way, every matrix entry is the test result for a test-fault combination.

To derive necessary stimuli's for the test pattern generation, the test stimuli optimization on the results available in the database is necessary. Several possible ways exist to optimize test stimuli: A brute force method would be to first gather fault coverage data for each of the circuit specifications. This data can then be used to find the best order of the test stimuli's, by trying all possible permutations of the test stimuli set. If the circuit has n specifications, $n!$ permutations would have to be tried, for example for 20 specifications 2.4×10^{18} permutations are needed, which is clearly not computationally feasible. The approach in [252] is more advanced. Here Dijkstra's Algorithm is applied to select and order analog specification tests. The computational complexity of the algorithm is $n \times 2^n$, so for $n = 20$, 1×10^6 probabilities need to be computed. This is a great improvement over the brute force method, but it is still too large. The problem with Dijkstra's algorithms is that it does not take into account fault coverage, which appears to be critical for test ordering. In other word, a not-costly test stimulus is useless if it does not detect any faults. In general it is best to perform the test stimuli's with high fault coverage's first, provided that they are not too expensive.

The algorithm in [225] and partially implemented here are based on these observations. First test stimuli's that have very low fault coverage's are eliminated from the test stimuli set, and then the remaining test stimuli's are ordered. Two approaches to test stimuli ordering are considered: in the first stage, the test stimuli are ordered so that the test stimuli detecting the most faulty parameters that are detected by no other test stimuli is performed first, and the least important test stimuli is performed last (the test stimuli are ordered in descending order of unique coverage values). In the second stage the more costly test stimuli are moved downwards and are therefore used on a smaller number of circuits thus reducing the total test stimuli cost. Going from top to bottom, test stimuli, which do not increase the cumulative coverage, are moved to the bottom of the list.

Because some test stimuli are eliminated from the test stimuli set before the test stimuli's are ordered, both algorithms are heuristic, and both can handle circuits with many more specifications at much less computation cost. The computational complexity of the algorithm is n^2, so for $n = 20$, 400 probabilities need to be computed. Because of the heuristic approach, the algorithm does not necessarily produce the real optimum value, but usually a value close to the optimum value. In order to find optimum test stimuli set, the algorithm tries various permutations of the test stimuli set. The more costly test stimuli are considered first, and the less costly test stimuli are considered last. Suppose a costly test stimulus is in the ith position. The algorithm considers moving it to each position j with $j > i$. When the ith test stimulus is moved to the j_{th} position, the test stimuli currently in the $i + 1$st to jth positions are moved forward one step. When a permutation is made, all of the test stimuli's yields, Y_i, change. The yield Y_i of test T_i is defined to be the fraction of devices passing test T_i, given all previous test stimuli's in the sequence. This means that the yield of test stimuli depends on its place in a test sequence. The optimal test sequence is the test sequence for which the test stimuli cost is a minimum. The average test stimuli cost is

$$T_c = w_1 + w_2 \frac{(D_T - N_1)}{D_T} + w_3 \frac{(D_T - N_1)}{D_T} \frac{(D_T - N_1 - N_2)}{(D_T - N_1)} + \dots$$

$$= \sum_{i=1}^{n} w_i \prod_{j=1}^{i-1} Y_j \qquad (4.19)$$

where w_i is cost of applying test stimuli performing test number i, N_i is defined to be the number of faults detected uniquely by test stimuli T_i and not by any of the previous test stimuli's and D_T is the total number of devices.

Optimization algorithm

Initialization
- Initialize the input test stimuli set T, w_i, $i \in T$ (test costs)
- Initialize the number of test stimuli's m

Main body
- Compute the average test stimuli cost T_c
- Compute the test stimuli cumulative coverage C_c
- Find min argument of w_i, $i \in T$

$j = i + 1$

While $j \leq m$

    {Compute (4.22) and (4.23)

    if (4.23) < (4.22)

        {Compute $T_c{}'$ (the new average test stimuli cost) using (4.19)

        If $T_c{}' < T_c$

            move the ith test stimuli to the jth position

            move the $i + 1^{st}$ to j^{th} test stimuli's up one position and update T_c

        }

    Compute $C_c{}'$ (the new average test stimuli cost) using (4.24)

    If $C_c{}' < C_c$,

        {move the ith test stimuli to the jth position

        move the $i + 1^{st}$ to j^{th} test stimuli's up one position and update C_c

        }

    $j = j + 1$

    }
- Stop when the optimum test stimuli set T is found

The algorithm avoids the calculation of all the new test stimuli's yields, Y_i, for each possible permutation, and instead it first estimates if the change is likely to improve average test stimuli cost. If the ith test is moved to the jth position then only the test yields of the ith to jth tests can change. In particular, the only change in previous equation is

$$\sum_{k=1}^{j} w_k \prod_{l=i}^{k-1} Y_l \tag{4.20}$$

Defining x as

$$x = \sum_{k=1+1}^{j} w_k \prod_{l=i+1}^{k-1} Y_l \tag{4.21}$$

Then before a permutation, (4.20) is equal to

$$w_i + Y_i x \tag{4.22}$$

If Y_i do not change significantly, after a permutation, (4.20) can be approximated by

$$x + w_i \prod_{l=i+1}^{j} Y_l \qquad (4.23)$$

So if (4.23) is smaller than (4.22), the permutation is likely to reduce test stimuli cost, and the true Y_i are computed. If test stimuli cost decreases, the move is accepted. The algorithm continues by trying more permutations until the shortest tests have been tried. The cumulative coverage of test stimuli is the number of faults detected by the test stimuli or a previous test stimuli divided by the total number of faults.

$$C_c = \frac{1}{N} \sum_{i-1}^{n} c_h(i) \qquad (4.24)$$

where $c_h(i)$ is the highest achieved normalized fault coverage and N is the total number of faults.

Following the steps described in this section, the total time required for fault injection $t_{injection}$, fault simulation $t_{simulation}$, discrimination analysis $t_{discrimination}$ and test stimuli optimization $t_{optimization}$ can be expressed as

$$
\begin{aligned}
t_{injection} &= N_{nodes} \cdot t_{swap} \\
t_{simulation} &= N_{bias} \cdot N_{supply} \cdot N_{input} \cdot N_{reference} \cdot t_{circuit} + t_{tolerance} \\
t_{discrimination} &= N_{faults} \cdot t_{analysis} \\
t_{optimization} &= N_{permutation} \cdot N_{faults}
\end{aligned}
\qquad (4.25)
$$

where t_{swap}, $t_{circuit}$, $t_{tolerance}$ and $t_{analysis}$ is a time required to introduce the fault into circuitry, simulate the circuit netlist, derive the boundaries of circuit response and perform the Neyman-Pearson test, respectively. N_{nodes} denote the number of the nodes in the circuit, N_{bias}, N_{supply}, N_{input} and $N_{reference}$ designate the number of bias, supply, input and reference nodes where a quasi-static stimulus is applied, respectively, N_{faults} indicate the number of the faults and $N_{permutation}$ designate the number of permutations of the test stimuli set.

4.2 Design for Testability Concept

Modern Systems-on-Chip (SoC) integrate digital, analog and mixed-mode modules, e.g. mixed-signal, or RF analog and digital, etc., on the same chip. This level of integration is further complicated by the use of third-party cores obtained from virtual library descriptions of the final IC block. Furthermore, the variety and number of cores and their nature type, e.g. analog, complicate the testing phase of individual blocks, of combinations of blocks and ultimately of the whole system. The problem in the analog domain is that it is much more difficult to scan signals over long distances in a chip and across its boundary to the outside world, since rapid signal degradation is very likely to occur. The IEEE 1149.4 [253] is a standard mixed-signal test bus that can be used at the device, sub-assembly, and system levels. It aims at improving the controllability and observability of mixed-signal designs and at supporting mixed-signal built-in test structures in order to reduce test development time and costs, and to improve test quality.

There are various known approaches for designing for testability of analog circuits. The most common approach is to partition a system into sub-blocks to have access to internal nodes such that each isolated sub-block receives the proper stimuli for testing [254]. DfT in [255] is oriented at testing sub-blocks of a filter such that each stage is tested by increasing the bandwidth of the other stages. This is done by adjusting the switching scheme in the case of switched-capacitor filters, or by bypassing the filter capacitors using additional MOS transistor switches. The problem with the latter approach is that MOS transistors are in the direct signal path of the filter degrading the performance. A switched op amp structure can overcome this limitation [256]. In essence, this switched op amp has basically two operational modes, test and normal, depending upon a digital control signal. The op amp is used at the interface between any two sub-blocks. An enhanced approach similar to [256] that makes use of an op amp with duplicated input stages is given in [257, 258]. In this approach every op amp in the initial filter design is replaced by the DfT op amp. In [259] a DfT scheme suitable to detect parametric faults in switched-capacitor circuits based on a circuit that can compute all the capacitor ratios that determine the transfer function of the filter is shown. Other DfT schemes include A/D converters with self-correction capability [260].

Unlike previous approaches that test the analog circuit for functionality, analog structural testing [261, 262] consists of exciting the circuit under test with a dc or low-frequency stimulus to sample the response at specified times to detect the presence of a fault. The dc-transient waveform can be formed from piecewise-linear segments that excite the circuit's power supply, biasing, and/or inputs. To facilitate this kind of testing, it is preferable to observe the current (or voltage) signatures of individual cores instead of observing the current (or voltage) signature of the whole analog SoC. Therefore, shown DfT method works like a power-scan chain aimed at turning on/off analog cores in an individual manner, at providing observability means at the core's power and output terminals, and at exciting the core under test [263]. Existent DfT techniques do not turn individual blocks on and off, but merely disconnect the observed block from the signal path; neither are they suitable for dc-transient testing such as V_{DD} ramping [264].

The supply-current monitoring technique has, so far, found no practical wide-spread application in analog circuits. This is mainly because analog faulty behaviours are not so pronounced as those in the digital case; special test vectors often have to be applied to increase fault coverage, in a special test mode. To apply the power-supply-current observation concept to analog fault diagnosis, major modifications should be made to the existing current testing techniques. This is because of two factors: (i) the method requires more than a simple observation of abnormal currents; (ii) unlike digital circuits, bias currents always exist between the power supplies in analog circuits and in most cases abnormal currents cannot be defined.

For analog circuits and systems, fault diagnosis techniques are more complex when compared to their counterparts in digital circuits for various reasons: (i) the requirement of measurements of current and voltage signals at the internal nodes; (ii) diagnosis errors caused by soft-faults, which are due to the tolerance of the components.

An advanced methodology for testing analog circuits that overcomes the majority of the obstacles encountered with the supply-current-monitoring technique was proposed in [264]. To obtain signatures rich in information for efficient testing, the transistors in the circuit are forced to operate in all possible regions of operation by applying a ramp signal to the supply terminals instead of the conventional constant *dc* signal or ground voltage. The power-supply-ramping technique can potentially detect and diagnose catastrophic [261] as well as parametric faults [262]. The application of a ramp signal to the power supply voltage nodes instead of the conventional step (or *dc* voltage) can force the majority of the transistors in the circuit to operate in all the regions of operation, and hence, provide bias currents rich in information. The method measures current signatures, not single values. Many faults have unique or near-unique signatures, easing the diagnosis process. Indeed it is independent of the linearity or nonlinearity of the systems, circuit or component.

4.2.1 Power-Scan Chain DfT

Conventional DfT methods can provide test access points to each input and output node in the signal path. The CMOS switch matrix is capable of three modes. First, it can pass the signal from one block to the next for normal operation. Second, it can disconnect the output from one block and connect the input of the following block to the analog test input bus. The tester can then inject a test signal into the input of the circuit under test. Finally, the switch matrix allows observation of the output of the circuit under test through the analog test output bus. However, when using CMOS switches in the signal path of sensitive analog circuits, some possible problems may occur such as crosstalk, capacitive loading, and increased noise and distortion.

An objective of the power scan-chain DfT [263] illustrated in Fig. 4.8 is to avoid (minimize) the number of switches in the signal path as well as to provide means to support current-based testing techniques. The main feature of the DfT is to selectively turn individual cores on and off such that the core(s) *dc*, *ac*, and/or transient characteristics can be tested in isolation or together with other cores of an analog SoC. Conceptual overview of power-scan chain implemented in the two-step

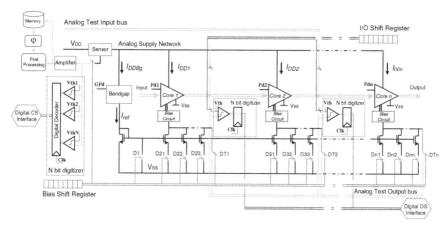

Fig. 4.8 Power-Scan Chain Analog DfT [263]

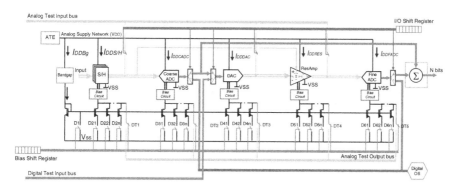

Fig. 4.9 Conceptual overview of power-scan chain DfT implemented in the two-step ADC

A/D converter is shown in Fig. 4.9. There are several ways of turning on/off analog cores. Preferred choice is by placing switches in the biasing network of the analog core, as this does not interfere with signals in the signal path. Another option is to place switches in the power path. However, these switches introduce a voltage drop that can have an impact on the core's performance. Additionally, the latter switches may be bulky. The DfT network shown in Fig. 4.8 has been designed to operate in all possible regions of operation by using a ramp signal at the supply instead of the conventional constant *dc* signal or ground voltage. The DfT consists of an Analog Test Input Bus to provide input stimuli to the core under test, an Analog Test Output Bus to read out stimuli response, a Digital CS and OS Interface to read out digitized current signatures (CS) and digitized output response stimuli (OS), an Analog Supply Network to read out currents in the power line and two Shift-register controllers to turn on/off individual cores and to select/deselect input/out test busses, respectively.

4.2.1.1 Analog Supply Network and Biasing Control

Only one sensor is inserted in the V_{DD} path of analog supply network, which is essentially an amp meter to measure the current flowing through the power supply terminal. An important property of any current sensor is to provide accurate measurements of extremely small current readings in the environment where the current changes are very fast and usually masked with high transients.

Conversely, under the condition that the ratio of the supply current to the background current is not sufficient to differentiate between a fault-free and a faulty circuit, the main limitation would be off-chip sensing. However, with development of built-in current monitor on-chip, these constraints have been largely overcome [265, 266]. To facilitate supply current readings of the individual cores (I_{DDBg}, I_{DD1}, I_{DD2}, ..., I_{DDn}), the biasing network of the cores under consideration are turned on/off in an individual manner. The supply currents of the individual cores are found from the difference between the supply currents found for the different codes, clocked in from left to right out of the bias shift register. The supply current readings are performed at the core's nominal operating conditions. The needed test time depends on the complexity of the analog blocks. ATPG results provide the optimum choice of the applied ramp and of the sequence of switching on/off individual analog core leading to the minimum test time under the restriction of a certain minimum allowed fault coverage.

To increase readability and avoid the rounding effects of the low-level power supply current, the quiescent power supply current is sensed across a sensing element and further amplified in a voltage amplifier with automatic-gain-control features to deal with wide dynamic ranges. After amplification, all individual supply current readings are stored in a memory unit. The φ operator allows any type of mathematical operation, such as addition, multiplication, convolutions, etc., for any combination of individual current signatures. Post-processing of the current signature is done by any post-processing means, such as for instance integration or FFT. Besides observing the analog waveform of the power supply current signature this approach allows observing its digitized version as well. Figure 4.8 shows a flash N-bit digitizer, with $N \geq 1$, whose V_{TH} references originate from the Analog Test Input bus, although, any N-bit digitizer architecture can be implemented. The Digital-Decoder clock signal is related to the system. The biasing network consists of a bandgap circuit providing an I_{ref} current, and of a current mirror with multiple legs, each feeding an analog core. The (bias) shift register controller is a digital circuit with V_{DD} (logical 1) and V_{SS} (logical 0) voltage levels. For the n-channel-based biasing network of this example, a bit value 1 in any position of the shift register turns on any of the switches $D_1, D_{21} \ldots D_{2N} \ldots D_{NN}$. Note that when node D_1 is switched on or when the bandgap is powered down by the global power-down signal (GP_d), I_{ref} will no longer flow and all bias currents will become zero irrespective of the $D_{21} \ldots D_{2N} \ldots D_{NN}$ signals. When node D_1 is switched on all bias currents will become zero irrespective of the $D_{21} \ldots D_{2N} \ldots D_{NN}$ signals, however, I_{ref} will still flow. Switches $D_{21} \ldots D_{2N} \ldots D_{NN}$ are used to turn off the core-under-test or to adjust the corresponding current biasing. One can regard switch D_{k1}, where $k = 2 \ldots N$, as the master switch that

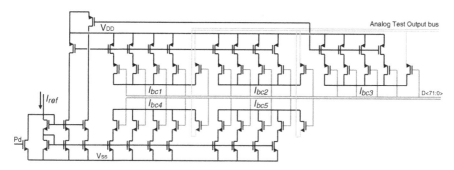

Fig. 4.10 Part of the switching matrix used to turn on and off the core's biasing circuits

enables the nominal biasing current to the core under normal operation. The remaining switches are used in test mode to allow the biasing current to be modified for testing purposes. In Fig. 4.8, it is assumed that switches in the biasing network are composed of n-channel transistors. However, actual implementation comprises both n-channel and p-channel transistors as shown in Fig. 4.10. Part of the switching matrix control circuitry is shown in Fig. 4.11.

The IC may consume a certain amount of current even if all blocks are off due to the leakage current of the digital circuitry used in mixed-signal circuits. It is important to note that by placing the switches at the ground nodes of the core's biasing circuit and not at the ground nodes of the analog core itself, an impact on the core's bias point due to voltage drop caused by switch on-resistance is limited. To ensure that individual core is totally off, e.g. that it does not have floating nodes, a local power-down signal, P_d, is made available. The local power-down signal is active, when switches $D_{21} \ldots D_{2N}$ connected to the biasing network of the core 1, $D_{31} \ldots D_{3N}$ for core 2, $D_{N1} \ldots D_{NN}$ for core N, are disconnected. The implementation for these settings is a logical *nor* port. Switches $D_{T1} \ldots D_{TN}$ are used to test the biasing network itself. They connect the bias nodes to the Analog Output Test bus. The core's biasing network is tested by turning on one core at a time and reading out the corresponding biasing voltage levels. Sensing of the bias network can be easily adjusted to detect the biasing current as well. As a final test provision is made to test the bandgap power-supply current.

4.2.1.2 Analog Test Input and Output Bus

The scheme provides an Analog Test Input Bus to control (excite) the cores that are enabled. When using CMOS switches in the signal path of sensitive analog circuits, some possible problems may occur, such as cross-talk and increased noise, distortion and capacitive loading. In this approach two switches are implemented instead of, conventionally used, four. Observe that there are no switches in the signal path from core to core, and that the Analog test bus is shunted with the signal path. This is possible when assuming that the core prior to the core under

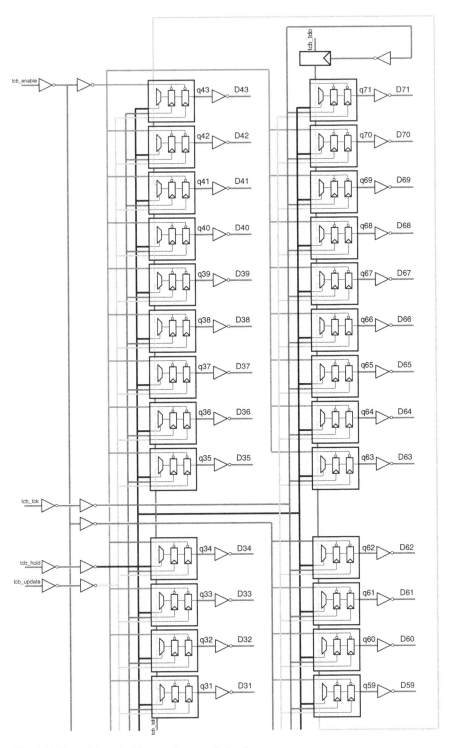

Fig. 4.11 Part of the switching matrix control circuitry

test is turned off through the I/O register-controlled switch and the local power-down signal. The Analog Test Output Bus provides observability of the core-under-test; associated with this bus is the Digital Output Interface, which digitizes the analog signal. The Analog Test Input bus can offer the input signal to the system, and the system output signal can be digitized and offered to the Digital Output Interface as well, to allow voltage read outs in Power-Scan chain mode in the whole system.

4.2.1.3   Power Scan–Chain Interface

The serial shift register provides the control of the various analog switches for bias-current control and connections to the analog test bus. In this implementation this register is a user register controlled by an IEEE Std 1149.1 TAP controller [267]. The analog test bus interface is provided by the IEEE 1149.4 analog test bus extension to 1149.1. Using an 1149.1 TAP allows access to the serial register, while the device is in functional mode. Other serial access interfaces, such as I^{2}C or a 3-wire bus can also be used to control the serial shift register, although they do not allow accessing the register in functional mode. The fact that one can gain access to the shift registers while the device is in functional mode opens unique opportunities for on-chip silicon debugging. For instance, it is possible to debug certain circuit specifications by properly tuning the biasing network of a core in operating mode, e.g., the gain of an amplifier, or the bandwidth of an active filter could be debugged by changing their biasing conditions. Similarly, a core can be isolated and its specification debugged with external input stimuli through the analog test input bus and by adjusting its biasing network. For instance, one can debug the bandwidth and insertion loss of a filter to known input signals, or debug the LSB sections of a data converter. Yet another example is to bypass a core in normal operating mode and to provide its neighboring cores with known input signals, or to analyze a core's output signal through the analog test output bus.

4.2.1.4   Dft with Integrated Current Sensors

The current sensors can also be placed between the V_{DD} pad and the core's V_{DD} terminal, as illustrated in Fig. 4.12. Although at the cost of the increased complexity, placing the sensors in individual power supply lines will enable parallel current-sensing operation, which as a consequence, results in significantly reduced test times. Compared to Fig. 4.8, the DfT now incorporates an Analog Power Output Bus to read out currents in the power line. The biasing network, Analog Test Input Bus, Analog Test Output Bus, Digital CS and OS Interface and two Shift-register controllers are similar. Note that band-gap sensor alterations can be made, so that the band-gap voltage can be measured and read out on the analog or digital power output bus. In tested systems with limited chip package pin-count, multiple individual blocks can be powered through one chip package pin. In those cases, on-chip current

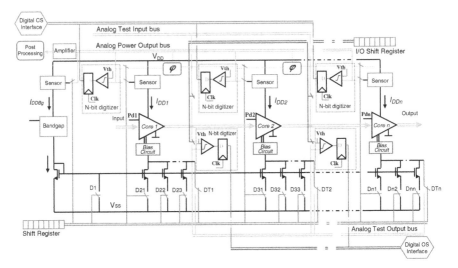

Fig. 4.12 DfT with integrated current sensors

sensors provide additional observability. The supply currents of the individual cores can be found by turning on the biasing network of the cores under consideration. All supply current readings are performed at the core's nominal operating conditions. The outputs of the sensors are chained into an amplifier, with automatic-gain-control features to deal with wide dynamic ranges. The amplifier and post-processing units are part of the Automatic Test Equipment (ATE). The measured current(s) of the core(s)-under-test are used for testing purposes. Similar to Fig. 4.8, the φ operator allows any type of mathematical function on the current signatures of any two consecutively enabled cores. Post-processing of the current signatures can be done by any post-processing means, such as for instance integration or FFT.

4.2.2 Application Example

The test generation methodology given in Section 4.1 can be employed for both fault detection (to determine whether the circuit is good or bad) and fault isolation (to identify the faulty sub-circuit). As mentioned before, the input to the test generator consists of the circuit description, test stimuli and test program giving the output consisting of test nodes, fault coverage and optimum test stimuli's that are sufficient to detect (or isolate) the failure modes (faults) in the circuit under test. To overcome system-test limitations of the structural current-based testing, the device-under-test is partitioned into smaller blocks with only limited additional hardware by means of the power-scan DfT technique.

The results shown in the next paragraphs were obtained with limited area overhead (approximately 5%), primarily from additional biasing transistors and routing, and at negligible extra power consumption since these bias transistors are not used in normal functional node. Performance loss is insignificant as no switches are placed in the signal path. The total test time required at wafer-level manufacturing test based on current-signature analysis is in 3–4 ms range per input stimuli. As the results indicate, most quasi-static failures in various blocks of the multi-step A/D converter from Chapter 3, depending on the degree of partitioning, are detectable through quasi-static structural test. For the entire A/D converter 11 input stimuli are required, which results in a total test time for the quasi-static test of at most 45 ms. This pales in comparison to around one second needed to perform histogram-based static or approximately one second for FFT-based dynamic A/D converter test offering more than 20-fold reduction in test time. Note that time required to perform these functional tests depends on the speed of the converter and available post-processing power.

4.2.2.1   Quasi-static Structural Test of Fine A/D Converter

Since the linearity of the fine A/D converter determines the overall achievable linearity of the overall converter, let's firstly examine the fine A/D converter into more details. As elaborately examined in Section 3.4.2 using folding and interpolation technique in the fine A/D converter (Fig. 4.13), the number of the required comparators is significantly reduced, however, at the cost of the reduced bandwidth by the number of folds. The nonlinear distribution of the reference voltages and resistor and preamplifier matching limits the linearity in fine A/D converter, since mismatch effects put a lower boundary on the smallest signal that can be processed

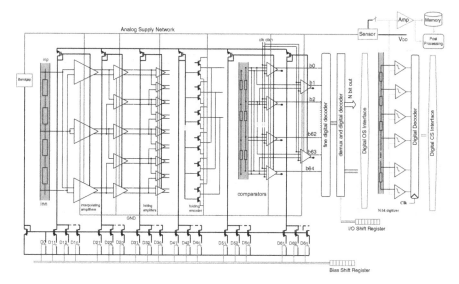

Fig. 4.13 Power Scan Chain DfT employed in the Fine A/D converter

in a system. As a result, their influence is most important at the first stage, where the signal levels are small. The offset voltage is major limiting factor of the comparator accuracy as well. If the gain mismatch of preamplifiers is present in the circuit, zero crossing of interpolated signal will be altered. Non-linearity of preamplifiers prevents perfectly linear interpolation as well by shifting the ideal zero crossings, while comparators non-linearity of comparators lead to the wrong comparator decision. Additionally, the error due to the tail-current mismatch of folding pre-amplifiers appears as if it is the offset of the differential pair. Since the zero crossing error due to the tail current mismatch is additive, the number of tail current that need to be matched is the same as the folding degree. As a result, it is beneficial to have a low folding degree since the error due to the tail current mismatch also decreases as the folding degree decreases. Each zero-crossing of the output is ideally determined only by one folding amplifier.

Detecting the faults in the first stage is essential since offset of the first stage preamplifiers determine the total referred input offset, providing that the gain of the preamplifiers is high enough. For illustration, two faults as defined by (4.12) are inserted (Fig. 4.14), at the output Ξ_{r1} and at the input biasing node Ξ_{r2} of first stage amplifier. This fault injection sets the node voltage outside the permitted node variation range characterized by (4.14) as illustrated in Fig. 4.15a and leads to easily spotted integral nonlinearity (INL) errors as shown in Fig. 4.15b.

Now, let's look at the influence of the modeled fault at the power supply current I_{DD}. As shown in Fig. 4.16a, inserting the fault at the observed nodes and simulating at nominal input values will lead to a deviation in the supply current, although, as

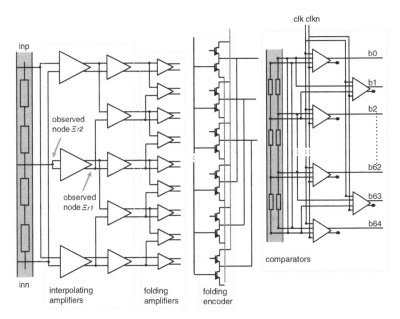

Fig. 4.14 Schematic of the fine A/D converter. Two observed nodes, at the input and output of the first stage preamplifier are shown

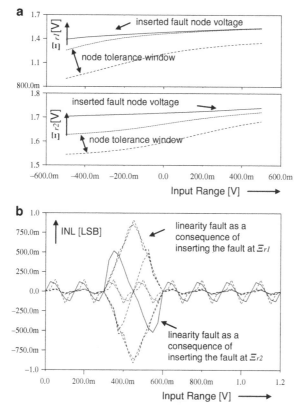

Fig. 4.15 (**a**) Fault insertion and their relation towards tolerance window defined in (4.14). (**b**) Integral Nonlinearity when faults are inserted in the first stage preamplifier (© IEEE 2009)

seen for the fault at the negative output node $\varXi_{r1}$ a large ambiguity region within error-free $P(G)$ and faulty $P(F)$ circuit probabilities make any decision subject to an error. However, by concurrently applying a linear combination of the inputs stimuli (the input signal, the biasing, reference and the power supply voltage) one can find the operating region where uncertainty due to the ambiguity regions for both modeled faults is reduced (Fig. 4.16b), and, thus, the corresponding probability of accepting the faulty circuit as a error-free is decreased. Next, the significance level of the test α is set and based on the distribution of the test statistics (4.18) the Neyman-Pearson critical (rejection) region C^* is formulated.

Continuing with the example illustrated in Fig. 4.16b, where $P(G)$: $I_{DDn}{\sim}N$ *(1.946mA, 0.23mA)* and $P(F)$: $I_{DDn}{\sim}N(2.16mA, 0.25mA)$ for the fault at node $\varXi_{r1}$, the critical region C^* for the test to be of significance level $\alpha = 0.1$

$$\alpha = P\left(Z \geqslant \frac{\lambda - \mu_G}{\sigma_G/\sqrt{n}}\right) = 1 - \Phi\left(\frac{\lambda - \mu_G}{\sigma_G/\sqrt{n}}\right) = 0.1 \qquad (4.26)$$

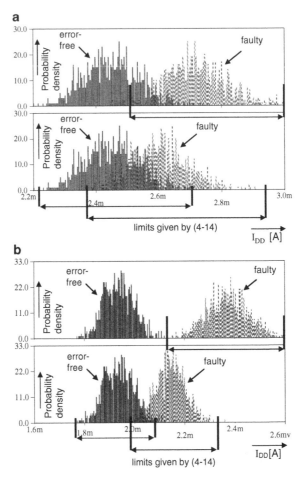

Fig. 4.16 (a) Preamplifier I_{DD} histogram plot at nominal input values for the error-free and for top) fault at node Ξ_{r2} and bottom) Ξ_{r1}. (b) Preamplifier I_{DD} histogram plot after ATPG optimization for the error-free and for top) fault at node Ξ_{r2} and bottom) Ξ_{r1} (© IEEE 2009)

Hence, from standard normal tables,

$$\Phi\left(\frac{\lambda - \mu_G}{\sigma_G/\sqrt{n}}\right) = 0.9 \Rightarrow \frac{\lambda - \mu_G}{\sigma_G/\sqrt{n}} = 1.282 \qquad (4.27)$$

leading to the critical region of

$$C^* = \left\{ (I_{DD1,...,}I_{DDn}) : \bar{I}_{DD} \geqslant \mu_G + 1.282\frac{\sigma_G}{\sqrt{n}} = 2.01mA \right\} \qquad (4.28)$$

where μ_G and σ_G are mean and variance of error-free power supply current. Thus, the circuit will be specified as a faulty if its power supply current value I_{DDn} is higher or equal to the threshold λ.

A comparable discrimination analysis is performed for all circuit's power supply current values generated as a consequence of inserting the faults at all nodes in the circuit in the entire range specified in the test program. The probabilities $P(G)$ and $P(F)$ as specified in (4.14) match the spread of more than 1,500 Monte Carlo iterations within 1%, while allowing multiple order of magnitude CPU time savings. Figure 4.17a displays the accuracy of using (4.14) for a power supply sweep. Influence of the inserting of the fault Ξ_{r1} on transfer function is illustrated in Fig. 4.17b.

In the entire fine A/D converter a total of 2,198 faults, corresponding to a similar number of nodes, are injected in the fault-free circuit netlist, and simulated according to

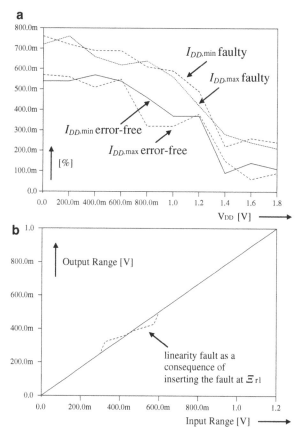

Fig. 4.17 (**a**) Difference in percentage (%) between Monte Carlo analysis and limits given by equation (4.14) for the supply current I_{DD} of the error-free and faulty circuit as function of the supply voltage V_{DD}. $I_{DD,max}$ and $I_{DD,min}$ denote the tails of the probability function; (**b**) influence of the inserting of the fault at Ξ_{r1} on transfer function (© IEEE 2009)

the test stimuli and the specified test program. The results of the test generator offer an indication for the required circuit partitioning through power-scan DfT, within the framework of circuit performance, area and testability. The tests are performed hierarchically and influenced by circuit architectural aspects, such as feedback among subblocks. At the first instance, the supply current of the entire fine A/D converter is found. Next, starting from the comparators and finishing at the first preamplifiers stage, one by one stage is switched off and the supply current is measured.

Implementing the DfT in such a way ensures that the reference voltages for the preamplifiers inputs of all stages are still provided. The supply currents of the individual cores can be found from the difference between the supply currents found for the different codes, clocked in from left to right out of the bias shift register. Notice that the use of DfT, as shown in Table 4.2, increases the fault coverage of the preamplifier stages and folding encoder to 100%. The undetected faults in the inactive parts of the comparator's decision stage and storage latch, expose the limitations of the quasi-static approach, due to the dynamic nature of the response. After test stimuli optimization, only three test stimuli (Table 4.3) are needed to achieve designated fault coverage from a given input stimuli ramps.

Table 4.2 Fine A/D converter test results

		Fault coverage (%)		
	Nr of faults	$\alpha = 0.05$	$\alpha = 0.1$	$\alpha = 0.2$
1st stage amplifiers	45	81.5	100	100
2nd stage amplifiers	119	56.3	68.9	70.6
1st and 2nd stage amplifiers	164	67.8	78.8	84.9
Folding stage amplifiers	102	76.5	82.4	84.3
Total without folding encoder	266	68.4	78.2	93.5
Folding encoder	96	66.7	85.4	85.4
Total without DfT	362	68.3	79.9	81.1
Total with DfT (43 transistors)	362	100	100	100
Comparators block	1836	70.4	80.1	93.2
Total fine ADC	2198	69.9	80.1	91.3

Table 4.3 Optimum test stimuli set for fine A/D converter

V_{DD} [1.8–0.0,0.1]	V_{RESA} [1.2–1.8,0.1]	V_{NRESA} [1.2–0.6,0.1]	V_{RESB} [1.8–1.2,0.1]	V_{NRESB} [0.6–1.2,0.1]	Fault coverage
[0.8]	[1.3]	[1.0]	[1.5]	[0.8]	61.3%
[0.0]	[1.8]	[0.6]	[1.4]	[1.0]	67.2%
[0.0]	[1.4]	[1.1]	[1.6]	[0.7]	69.9%
. . .	. . .	. . .	. . .	. . .	. . .
[0.0]	[1.6]	[0.9]	[1.6]	[0.6]	69.9%

4.2.2.2   Quasi-static Structural Test of S/H, Coarse ADC, Sub DAC and Residue Amplifier

The rest of the observed A/D converter has been evaluated following the similar principles. To enable fault detection based on differences of the power-supply current and prevent fault masking due to the feedback and common-mode regulation the sample-and-hold units are monitored in open-loop. The voltage deviation of the biasing transistors translates to voltage deviation of all biasing levels, which in itself leads to, easily detectable, variations of the power supply current. The faults on the output switches are relatively easy to detect since they have direct impact on the power supply current. A total of 244 faults are injected in the fault-free circuit netlist and simulated according to test stimuli and the test program. The low-frequency test stimuli offered are ramp for bias current, triangle for differential input voltage and adapted pulse for power-supply voltage. The results with the DfT in place are shown in Table 4.4.

The total fault coverage for the whole sample-and-hold circuit reaches 95.5%, 97.9% and 100% for the probability $\alpha = 0.05, 0.1, 0.2$, respectively, that the fault-free circuit is rejected as a faulty. After test stimuli optimization, five test stimuli are needed to achieve 95.5% fault coverage from the given input stimuli's ramps (Table 4.5).

Table 4.4 S/H test results

	Nr of faults	Fault coverage (%)		
		$\alpha = 0.05$	$\alpha = 0.1$	$\alpha = 0.2$
S/H 1	74	94.6	97.3	100
S/H 2	74	98.6	100	100
S/H 3	74	93.2	95.9	100
Bias circuit	10	100	100	100
Output switches	12	91.7	100	100
Total	244	95.5	97.9	100

Table 4.5 Optimum test stimuli set for S/H

V_{DD} [0.0–1.8,0.1]	V_{IN} [0.5–0.5,0.1]	V_{BIAS} [0.4–0.6,0.1]	C_{LK1} [0.0–1.8,0.1]	C_{LK2} [0.0–1.8,0.1]	C_{LK2} [0.0–1.8,0.1]	Fault coverage
[0.5]	[0.2]	[0.4]	[1.8]	[1.8]	[0.0]	60.7%
[1.0]	[0.1]	[0.4]	[1.8]	[0.0]	[1.8]	78.3%
[1.0]	[− 0.4]	[0.5]	[1.8]	[1.8]	[0.0]	87.3%
[0.5]	[0.5]	[0.5]	[0.0]	[1.8]	[1.8]	90.6%
[0.7]	[0.4]	[0.4]	[1.8]	[1.8]	[0.0]	95.5%
...	...	...	...	...		...
[0.5]	[− 0.2]	[0.4]	[1.8]	[0.0]	[1.8]	95.5%

The offered stimuli for coarse A/D converter quasi-static test are adapted ramp-down and ramp-up for the power-supply and input voltage. The fault coverage of 69.5, 79.0 and 88.3% for the probability $\alpha = 0.05$, 0.1, 0.2, respectively, is accomplished with only two optimized test stimuli for the 996 injected faults in the five-bit coarse A/D converter as shown in Table 4.6.

Once again the faults responsible for the fast dynamic behavior of the comparators were not entirely captured by quasi-static approach. After test stimuli optimization, only two test stimuli's (Table 4.7) are needed to achieve the previously indicated fault coverage.

Adapted ramp-down and ramp-up at the top and bottom of the reference ladder and adapted ramp-down (digital code from $2^N - 1$ to 0) and ramp-up (digital code from 0 to $2^N - 1$) at the input of the D/A converter have been offered and the current through the resistor ladder was measured. When inserting the two modeled parametric faults (resistor in the reference ladder and transistor switch) D/A converter differential outputs will be shifted by the ΔV, as illustrated in Fig. 4.18. The fault in the resistor will shift entire output signal by the ΔV, due to the shifting of the all reference voltages, while the fault at the switch will shift the output value only for the code corresponding to the digital value of that signal. The fault coverage obtained (less than 3%) shows that the resistor-based D/A converter is not suitable for current signature-based testing without additional, application specific, adjustments.

Inserting the couple of the parametric faults at the input and output of the residue amplifier shifts selected range by the ΔV as illustrated in Fig. 4.19. Fault coverage of the residue amplifier reaches 100% with the adapted ramp-down and ramp-up for the power-supply and input voltage.

Table 4.6 Coarse A/D converter test results

	Nr of faults	Fault coverage (%)		
		$\alpha = 0.05$	$\alpha = 0.1$	$\alpha = 0.2$
Preamplifier 1st stage block	54	25.9	57.4	81.5
Preamplifier 2nd stage block	51	100	100	100
Total without DfT	105	61.9	78.1	90.5
Total with DfT (21 transistors)	105	100	100	100
Comparators block	891	70.3	79.1	89.5
Total coarse ADC	996	69.5	79.0	88.3

Table 4.7 Optimum test stimuli set for coarse A/D converter

V_{DD} [1.8–0.0,0.1]	V_{INP} [0.4–0.9,0.1]	V_{INN} [0.9–0.4,0.1]	Fault coverage
[0.1]	[0.8]	[0.5]	32.1%
[0.2]	[0.5]	[0.8]	69.5%
. . .	. . .	. . .	. . .
[1.4]	[0.7]	[0.5]	69.5%

Fig. 4.18 (a) D/A converter differential outputs, (b) influence of the resistors and switches parametric faults on the D/A converter outputs

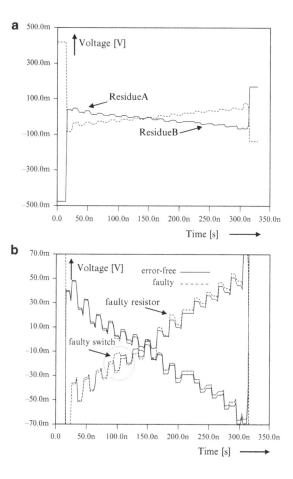

4.3 On-Chip Stimulus Generation for BIST Applications

The constant evolution of the integration capability in CMOS technologies is making possible the development of complex mixed-signal SoC (System on Chip). However, this increasing complexity has the associated issue of more complex and hence more expensive test, which is identified in the SIA Roadmap for Semiconductors as one of the key problems for present and future mixed-signal SoCs. Usually the test of the analog parts represents the main bottleneck in this line. These circuits are usually tested using functional approaches, often requiring a large data volume processing, high accuracy and high speed ATEs (Automatic Test Equipment). In addition, these analog cores are normally very sensitive to noise and loading effects, which limit the external monitoring and make their test a difficult task. BIST (Built-in Self-Test) schemes offer one of the most promising solutions to the problems cited above. In general, these schemes consist on moving

Fig. 4.19 (a) Output of the residue amplifier, (b) output of the residue amplifier in the presence of the parametric fault at the input

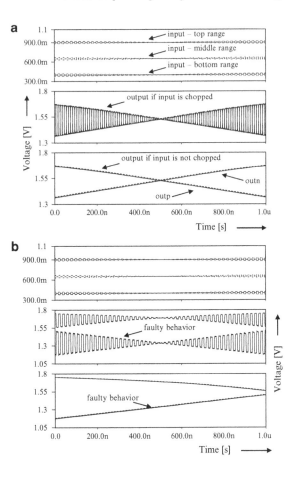

part of the required test resources (test stimuli generation, response evaluation, test control circuitry, etc.) from the ATE to the chip. However, the diversity of analog circuit designs, the multitude of their performance parameters and their limited observability, make analog and mixed-signal circuit BIST a very challenging problem compared to pure digital circuit BIST. Performing the built-in characterization of all the possible parameters would completely avoid the need of external testing, but the required design time and silicon area overhead would often make that option unaffordable. Nevertheless, a reduction of the testing time, through the built-in aided test of a sub-set of the performance parameters of a mixed-signal IC, can positively influence the final cost of the chip. On-chip evaluation and generation of periodic signals are of undoubted interest from this point of view. They have a wide potential applications in the field of mixed-signal testing as most of these systems (filters, A/D converters, D/A converters, signal conditioners, etc.) can be characterized and tested (frequency domain specifications, linearity, etc.) using this kind of stimuli.

Conventional sine-wave signal generation methods relay on an analog oscillator consisting on a filtering section and a non-linear feedback mechanism [268] or by adapting digital techniques, which facilitates a digital interface for control and programming tasks [269–272] or by employing the programmable integrator followed by filtering segment [273]. In an analog oscillator based techniques a non-linear feedback mechanism forces the oscillation while the filtering section removes the unwanted harmonics. However, the quality of the generated signal depends on the linearity and selectivity of the filter and the shape of the nonlinear function (smooth functions are needed for low distortion), which requires a lot of area and power. In digital techniques, by exploiting the noise shaping characteristics of $\Sigma\Delta$ encoding schemes use of the D/A converter is avoided [269–272]. In essence, a one-bit stream $\Sigma\Delta$ encoded version of an N-bit digital signal is generated and the shape of a filter is matched with the noise shaping characteristics of the encoded bit stream. Although, these schemes are valid for single and multi-tone signals they require large bit-stream lengths and a highly selective filter to remove the noise. In addition, these approaches are frequency limited due to the need of very high over-sampling ratios. On the other hand, the programmable integrator allows discrete-time or continuous-time periodic analog signal generation and in essence, fulfils the function of the D/A converter. The method has the attributes of digital programming and control capability, robustness and reduced area overhead, which make it suitable for BIST applications.

4.3.1 Continuous- and Discrete-Time Circuit Topologies

4.3.1.1 Continuous-Time Integrators

The difference between a continuous-time and a discrete-time sampled-signal, i.e. switched capacitor or switched current, integrator, is that the former processes the whole continuous time signal while the latter performs calculations using discrete-time analog samples and state variables. Figure 4.20 illustrates four circuit techniques to implement a continuous-time integrator: opamp-RC, MOSFET-C,

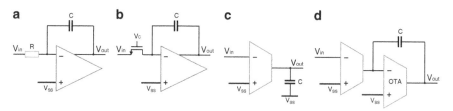

Fig. 4.20 (a) Opamp-RC integrator, (b) MOSFET-C integrator, (c) G_m-C integrator, (d) G_m-C-OTA integrator

G_m-C and G_m-C-OTA. For simplicity, single-ended configurations are shown; however in circuit implementation all would by full differential.

In an opamp-RC integrator, the opamp input forms a virtual ground and therefore the resistor transfers the input voltage into current. The feedback capacitor integrates the input current, since no current flows to the opamp input. In an inverting opamp-RC integrator (Miller integrator) such as shown in Fig. 4.20a, input signals can be summed by adding resistors to the opamp input without the need for additional amplifiers for summing. The opamp-RC technique is relatively insensitive to parasitic capacitances. In an ideal opamp the parasitic capacitances at the negative input and output of the opamp are driven by a voltage source or are connected to a virtual ground, and therefore do not affect the integrator transfer function. However, due to the limited *dc* gain, unity gain bandwidth and output resistance of the opamp, the parasitic capacitances, in practice, slightly affect the transfer function of the integrator. The time constant determined by R and C varies considerably as a result of variations in temperature and process, since the variations in the resistors and capacitors do not correlate. As a result, controllable time constants, which are handled by a separate tuning circuit, are necessary, such as tuning circuitry using series or parallel capacitor or resistor matrices. The series resistor and parallel capacitor matrices occupy less area than the other two options [274, 275]. In practice, parallel capacitors are usually employed, since the switch on-resistance produces a left-half-plane zero without shifting the integrator time-constant (an ideal opamp is assumed), as in the series-resistor tuning scheme. Since the temperature dependency of the capacitors is much smaller than that of resistors, it may be sufficient to calibrate the time-constants during the testing phase if resistors with sufficiently low temperature coefficients are available [276]. The clock signal having an accurate frequency can be used to derive an accurate integration time for a time-domain test integrator, which integrates a reference voltage for a fixed time (master-slave tuning scheme) [277, 278]. If the circuit can be disconnected from the signal path, it can be tuned by feeding a test signal to the input and changing the time-constants according to the measured output signal [279–281].

In general, the opamp-RC technique is largely insensitive to parasitics, is suitable for low supply voltages and has excellent linearity and large dynamic range [282]. The technique is appropriate for applications where high performance and moderate speed are required.

The resistors in an opamp-RC integrator can be replaced with a triode-region MOSFET as shown in Fig. 4.20b. In a MOSFET-C integrator, the time constant can be tuned by controlling the gate voltage V_C. Since no capacitor matrices are required for tuning, the area can be reduced in comparison to its opamp-RC counterparts. However, triode-region MOSFETs are more nonlinear than passive resistors and have a significant second-order term implying balanced design requirements to suppress even-order distortion. Even-order distortion can be suppressed further by using four cross-coupled MOSFETs [283] and a higher linearity can be achieved if a combination of resistors and cross-coupled MOS-FETs are used [284]. A large gate voltage necessary for high linearity and to keep the MOSFETs in the triode region entail requirement of techniques such as bootstrapping (Section 3.3.3.2). The control voltage can be driven over the positive supply with a charge-pump to improve the dynamic range of a MOSFET-C filter and to mitigate the sensitivity of the transfer function to dc offsets and variations in the common-mode level [274, 285, 286]. In principle, the noise performance of a MOSFET-C integrator equals that of an equivalent opamp-RC structure. The speed of MOSFET-C technique corresponds to that of the opamp-RC and technique's insensitivity to parasitic capacitances equal to that of the opamp-RC technique.

In a G_m-C integrator such as that shown in Fig. 4.20c, the transconductance g_m and integration capacitor C determine the unity-gain frequency of the integrator. g_m and the transconductor output resistance r_0 determine the dc gain of the integrator. The internal time constants of the transconductor form high-frequency poles and zeros, while, the quality factor is determined by the dc gain and the high-frequency poles and zeros. If the dc gain is sufficiently high due to a cascoded output stage or gain-boosting techniques, for example, the integrator quality factor becomes $Q_{int} = \omega_p/\omega_{int}$, where ω_p is the non-dominant pole. As this pole can be at a much higher frequency than the ω_{GBW} of an opamp, the bandwidth of a G_m-C filter can be much higher than that of an opamp-RC structure. To avoid instability, the opamp gain-bandwidth has to be less than the first parasitic pole determined by the load capacitance, which limits the usable frequency range of the technique. In the G_m-C technique, the bandwidth is only limited by the internal poles of the trans-conductor, which can approach that of the transistor f_T [274]. Additionally, the bandwidth can be maximized using transconductors without internal poles [287]. Parasitic wiring capacitances, the output capacitance of the transconductor and the input capacitances of the following transconductors augment the integrating capacitors and thus shift the integrator time-constant from the desired value. The integrating capacitances can be slightly decreased to take into account the parasitic capacitances. However, the accurate value of the parasitics may not be known, they degrade capacitance matching accuracy, do not track the actual capacitor values, and are nonlinear. A G_m-C filter has an additional source of inaccuracy in the frequency response when compared to the opamp-RC technique as a cascoded output stage is typically used to enhance the dc gain to a sufficiently high value limiting the output swing. G_m-C integrators typically require a folded cascode output stage, which can make the output current sources the dominant noise source [288]. Typically, the time-constant of a G_m-C integrator is tuned by changing the bias current of the transconductor or the gate-source voltage of triode-region MOSFET in an analog manner. The tuning accuracy can be higher than in a digitally controlled opamp-RC integrator. The silicon area is minimized since matrices of passive elements are not required. The continuous tuning of transconductors is performed at the expense of large signal handling capability [282]. The correct control signal, a current or voltage, can be derived using the master-slave tuning scheme [274]. The open-loop nature of the G_m-C integrator means higher nonlinearity than in opamp-RC integrator [162, 290–292]. Typically, G_m-C circuits consume less power and silicon area than do the corresponding opamp-RC circuits, although dynamic ranges achieved are higher in the opamp-RC technique.

In a G_m-C-OTA integrator, the output of the transconductor is buffered with a Miller integrator, as shown in Fig. 4.20d [274, 293], thereby reducing the sensitivity to parasitic capacitances. The latter amplifier can be a simple, wide-band OTA, like a common-source stage or differential pair, as it does not drive resistive loads. The dc gain of the structure is the sum of the dc gains of the two amplifiers meaning that cascoding in the OTA is not required to achieve a sufficient dc gain [288]. If both amplifiers are differential structures, however, two common-mode feedbacks are required. The G_m-C-OTA filters have low complexity and are power- and area efficient. As the transconductor limits the linearity of a G_m-C-OTA integrator distortion level is comparable to that of G_m-C structures using equivalent transconductors. The G_m-C topology suffers from low dc gain and, more importantly, high sensitivity to parasitic capacitance at the output node. However, since the G_m-C-OTA structure does not suffer from the aforementioned problems as the transconductor does not have to swing a large voltage range; it can tolerate larger and more nonlinear parasitic capacitance and it can be less sensitive to noise, G_m-C-OTA structure is chosen for continuous-time implementation. Furthermore, either a low- or high-output-impedance Miller integrator can be employed as it only needs to drive capacitive loads, although, typically, high-output-impedance amplifier is used as it tends to have higher unity-gain frequencies, be simpler and have less power dissipation.

4.3.1.2 Discrete-Time Integrators

A switched-capacitor circuit, although its signal remains continuous in voltage is, in fact, a discrete-time circuit since it requires sampling in the time domain. Because of this time-domain sampling, the clock rate must always be at least twice that of the highest frequency being processed to eliminate aliasing and as a result these circuits are limited in their ability to process high-frequency signals. By replacing the resistor from opamp-RC integrator shown in Fig. 4.20a with capacitor whose resistive value is equal to $(C_H f_{clk})^{-1}$ its discrete-time equivalent can be derived as shown in Fig. 4.21a. In every clock cycle, C_H absorbs charge equal to $C_H V_{in}$ when S_1 is on and deposits the charge on C_i when S_2 is on. If V_{in} is constant, the output changes by $V_{in} C_H / C_i$ every clock cycle. The sampling instants by the end of clock phase φ_1 are defined as $(k-1)/f_{clk}, k/f_{clk}, (k+1)/f_{clk}, \ldots$, whereas those by the end of clock phase φ_2 are deemed to be $(k-3/2)/f_{clk}, (k-1/2)/f_{clk}, (k+1/2)/f_{clk}, \ldots$ Even if the falling edge of φ_1 is not precisely half clock period apart from that of φ_2, effect on the circuit operation is limited as long as the duration of each clock phase is long enough for the signal to settle properly. The final value of V_{out} after every clock cycle can be written as $V_{out}(k/f_{clk}) = V_{out}[(k-1)/f_{clk}] - V_{in}[(k-1)/f_{clk}] C_H/C_i$.

Integrator performance is specified in terms of non-idealities which limit the overall modulator dynamic range and speed of operation. Parameters such as finite dc gain, linear settling and slew rate can raise the quantization noise floor

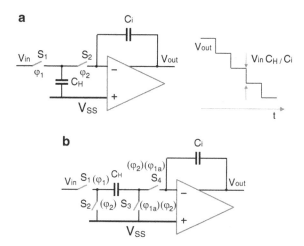

Fig. 4.21 (a) Discrete-time integrator with response of circuit to a constant input voltage, (b) parasitic-insensitive discrete-time integrator

or introduce distortion depending on the circumstances. Thermal noise introduced by the sampling process as well as by the amplifiers adds directly to the quantization noise. Finite amplifier *dc* gain can degrade both the distortion and noise performance and change the positions of the poles in a switched-capacitor integrator. The pole shift has an impact on the choice of operational amplifier topology in practical designs. Since the effects of the pole shifting are difficult to calibrate due to its small values, the amplifiers with large *dc* gains have to be employed. Linear settling and slew rate specify the small signal and large signal speed performance of an integrator, respectively. A linear settling error results in an integrator gain error while slew rate results in harmonic distortion. Harmonic distortion due to amplifier slew rate can directly degrade the large signal performance of the integrator. Due to the low g_m/I ratio of short channel CMOS devices and the required high speed operation, this distortion constraint will be satisfied if the amplifier slews for a small fraction of the settling period and spends the majority of its time in a linear settling regime. Output swing defines the maximum signal handling capability of an operational amplifier and is directly related to the integrator input overload level. Maximizing the output swing will increase the maximum signal handling capability. For a *kT/C* noise limited design, this will minimize the required sampling capacitance and power dissipation as described in Chapter 3. Output swing is ultimately limited by the power supply voltage, but in practical designs the swing will be lower due to the requirement that the output devices remain in saturation. In practice, saturation voltage of *n*-channel devices will need to be large enough to bias the device at sufficient f_T that settling requirements can be met. The saturation voltage of *p*-channel devices will need to be large enough that the device parasitics do not appreciably load the amplifier output. As a result, output swing will trade-off with amplifier settling requirements.

The integrator in Fig. 4.21a suffers from two important drawbacks. First, the input-dependent charge injection of S_1 introduces nonlinearity in the charge stored on C_H. Second, the nonlinear capacitance resulting from the source/drain junctions of S_1 and S_2 leads to a nonlinear charge-to-voltage conversion as explained in Section 3.3. A parasitic-insensitive SC integrator [162, 294] illustrated in Fig. 4.21b resolves both of these issues. The clock phases for the inverting integrator are in parentheses. Bottom plate sampling [125] as discussed in Section 3.3, is implemented with the clock phase φ_{1a}, which is a slightly more advanced version with respect to the clock phase φ_1. The parasitic capacitances at the top plate of the capacitor C_H and at the input of the operational amplifier are always connected to a fixed potential through the switches S_3 and S_4, and therefore do not affect the operation of the circuit. Parasitic capacitances at the input and output of the operational amplifier only have an effect on the settling speed of the operational amplifier, and do not introduce error, if taken into account in the characterization of the settling time of the operational amplifier. The effect of the parasitic capacitances at the bottom plate of the capacitor C_H is also canceled. Even though the parasitic capacitances are charged to the input voltage V_{in} in phase φ_1, the parasitic capacitances are discharged to a fixed potential in the clock phase φ_2 through the switch S_2, and no discharge current flows through the capacitor C_H.

Probably the most common non-filtering analog function is a gain circuit where the output signal is a scaled version of the input. In continuous-time opamp-RC circuits, a gain circuit can be realized as parallel combinations of resistors and capacitors in both the feedback and feed-on path as illustrated in Fig. 4.22a. By replacing resistors with their switched-capacitor counterparts, equivalent discrete-time gain circuit can be derived (Fig. 4.22b), whose gain G is given by the ratio of $V_{out}(k)/V_{in}(k)$. However, although its output is a continuous waveform that does not incur any large slew-rate requirement, such a circuit will also amplify the $1/f$ noise and opamp offset voltage with the gain G. To suppress finite offset and opamp gain as well as $1/f$ noise, commonly applicable technique for various switched-capacitor circuit types, e.g. gain amplifiers, sample and holds, integrators, is correlated double sampling such as the one illustrated in Fig. 4.22c. A thorough-full discussion on this technique is given in Section 3.3. Here, just a short reminder, during the clock phase φ_2, the finite opamp input voltage caused by above mentioned limitations is sampled and stored across C_1 and C_2. Next, during φ_1, this input error voltage is subtracted from the signal (applied to the opamp input) at that time. Assuming that the input voltage and the opamp error voltages did not change appreciably from φ_2 to φ_1, error due to them will be significantly reduced.

4.3.1.3 Discrete- and Continuous-Time Filters

The realizations of discrete, switched-capacitor filters can be categorized into three basic groups: (i) continuous-time filter emulations, (ii) SC ladder filters, and (iii) cascade SC filter realizations. (i) In continuous-time filter emulations, SC passive elements replace the resistors in a classical continuous-time active RC filter

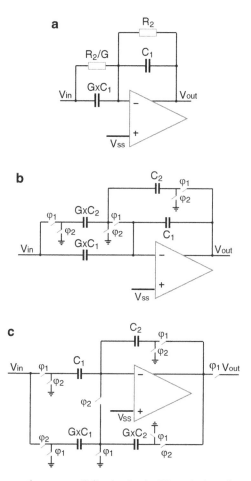

Fig. 4.22 (**a**) Continuous-time opamp-RC gain circuit, (**b**) equivalent discrete-time gain circuit, (**c**) discrete-time gain circuit with correlated double sampling

such as Tow-Thomas filter. However, the limited number of filter types and inappropriateness for high-Q applications provide the foremost restrictions. (ii) In SC ladder filters, SC circuits simulate the low-sensitivity response of a high-Q doubly terminated passive reactance two-port, or, equivalently, an RLC ladder network for realizing a desired high-order transfer function with high-Q poles [297]. In general, three different approaches to the realization of an SC ladder filter exist: first, the ladder component substitution, where an equivalent SC element substitute each R, L, and C component of the original continuous-time ladder network maintaining the original filter's low sensitivity to component variations; second, the voltage-inverter-switch approach [298, 299], which, although it does not require complex design, due to its inherent sensitivity to the top-plate parasitic capacitance and multi-phase operation provide severe limitation and third approach based on the signal-flow graph ladder realization. Here, the s-domain transfer function of the *RLC* ladder filter is converted into its *z*-domain counterpart using either lossless discrete integrators (LDI) (i.e., approximate design) or bilinear (i.e., exact design) *s*-to-*z* transformations, depending on the accuracy requirements and the ratio of clock frequency over signal bandwidth (i.e., f_{clk}/f_0). (iii) The cascade SC filter realizations adopt a direct building block approach to determine the filter's transfer function in the *z*-domain. However, the main transfer function is broken into products of first- and second-order terms. In other words, the numerator and denominator of a high-order transfer function are factored into first- and second-order sub-functions. Each sub-function can be realized by one of the low-order (first-order or biquad) filters. Since each of the low-order filters is individually buffered and capable of functioning independently, cascading them together will not affect their own transfer functions [300].

A generic first-order active switched-capacitor filter is shown in Fig. 4.23a. The circuit originates from a circuit similar to one shown in Fig. 4.22a, where to obtain a switched-capacitor filter having the same low-frequency behavior, the resistors are replaced with delay-free switched capacitors, while the non-switched capacitor feed in is left unchanged. In essence, this is a three-in-one filtering circuit as it incorporates three different filtering types, e.g. low-pass, all-pass and high-pass. Note that there are three specially labeled switches: for instance, the switch labeled $\varphi_2\varphi_{LP}$ will be turned on when both φ_2 and φ_{LP} go to one. Here, φ_{LP}, φ_{AP} and φ_{HP} stand for low-pass, all-pass and high-pass, respectively. The clock phase φ_{AP} takes up a portion of φ_{LP}, whereas φ_{HP} is not overlapping with φ_{LP}. In other

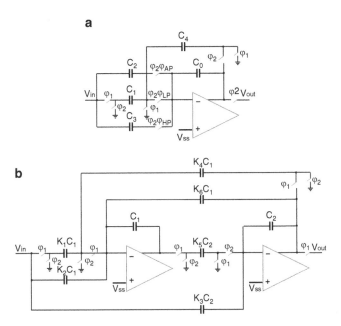

Fig. 4.23 (a) First-order switched-capacitor filter without switch sharing, (b) a high-Q switched-capacitor biquad filter without switch sharing

words, when the circuit is employed to realize a first-order all-pass filter, both C_1 and C_2 branches are activated. By contrast, when the circuit is used as a first-order high-pass filter, only the C_3 branch is activated. Similarly, when the circuit is employed as a first-order low-pass filter, only the C_1 branch is activated. A second-order SC filter or biquad may be realized using one of many approaches, depending on the application's requirements and the arrangement of the switches. An SC biquad, which is shown in Fig. 4.23b [301], is a three-in-one filtering system as well capable of realizing second-order low-pass (all-pole) band-pass, and high-pass filters. Specifically, the input should be sent through the K_1C_1, K_2C_2 and K_3C_2 signal paths into the core circuitry when realizing low-pass, band-pass, and high-pass filters, respectively. Other possibilities for realizing different types of functions exist as well. For example, a non-delayed switched capacitor going from the input to the second integrator could also be used to realize an inverting bandpass function. If the phases on the input switches of such a switched capacitor were interchanged a non-inverting band-pass function would result, although with an additional period delay.

The major bottleneck of realizing a high-frequency CMOS SC filters is the requirement for high-gain and large-bandwidth op-amps, which provide virtual grounds for accurate charge transfer [302]. Although it is possible to realize CMOS op-amps that can provide the required gain and bandwidth values, the resultant high power consumption often hinders the practical realization [302]. Besides the large-gain-bandwidth approach, two techniques, which are meant to enhance a CMOS SC filter's ability to operate in the 100 MHz range, can be roughly classified into two groups: op-amp based and unity-gain-buffer based. The large-bandwidth or high-speed op-amps tend to have low dc gains due to the fundamental tradeoff between RC time constant and $g_m R_{out}$. A low op-amp dc gain tends to introduce nonlinearity errors to the SC integrator's output, hence compromising the filter's accuracy performance. As a response to this problem, op-amp-based techniques typically emphasize modifying the traditional op-amp structures so that the resultant new op-amps are capable of fulfilling both the speed and accuracy requirements of high-frequency SC filters, despite the fact that they typically have low dc gains. In the gain regulating approach [303], the low dc gain of each large-bandwidth op-amp in the high-frequency SC filter is precisely controlled and the regulated gain value is used as a reference to scale the capacitances in the circuit. As the alternative to the op-amp-based approach, the unity-gain-buffer-based technique makes use of unity-gain buffers to build SC integrators [304, 305]. A typical unity-gain buffer is able to operate over a much wider signal bandwidth than the conventional op-amp. In addition, this buffer can be realized using simpler circuitry; as a result, it occupies less silicon area and consumes less power than a conventional large-gain-bandwidth op-amp does. However, unity-gain buffer -based SC integrators suffer from parasitic capacitances, which are mainly caused by the source-gate diffusions in the unity-gain buffer's input transistors, whose values tend to vary with process and temperature. Similarly, the parasitic-insensitive techniques introduced in Chapter 3 are not applicable on unity-gain buffer-based SC integrators, due to its inherent low gain.

The most widely known continuous-time filter is the active RC filter, which is characterized with high dynamic range, low distortion features. However, restricted speed due to the negative feedbacks limits the active RC filter to low speed applications. Similarly, the gain-bandwidth product of the filter's amplifier has to be much higher than the filter cut-off frequency to minimize phase error and any other non-idealities which occur near the gain-bandwidth product [306], which in itself requires excessive power consumption. The main advantage of the MOSFET-C filter, which is in essence active RC filters where resistors are replaced with tunable triode-region transistors as elaborated in previous section, in comparison to its active RC counterpart is that by controlling the resistance by a control voltage V_C results in an extended tuning range as the resistance increases to infinity [274, 307]. However, device mismatches and noise limit the implementation of very high resistor values. In addition, rather high control voltage V_C is required to ensure the triode region operation of the transistor. Similarly, achievable signal swing in low voltage application is limited. On the other hand, the tunable CMOS resistors eliminate the use of capacitor matrices for time constant tuning. Many different active RC filters structures exist; among them, Tow-Thomas, Ackerberg-Mossberg and Sallen-Key filter are widely known. Usually, MOSFET-C filter can be realized when passive resistors in the filter are replaced with triode region MOSFETs. However, parasitic capacitances of the triode region transistor affect the performance of the filter. As parasitic capacitance can be minimized when it is connected to virtual ground or output of opamp, consequently, filter architecture insensitive to parasitics, such as a Tow-Thomas or a Sallen-Key filter can be employed [308].

The G_m-C filters often make use of open-loop rather than closed-loop operational transconductance amplifiers and thus need not be constrained by the stability requirement [162, 274]. As a result, G_m-C filters typically have a speed advantage over both SC and RC filters, particularly for applications in the hundreds of megahertz range. However, the drawback of using G_m stages in an open loop configuration is that the circuit is limited to small input levels in order to operate the transconductor in linear region [274, 309]. Even though many different techniques have been reported to increase the input range while maintaining linearity, these techniques often degrade the frequency response due to additional parasitics [309, 310]. Even with the linearization technique, the input signal swing range is still small compared to the active RC filters. Another drawback of G_m-C filters is their dependence on the transconductance g_m, which makes them highly susceptible to process variations. A single-ended G_m-C filter that realizes a first-order filter is shown in Fig. 4.24a. One advantage of the G_m-C filter is that passive components like the resistor and inductor can be realized with OTA and capacitor. The resistor is realized by connecting negative input of OTA to positive output and the function of inductor can be realized at certain frequency by using two OTAs and a capacitor. However, linearity of the resistors and the inductors is highly dependent on the performance of OTA. If the OTA has finite bandwidth characteristics, the gain and phase of the filter will deviate from its ideal frequency response.

The specific nature of the deviation will be a function of the OTA's open-loop characteristic and the filter's desired response. However, in comparison to the

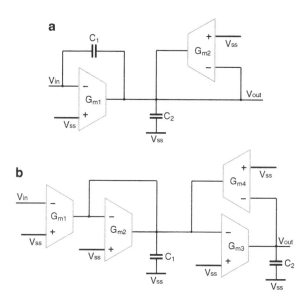

Fig. 4.24 (a) First-order G_m-C filter, (b) a G_m-C biquad filter

active RC realization where full opamp is employed, due the simplicity of the OTA, active G_m-C filters can be pushed to higher frequencies before OTA gain-bandwidth product impacts the filter's response. In opamp RC filters, linearity is mainly a function of opamp linearity since resistors exhibit quite linear behavior. Opamp is characterized by its low output impedance, meaning that it can drive resistive loads, making it applicable for RC filters and until certain degree for SC filters as well. Their output is in the form of voltage and extra buffer adds complexity and power dissipation. OTA's on the other hand have high output impedance and cannot drive resistive loads. Their output is in the form of current and they are applicable for SC and G_m-C filters. Typically, they are less complex, lower power solution in comparison to opamp with higher frequency potential as well. Figure 4.24b shows the G_m-C biquad filter. In this filter, the passive resistor is replaced with OTA by negative feedback connection. Similar to SC filters, this biquad filter can be cascaded in order to realize higher order filters. For easy and effective tuning of analog filter, there are several things which need to be considered. First, the frequency and quality factor needs to be tuned independently. Sometimes the tuning of frequency causes the change of quality factor which degrades the performance of filter. Usually, in a biquad filter, the frequency and quality factor can be tuned separately. Second, it is desirable to use the same value of R and C in the filter. If this is impossible, it is recommended to choose an integer ratio of resistor or capacitor values for better matching. Instead of resistor matrices, capacitor matrices are normally used for tuning since they are better in matching the components. Third, parasitic insensitive filter architecture is desired. Usually, a filter based on the signal flow graph is insensitive to parasitics. For the tuning, the output of the filter must be compared with another reference frequency. The reference frequency or signal usually comes from the outside of the chip [162, 313]. There are many different tuning systems such as adaptive tuning circuit, direct tuning strategy, master-slave tuning circuit, etc. Among them, master-slave filter tuning is a widely used tuning method as it is relatively easy to build and demonstrates acceptable accuracy if master and slave filter are well matched [313].

4.3.2 Design of Continuous- and Discrete-Time Waveform Generator

4.3.2.1 Discrete-Time Waveform Generator

The block diagram of the system for the on-chip waveform generation intended for BIST applications is shown in Fig. 4.25. It consists of a non-overlapped clock generator, a programmable (multi-) gain stage combined with linear time-variant filter forming a programmable sine-wave generator, a clock mapping blocks, gain decoders, a digital control unit, a band-pass filter, an additional programmable gain amplifier at the output of the band-pass filter to further improve the dynamic range of the system and a final stage for the amplitude detection and/or digitization of the PGA output. Based on an external master clock, the non-overlapped clock generator generates the appropriate clock signals for the signal generator and the programmable bandpass filter. The amplitude of the generated signal is adjustable to make it suitable for the input range of the device-under-test. The signal generator has two cascaded gain stages: a first gain stage (FGS) and a second gain stage (SGS), where the overall gain is the sum of the gains, of the FGS and SGS. The amount of gain realized by the first- and second gain stages is controlled by an array of non-overlapped clocks coming from the FGS and SGS gain mapping block, respectively. The select signal serves as a sequential selection signal feeding reference values to the programmable gain amplifier gain stage. Consequently, and in order to

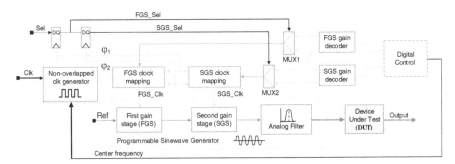

Fig. 4.25 Overview of the system for on-chip analog signal generation

have independent control of the gain stages, the first- and second gain clock mapping units generate a different array of clocks. Gain decoders generate combinational logic information on how a particular gain is realized in the first and second gain stages. Two decoders are therefore needed, an FGS gain decoder, and a SGS gain decoder for independent gain control. A state machine and two control muxes control the manner in which the information from the digital decoder is fed to the first- and second gain clock mapping unit, respectively. The programmable, high-Q bandpass filter selects the proper harmonic component at the output of the device-under-test for magnitude response or harmonic distortion characterization.

Depending on the characteristics of the integrated system in which the DUT is embedded and of the external test system, different options can be employed for the output building block. If an A/D converter is available on-chip, the output of the programmable gain amplifier could be directly digitized. One of the main advantages of the system is its inherent synchronization; both the stimuli frequency and the filter center frequency are controlled by the master clock; when it is swept, both the signal generator and the filter follow these adjustments. The testing strategy does not require any device-under-test re-configuration and is able to directly test frequency response related specifications. Before the characterization of the DUT, the functionality of the method can be easily verified by bypassing the output of the signal generator to the band-pass filter as illustrated with a dashed-line arrow in Fig. 4.25. A timing diagram showing the processing of data through the gain stages is illustrated in Fig. 4.26a. When FGS or SGS select is high, first or second gain stage is processing data, respectively. At the clock rising edge, the first gain stage starts acquiring data. When the FGS starts holding, the second gain stage starts attaining the FGS data. When the first gain stage finishes holding the SGS has fully acquired the signal from the first gain stage. It should be noted that the FGS_Sel and SGS_Sel signals are 90° out of phase due to the reversal of roles between φ_1 and φ_2 in the two stages.

With the exploit of a four step capacitor array as the analog up-sampler as shown in Fig. 4.27a programmable gain amplifier (PGA) whose preset gain stages

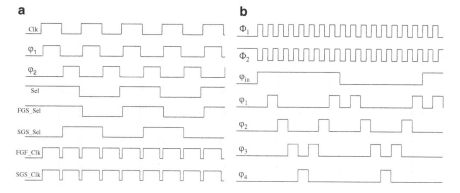

Fig. 4.26 (a) Timing diagram of the overall system, (b) timing diagram of the waveform generation

a   **b**

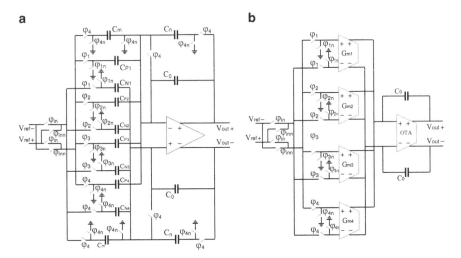

Fig. 4.27 Conceptual view of the waveform generation in (**a**) switched-capacitor technique and (**b**) G_m-C-OTA technique

correspond to the values of an ideally sampled and held sine wave can be constructed. Corresponding circuit implementation in G_m-C-OTA technique is illustrated in Fig. 4.27b. To avoid inaccurate waveform generation matching between two or more capacitors is of the utmost importance. However, various layout techniques such as use of integer ratios of unit capacitors, deployment of unit elements with the same shape or area to perimeter ratio and unit elements centroide form layout make matching of <0.05% feasible. To realize the desired capacitance with adequate accuracy, the size of each unit element is, thus, not to small, otherwise its desired value would be submerged by normal fabrication tolerances.

The SC signal generator has four different gain settings, which generate a sinusoidal signal output with 16 steps per period. When the gain is changed in discrete steps, there may be a transient in the output signal. There are two different causes of transients when the gain of a programmable gain amplifier is changed. The first is the amplification of a *dc* offset with a programmable gain, which produces a step in the output signal even when the PGA has no internal *dc* offsets or device mismatches. Second, when the gain of a programmable gain amplifier is changed in a device, in which *dc* current flows, the *dc* offset at the output may be changed due to device mismatches, even when there is no *dc* offset at the input of the PGA. In the first case, the cause of a transient is in the input signal, which contains a *dc* offset. In the latter case, the output *dc* offset of the programmable gain amplifier depends on the gain setting because of changes in the biasing, i.e. the topology of the PGA and mismatches cause the transients. The step caused by a change in the programmable gain may be a combination of both effects, although if properly deployed, the following high-frequency low-pass filtering stage will filter out this step if sufficient small time constant is deployed. The switches φ_1 through

φ_4 are closed sequentially for one clock period to generate the four steps of the sine wave. Once the maximum value of generated waveform is obtained, the switches close sequentially for one clock period in the opposite direction (from φ_4 to φ_1) to generate the second quarter-period as illustrated in Fig. 4.26b; in this case the polarity of the reference voltage is changed through φ_{in} to generate negative and positive integration, respectively. The charge injected by the input capacitors is integrated on capacitor C_0 to generate the first quarter-period of the sinusoidal waveform. Capacitors C_m and C_n are added to subtract error voltage from the input signal (applied to the opamp input) during φ_4, when this error is the largest. By cross-inverting the inputs, the circuit capability of realizing both parasitic-insensitive bilinear and forward-Euler integrators increase [314]. Specifically, if both the input and output are sampled when φ_1 is on, the circuit realizes a bilinear SC integrator. On the other hand, if the input and output are sampled when both φ_{inp} and φ_{inn} are on, the circuit realizes the forward-Euler integrating function (inverting or noninverting).

The opamps in the first and second gain stages were implemented as gain-boosted folded cascode amplifiers as shown in Fig. 4.28. To reduce the power consumption, the opamp used in the first gain stage was implemented with a higher gain-bandwidth product than the second gain stage opamp due to the larger range of feedback factors required to realize the different FGS gains [140]. Since first gain stage amplifier experiences no large signal swings, an input pair is formed with only p-channel transistors due to their superior $1/f$ noise properties. In a second gain stage amplifier, a n-channel input pair T_{1-2} is placed in parallel with a p-channel input pair T_{3-4} to process signals from rail to rail. To make the transconductance as a function of the common-mode input voltage constant a simple feed-forward method is applying current switches T_{5-8} [154]. The current switches are divided into two transistors of which the drains are connected to the drains of the

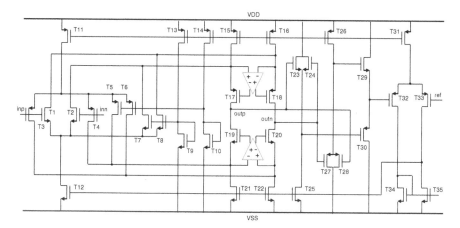

Fig. 4.28 Opamp schematic

corresponding input-stage transistors. By adding the currents to the outputs of the input-stage transistors the output current of the input stage does not change as a function of the common-mode input voltage. Since relatively small current-switch transistors can be used, their noise contribution to the noise of the amplifier can be made relatively small. With a common-mode feedback circuit comprising of T_{23-35} [315], where the resistor/capacitor network in balanced resistor/capacitor differential-difference amplifier common-mode feedback structure is replaced by a transistor network consisting of T_{23-28}, the output common mode level can be sensed without changing the impedance at the system output. The signal path from the opamp output to the gate of T_{32} is built exclusively by source-follower stages, so the gate voltage $V_{G(T32)}$ is a monotonically increasing function of the output common-mode level. Similarly, since T_{23-28} are complementary types of transistors, the output common-mode level is guaranteed to be detected in full-swing range without pulling any transistors of T_{31-35} away from the saturation region. Additionally, since all nodes in the network T_{23-28} are low impedance nodes, no additional stability problems will occur.

The major bottleneck of realizing a high-frequency CMOS SC filters is the requirement for high-gain and large-bandwidth op-amps, which provide virtual grounds for accurate charge transfer [302]. Although it is possible to realize CMOS op-amps that can provide the required gain and bandwidth values, the resultant high power consumption often hinders the practical realization [302]. Besides the large-gain-bandwidth approach, two techniques, which are meant to enhance a CMOS SC filter's ability to operate in the 100 MHz range, can be roughly classified into two groups: op-amp based and unity-gain-buffer based. The unity-gain-buffer-based technique makes use of unity-gain buffers to build SC integrators [304, 305]. A typical unity-gain buffer is able to operate over a much wider signal bandwidth than the conventional op-amp. In addition, this buffer can be realized using simpler circuitry; as a result, it occupies less silicon area and consumes less power than a conventional large-gain-bandwidth op-amp does. However, unity-gain buffer -based SC integrators suffer from parasitic capacitances, which are mainly caused by the source-gate diffusions in the unity-gain buffer's input transistors, whose values tend to vary with process and temperature. Similarly, the parasitic-insensitive techniques introduced in Chapter 3 are not applicable on unity-gain buffer-based SC integrators, due to its inherent low gain.

The large-bandwidth or high-speed op-amps tend to have low dc gains due to the fundamental tradeoff between RC time constant and $g_m R_{out}$. A low op-amp dc gain tends to introduce nonlinearity errors to the SC integrator's output, hence compromising the filter's accuracy performance. As a response to this problem, op-amp-based techniques typically emphasize modifying the traditional op-amp structures so that the resultant new op-amps are capable of fulfilling both the speed and accuracy requirements of high-frequency SC filters, despite the fact that they typically have low dc gains. In this implementation, the gain regulating approach [303] is followed, where the low dc gain of each large-bandwidth op-amp in the high-frequency SC filter is precisely controlled and the regulated gain value is used as a reference to scale the capacitances in the circuit.

4.3.2.2 Continuous-Time Waveform Generator

To realize analog signal generation in G_m-C technique shown in Fig. 4.27b and again illustrated in Fig. 4.29 multiple transconductors are connected in parallel to sum all the output currents. Scaling the input transconductance by a factor K introduces only a gain factor in the overall transfer function. The transconductor is based on a linearized voltage follower driving a resistor. The transconductance stems primarily from the resistor and the linearity is therefore limited only by the material of the resistor and the amount of loop gain linearizing the voltage follower. The operating principle of the G_m stage can be explained from Fig. 4.30. A differential input voltage causes a current to flow through the resistor R_{DEG} to the

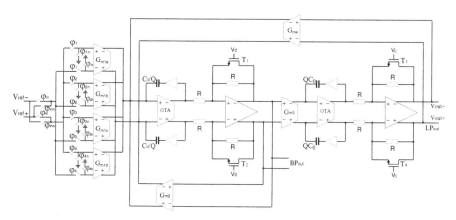

Fig. 4.29 G_m-C-OTA realization

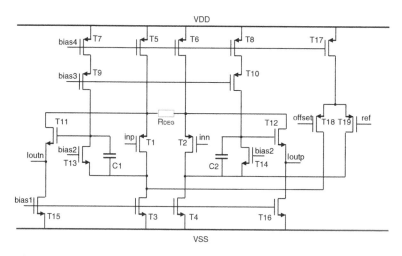

Fig. 4.30 Transconductor G_m realization

source node of T_1. The feedback loop formed by the transimpedance stage T_3 and T_{13}, and transconductance stage T_{11} and T_{15}, forces all of this current to flow to the other input, thus making the drain current and V_{GS}-voltage of T_1 constant and thereby linearizing the voltage follower T_1. The linear voltage follower gives a linear relationship between the input voltage and the current through the resistor R_{DEG} supplied by the output transconductance stage. Since the sum of the currents from the two outputs of the transconductance stage is constant it follows that the current from the other output is also a linear function of the input voltage, and that this current may be used as an output of the whole transconductor -structure. The transimpedance stage is a single transistor T_{13} in common-gate configuration. This stage provides a low impedance input for the voltage follower T_1 and a suitable bias level for the transconductance stage T_{11} extending the swing capability of the input. Similarly, it increases gain in the feedback loop reducing the gain error and linearizes the voltage follower. However, source degeneration lowers the value of the effective transconductance and, in this way, the dc gain of the integrator. To solve this, the output impedance is increased by using nonminimal transistor lengths in the output branches. Cascode devices are not needed in the output of the transconductor G_m since the virtual ground of the OTA in the G_m-C-OTA construction is connected to the output of the G_m stage. The capacitors C_1 and C_2 stabilize the feedback loops of G_m. To avoid significant reductions in the signal handling capability of G_m stage, the input-referred dc offset voltage has to be canceled in the input stage of G_m stage. This has been implemented with a p-channel differential pair T_{18-19}. The output signals from this differential pair are connected to the drains of the input devices of G_m so as to be able to control the gate-source voltages of the input devices and thus the input-referred dc offset voltage of G_m. The dc offsets are canceled with an off-chip control signal V_{offset}.

The OTA used in the integrator shown in Fig. 4.29 is a simple n-channel differential pair with an active load. Applying a capacitive feedback around it leads to a parasitic right half-plane zero in the closed-loop transfer function, causing a serious degradation in the phase response of the integrator near its unity-gain frequency. An ideal unity-gain buffer, inserted in the feedback path eliminates this right half-plane zero. However, the buffer, implemented here as a simple source-follower, has a finite output impedance. This creates a high-frequency left half plane zero which causes an unwanted phase-lead error. To create a parasitic phase-lag that nominally cancels the phase lead created by the source follower and provide additional gain a variable-gain voltage amplifier is connected in cascade. The gain of this amplifier can be varied by shunting the feedback resistor with a MOSFET in triode region (T_{1-4}). Changing the gate voltage of this MOSFET varies the filter unity-gain frequency.

To keep filter bandwidth relatively constant with temperature and process variations, tuning of the characteristic frequency and a quality factor Q of the filter is required. The most widely known continuous-time filter is the active RC filter, which is characterized with high dynamic range, low distortion features. However, restricted speed due to the negative feedbacks limits the active RC filter to low speed applications. Similarly, the gain-bandwidth product of the filter's amplifier has to be much higher than the filter cut-off frequency to minimize phase error and any other non-idealities which occur near the gain-bandwidth product [306], which in itself requires excessive power consumption. Similarly, device mismatches and noise limit the implementation of very high resistor values in the MOSFET-C filter. In addition, rather high control voltage V_C is required to ensure the triode region operation of the transistor. Similarly, achievable signal swing in low voltage application is limited. The G_m-C filters often make use of open-loop rather than closed-loop operational transconductance amplifiers and thus need not be constrained by the stability requirement [162, 274]. As a result, G_m-C filters typically have a speed advantage over both SC and RC filters, particularly for applications in the hundreds of megahertz range. However, the drawback of using G_m stages in an open loop configuration is that the circuit is limited to small input levels in order to operate the transconductor in linear region [274, 309]. Even though many different techniques have been reported to increase the input range while maintaining linearity, these techniques often degrade the frequency response due to additional parasitics [309, 310]. Even with the linearization technique, the input signal swing range is still small compared to the active RC filters. Another drawback of G_m-C filters is their dependence on the transconductance g_m, which makes them highly susceptible to process variations. One advantage of the G_m-C filter is that passive components like the resistor and inductor can be realized with OTA and capacitor. The resistor is realized by connecting negative input of OTA to positive output and the function of inductor can be realized at certain frequency by using two OTAs and a capacitor. However, linearity of the resistors and the inductors is highly dependent on the performance of OTA. If the OTA has finite bandwidth characteristics, the gain and phase of the filter will deviate from its ideal frequency response. The specific nature of the deviation will be a function of the OTA's open-loop characteristic and the filter's desired response. However, in comparison to the active RC realization where full opamp is employed, due the simplicity of the OTA, active G_m-C filters can be pushed to higher frequencies before OTA gain-bandwidth product impacts the filter's response.

The system coefficients of monolithic continuous-time filters are primarily determined by the products of resistors and capacitors in active RC filters or transconductances and capacitors in G_m-C implementations. Although the absolute value of integrated resistors, capacitors and transconductances are quite variable; resistors vary due to doping and etching non-uniformities (could vary by as much as ~40% due to process and temperature variations), capacitors due oxide thickness variations and etching inaccuracies (~15%) and transconductances due to mobility, oxide thickness, current, device geometry variations (could vary by more than 40% with process, temperature and supply voltage variations), their ratios can be very accurate and stable over time and temperature if special attention to layout (e.g. interleaving, use of dummy devices, common-centroid geometries ...) is provided, e.g. capacitor ratio matching with <0.05%, resistor ratio matching <0.1% and G_m ratio matching <0.5%, providing well-preserved relative amplitude and phase versus frequency characteristics with the need to adjust only continuous-time filter critical parameters. It is clear however, that besides the characteristic frequency, sometimes tuning is required of a quality factor Q as well, to keep filter bandwidth relatively constant with temperature and process variations. There are many different tuning systems such as adaptive tuning circuit, direct tuning strategy, master-slave tuning circuit, etc. In PLL tuning, the output of the VCO or master filter is compared with the reference signal through a phase detector. If there is any difference between reference signal and the output, a certain level of voltage, which is proportional to phase difference between reference signal and VCO or master filter output, is produced and the voltage is filtered through low-pass filter to eliminate high frequency component. Then only dc control voltage is applied to both master and main filter to correct the frequency difference. For Q factor control, amplitude detector or peak detector is often used to compare the magnitude at the certain frequency. Like frequency tuning, the peak of the master filter is compared with the reference signal at the certain frequency. Then peak difference is usually amplified and low pass filtered to apply Q factor control voltage to the main filter. However, to utilize such a tuning, it is assumed that the filter's bandpass gain at center frequency is equal (or proportional) to the quality factor Q. Nevertheless, in the presence of dominant parasitics this relation may not hold, especially at high-Q values. Also, nonlinearities of the filter affect the gain versus Q relationship, which may result in further Q-tuning errors. Here, the technique applied utilizes the phase response of a second-order block rather than the magnitude. In this case, Q-tuning, as well as center frequency tuning, is independent of the filter gain. By fitting the output phase at two reference frequencies to the known values, which can be calculated from the desired response, both center frequency and Q are tuned accurately. Utilizing a digital technique, only one tuning loop is allowed to operate at a given time, which improves the stability of the tuning circuit and eliminates the need for slow loops.

Figure 4.31 shows the complete tuning system. The low-frequency clock signal Clk determines the tuning cycles in which the filter's frequency is switched between referenced frequencies. The signals V_{high} and V_{low} are delayed clocks, whereas V_{up} and V_{down} are the control voltages. The filter is calibrated when the $Tune$ signal is high. Normal operation of the filter (i.e., processing the signal) is resumed when the $Tune$ signal is set to low, where the tuning voltages V_{up} and V_{down} are hold at their proper values. Since the two reference frequencies cannot be applied simultaneously, the frequency of the filter changes periodically between referenced frequencies. When one reference frequency is applied to the filter, the low-pass output is compared with V_{low}. The low-pass output is connected to the clock input of the flip-flop whereas the reference is applied to the D input.

Assuming that transition occurs at the rising edge of the clock, the flip-flop stores high output level if the low-pass delay is more than V_{low}, otherwise the output is zero. In the next phase, when the filter input is at different reference frequency, delay of the low-pass output is compared with V_{high}. The first set of D flip-flops compares the low-pass phase with the appropriate references and the output is stored in the following D flip-flops at the end of each cycle. The counters are updated by the Clk signal and are enabled only when the $Tune$ signal is high. Note that V_{up} and V_{down} do not change simultaneously (in the same tuning cycle), which ensures the stability of the tuning loops. Similarly, $DACs$ operate at very low

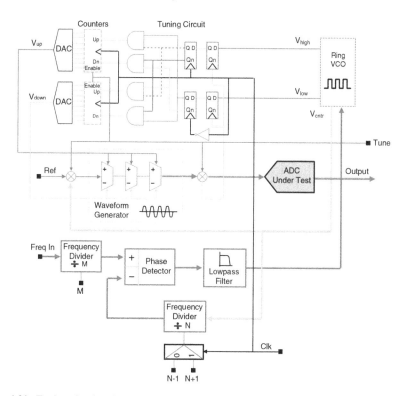

Fig. 4.31 Tuning circuitry for the Gm-C-OTA realization

frequency and can be avoided in case of a digitally tunable filter [317]. The simulation results shown in Fig. 4.32, illustrate the feasibility of both, the discrete- and continuous-time realizations.

4.4 Remarks on Built-In Self-Test Concepts

In the past, code transition levels of analog-to-digital converters were determined according to a step-wise statistical algorithm [333]. However, the necessity for each transition level of about five input changes, and, for each of them, of a significant calibrator settling time, makes the test duration prohibitive for high-resolution A/D converters. The histogram procedure [333–336] is adopted in order to reduce the sample number and the test duration in comparison to the step-wise statistical algorithm. The histogram or output code density is the number of times every individual code has occurred. For an ideal A/D converter with a full scale ramp input and random sampling, an equal number of codes is expected in each bin. The number of counts in the ith bin $H(i)$ divided by the total number of samples N_t, is the width of the bin as a fraction of full scale. By compiling a cumulative

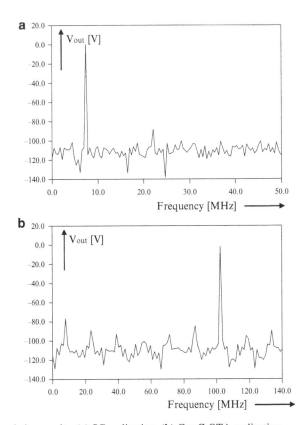

Fig. 4.32 Simulation results: (**a**) SC realization, (**b**) Gm-C-OTA realization

histogram, the cumulative bin widths are the transition levels. A linear signal (triangular wave) is used for achieving a uniform stimulus condition over the A/D converter range [335–337]. Conversely, the standard histogram test exploits a sinusoidal stimulus signal, which is easier to generate with sufficiently low distortion than a triangular one [333–336].

The viability of a histogram-based BIST approach in case of a sinewave input test signal is explored in [327]. Applying the sequential decomposition of the test procedure, although reducing the additional circuitry implies that a high number of input test patterns are required to complete the test. In histogram-based A/D converter testing, code frequency statistics are collected based on an input signal (ramp, sinusoidal) and analyzed to derive the A/D converter's static measures. To collect the data, most histogram techniques require access to an on-chip memory and a DSP. In the absence of such access either due to the lack of on-chip memory or due to layout constraints, the histogram of each code can be collected in a sequential manner, however, at the cost of appreciably increased the test time. Rather than relying on measuring the code frequency as in the histogram technique,

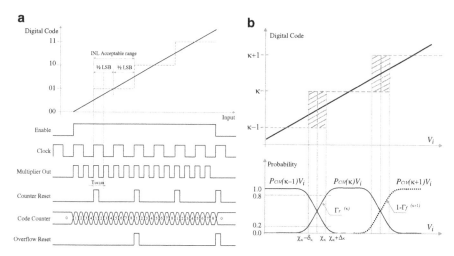

Fig. 4.33 (a) Timing diagram of the testing strategy, (b) probability of code transition

employing a counter along with a code change detector can directly measure the code width (Fig. 4.33). However, as shown in following discussion, linearity of the input signal is the main limiting factor.

The test time is determined by the A/D converter properties, such as the resolution and the sampling frequency f_s as well as of the samples per code ζ. The total test time can be calculated by $T_{total} = (\zeta \times 2^N)/f_s$. The required slop of the ramp signal δ_{ramp} for given full-scale-range V_{FS} of the A/D converter can also be calculated with the test parameters as $\delta_{ramp} = (f_s \times V_{FS})/[(\zeta/N) \times 2^N]$. In an ideal N bits A/D converter code transitions occur for input values given by $\lambda_\kappa = [(\kappa - \frac{1}{2}) \times V_{FS}]/2^N$. However, the input value, for which a change of a code to its adjacent code occurs, is not unique. That is, the code limits in the quantization stair are lost in a transition band. These areas of uncertainty in converters lead to parameter tolerance that characterizes the real converter and limits its resolution.

$$P_{c|V=v}(\kappa|v) = \Gamma_r^{(\kappa)}(v - \chi_\kappa) - \Gamma_f^{(\kappa)}(v - \chi_{\kappa+1}) \tag{4.29}$$

where $\Gamma_{r,f}[.]$ are probability functions that gradually rise from 0 to 1 in the zero neighborhood. They can be taken as ramp functions or as the sigmoid of the Normal Accumulation Probability function $Q(x)$. Thus, we have two functions relate to each code $\Gamma_r^{(k)}(v-\lambda_\kappa)$ representing the disappearance of code κ, when the input decreases, and $1-\Gamma_f^{(k)}(v-\lambda_\kappa)$ representing the slowly disappearance of code κ when the input is increased. This is illustrated in Fig. 4.33b, where each transition is formed by the overlapping of functions corresponding to adjacent codes. As overlapping reduces to two codes, it is evident from the concept of probability that $\Gamma_f^{(\kappa)}[.] = \Gamma_\gamma^{(\kappa)}[.]$. If the transition band is considered such that all input values forming it lead equally to effective transition between codes, the functions $\Gamma_r[.]$ are ramps and the values λ_κ are the central value of the transition band. On the other hand, if there is some value in the band that is most probably identified as representative of the real transition, and this probability decreases asymmetrically as it moves away, then a good estimate of the functions $\Gamma_r[.]$ are the Quasi-Gaussian density distribution functions. When the transition is considered symmetrical the Normal function can be used

$$\Gamma_r^{(\kappa)}(v - \chi_\kappa) = Q[(v - \chi_\kappa)/\sigma] \tag{4.30}$$

Figure 4.33b shows the application of this definition for a probability interval of 20–80%. For example, applying previous equation as the model in the figure obtains the parameter σ from the relation $0.8416 \times \sigma = \Delta_\kappa$, since $Q(0.8416) = 0.8$.

Beside the thermal noise, the accuracy of the test scheme is impacted by number of the samples per code ζ and the linearity of the ramp Λ. The impact of thermal noise on the accuracy can be evaluated assuming Gaussian distribution for the amplitude of the noise spikes having an average of 0 and standard deviation of σ_n. Denoting the fractional part of the analog input signal as v_f, which has a uniform distribution, $p_v(v_f)$ from 0 to 1 LSB, the amplitude of the noise spike should be either smaller than $-v_f$ or greater than $LSB-v_f$ to change the digital output code. It can be shown that using the probability obtained for incorrect conversion, the error in the DNL measurement due to the thermal noise effects can be calculated as

$$\varepsilon_n = \sum_{i=1}^{\zeta} C(\zeta, i) P_n^{(i)} (1 - P_n)^{(\zeta - i)} \frac{i}{\zeta} \tag{4.31}$$

Since error components due to the thermal noise, number of the samples per code ζ and the linearity of the ramp Λ are uncorrelated, the total error θ (in LSB's) will be the sum

$$\varepsilon_{total} = \varepsilon_\zeta + \varepsilon_\Lambda + \varepsilon_n, \quad \varepsilon_\zeta = \frac{1}{\zeta} \quad \varepsilon_\Lambda = \frac{1}{2^{(N_{ramp} - N_{ADC})}} \quad \varepsilon_n = \sum_{i=1}^{\zeta} C(\zeta, i) P_n^{(i)} (1 - P_n)^{(\zeta - i)} \frac{i}{\zeta} \tag{4.32}$$

where N_{ramp} represents the linearity of the ramp signal. When examining the typical data, $\sigma_n = 0.1LSB$, $(N_{ramp} - N_{ADC}) = 3$ and $\zeta = 100$ it is clear that error due to the linearity of ramp signal is the main limiting factor of the method (i.e. $\varepsilon_\Lambda > 10\varepsilon_n$ and $\varepsilon_\Lambda > 10\,\varepsilon_\zeta$).

One of the possibilities to relax the requirement on the source linearity for precision A/D converters test is to employ stimulus error identification and removal (SEIR) algorithm [339]. Here, two nonlinear ramp signals with a constant offset are applied to the A/D converter under test and two sets of histogram data collected from where two estimates for transition levels are calculated. By finding the difference between those estimates through least squares error (LSE) method, source generator ramp nonlinearity can be established. With the knowledge of ramp nonlinearity, their effects on the histogram data can be than removed and A/D converter transition levels accurately identified. However, for the algorithm to be practically applicable severe limitations are set on the test environment, e.g. the exactly the same input signal have to be repeated twice and the offsets have to be constant.

The error introduced by test environment non-idealities, similar to the device offsets as shown in Sections 3.4 and 3.5, will lead to test error whose maximum absolute value occurs at the middle of the A/D converter input range. As practical test solutions usually require the test accuracy to remain within 10% of the device specification, this error will severely limit linearity test precision. However, as the outcome of these environment non-idealities is similar to offset in matching-sensitive circuits that are susceptible to gradient effects, techniques such as inter-leaving (Section 3.2) can relax requirements of the SEIR algorithm [340]. In essence, to get two or more matched electrical quantities, the circuit components are divided into many small unit cells and evenly placed on silicon such that gradient effects on the electrical parameters of these components are averaged out. Similarly, instead of repeating a signal twice and adding an offset to the second one, to relax requirements of the SEIR algorithm many copies of the same signal are generated and an offset to some of them is added according to a given pattern. Generally speaking, the two desired input signals are generated as triangular waves and interleaved with each other in the time domain. When the A/D converter under test is converting the signal, output codes generated will be split into two histograms defined by specific time windows intervals [340]. To cancel out the low-order nonstationary environments error terms, a pattern of one single element is symmetrically extended in center-symmetric interleaving fashion.

To establish source generator ramp nonlinearity in SEIR algorithm, in [341] simplified stimulus identification algorithm calculates the differences and average of two obtained histograms from two offset-shifted stimuli to find the effect of the slope of the stimuli on the heights of the histogram bins. By using numerical differentiation and the measured histograms, derivative of the ideal histogram is approximated. After integrating this derivative through trapezoidal integration rule, code density ideal histogram is obtained, which corresponds to the histogram of an ideal A/D converter for that stimulus. The non-linearity of the test stimuli expressed in code widths is then extracted from the ratio of the measured and ideal histogram. Similarly, the linearity of the A/D converter is determinated through obtained code widths.

One effective approach to reduce production test time is by increasing the parallel efficiency of testing multiple devices-under-test on one tester is the multi-site testing. Nevertheless, for an A/D converter test, the increasing number of devices-under-test in parallel usually requires more high-quality analog signal sources weakening the gains acquired with multi-site testing.

One test technique, which does not need external analog instrumentation, to measure the specifications of DAC-ADC loopback combinations, is a small-signal method [338]. Here, in loopback fashion, digital sinewave stimuli are applied to the input of a D/A converter on-chip. Its output, the analog sinewave, stimulates the on-chip A/D converter whose digital output is analyzed on-chip or off-chip to measure the test specifications. When a ramp or triangle wave is used for histogram tests, additive noise has no effect on the results; however, due to the distortion or nonlinearity in the ramp, it is difficult to guarantee the accuracy. For a differential nonlinearity test, a one percent change in the slope of the ramp would change the expected number of, codes by one percent. Since these errors would quickly accumulate, the integral nonlinearity test would become unfeasible. From consideration above, it is clear that to accurately characterize an A/D converter's linearity the input source should have better precision than the converter being tested. When a sine wave is used, an error is produced, which becomes larger near the peaks. This error can be made as small and desired by sufficiently overdriving the A/D converter. However, in most system-on-chip applications the D/A converter have usually smaller dynamic range than the A/D converter being tested, and hence, cannot be used to generate a signal that stimulates all the ADC's codes.

In the small-signal loopback approach, the constraint on the linearity distortion of the sinewave generator is relaxed by using a signal of amplitude much lower than the A/D converter full scale. The converter input range is stimulated by small-amplitude sinewaves superimposed to a progressively increased dc level. The input range is stimulated entirely by acquiring the samples for the histogram in N_s steps, with the same small-amplitude sinewave, but with different offset levels C_j, $j = 0, 1, \ldots, N_S - 1$. In the test mode, after the conversion from the on-chip digital sinewave generator stimuli in the D/A converter, an analog sine wave with amplitude at least eight times smaller, which corresponds to extra three bit precision in comparison to the dynamic range of the ADC under test (i.e. $N_s > 8$), is applied to the aforementioned A/D converter. At the output of the D/A converter, the analog small sine wave stimulus is level shifted so that the dynamic range of the A/D converter is scanned by increasing the offset Cj step by step as shown in Fig. 4.34. In each of the N_s steps, the A/D converter acquires records of samples of the small wave with amplitude A such that $A = \Delta_s/2 + \Delta_{overdrive}$, where Δ_s is the offset increment value and $\Delta_{overdrive}$ is an extra amplitude to guarantee that all the codes are stimulated. From the samples acquired in each step, a cumulative histogram $CH_j[k]$, which is equal to sum of the number of counts in the ith bin $H(i)$, where $i = 0, ..k$, is built [336]. After all the steps, the obtained N_s transition level arrays have to be combined into a single one. However, the need for overdrive in each of the N_s steps and the inaccuracy of the stimulus signal, give rise to some transition levels having two values computed in two successive steps, which must be considered when combining the transition level arrays. Once the transfer curve is formed, all the static specifications of the A/D converter can be computed.

As argued above, if both converter types are present on the same chip, DAC-ADC loopback combination allows an all-digital test configuration and avoids the need for expensive external analog instrumentation. However, loopback method, suffers from fault masking caused by the uncorrelated interaction between non-functionally related components in loopback mode. In particular, unlike

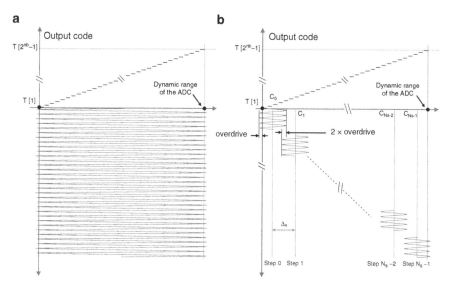

Fig. 4.34 (a) Full-scale and, (b) small-signal sinewave signals applied to the A/D converter

dynamic parameter testing where noise and distortion parameters are additive, static parameters such as INL and DNL tend to cancel, making standard test approaches such as a histogram-based method impractical. Similar to reconstruction and compensation technique, the SEIR algorithm implemented in loopback configuration can estimate linearity of both converters simultaneously without fault masking problems. In an ideal A/D converter, the code widths in the code density histogram are equal. Thus, the code density histogram in ideal case is similar to the probability density function (PDF) of the input signal. If the D/A converter input is uniformly distributed over all the codes, the D/A converter output is uniformly distributed as well; thus, the A/D converter code density histograms can be exploited to calculate ideal histogram. As the last step, the D/A converter linearity estimate can be found from ideal histogram with linear histogram method.

Another possibility to enhance multi-site testing is the possibility of inexpensive on-chip digital waveform generation in its pure or slightly adapted form. As mentioned previously, dynamic parameters of the A/D converter are evaluated through the conventional post-processing methods, such as histogram or FFT analysis, which exploit sine wave stimulus efficiency, i.e., negligible distortion and highly accurate and stable frequency. Employing similar methods to adapted pulse wave diminish advantages of these methods; spectrum of a pulse wave is distorted with harmonics related to the pulse rise and fall times, making accurate determination of A/D converter parametric faults complex and time excessive [344]. The spectral representation of adapted pulse wave is not only a function of sampling frequency and amplitude of the signal, as for sine wave stimuli, but a periodic function of a pulse rise and fall times as well. Additionally, the amplitudes of the expansion coefficients fall have the form of $sinx/x$ with a slope of -20 dB/decade demonstrating low-pass behavior. However, the two algorithms in [344] based on time-modulo reconstruction methods [345] simultaneously relief the accuracy requirement of excitation source in line with recent efforts and offer possibilities for characterization of analog-to-digital converter error mechanisms as a consequence of parametric faults. Additionally, the reconstructed waveform shows errors of the A/D converter more visibly and intuitively [345].

4.5 Stochastic Analysis of Deep-Submicron CMOS Process for Reliable Circuits Designs

One of the most notable features of nanometer scale CMOS technology is the increasing magnitude of variability of the key parameters affecting performance of integrated circuits [346]. Although scaling made controlling extrinsic variability more complex, nonetheless, the most profound reason for the future increase in parameter variability is that the technology is approaching the regime of fundamental randomness in the behavior of silicon structures where device operation must be described as a stochastic process. Since placement of dopant atoms introduced into silicon crystal is random, the final number and location of atoms in the channel of each transistor is a random variable. In addition to device variability, which sets the limitations of circuit designs in terms of accuracy, linearity and timing, existence of electrical noise associated with fundamental processes in integrated-circuit devices represents an elementary limit on the performance of electronic circuits.

Device variability effect and noise limitations as a rudimentary issue for the robust circuit design and evaluation have been subject of numerous studies. Several models have been suggested for device variability [347] and for noise [348, 349], and correspondingly, a number of CAD tools [350, 351] for statistical circuit simulation and noise analysis [349]. Monte Carlo analysis is a widespread approach for statistical analysis of circuits affected by technological variations and/or noise simulation in time-domain. The Monte Carlo algorithm takes random combinations of values chosen from within the range of each process parameter and repeatedly performs circuit simulations. The result is an ensemble of responses from which the statistical characteristics are estimated. Unfortunately, if the number of iterations for the simulation is not very large, the Monte Carlo simulation always underestimates the tolerance window. Accurately determining the bounds on the response requires a large number of simulations, so consequently, Monte Carlo method becomes very cpu-time consuming if the chip becomes large. Other approaches for statistical analysis of variation-affected circuits, such as response surface methodology, are able to perform much faster than Monte Carlo method at the expense of a design of experiments preprocessing stage [351]. The noise performance of a circuit can be analyzed in terms of the small-signal equivalent circuits by considering each of the uncorrelated noise sources in turn and separately computing its contribution at the output. Unfortunately, this method is only applicable to circuits with fixed operating points and is not appropriate for noise simulation of circuits with changing bias conditions.

In this section, a direct approach to statistical simulation based on solving the equations (necessarily stochastic), which describe the statistical behavior of the circuit, rather than estimating it by a population of realizations is shown. The circuits are described as a set of stochastic differential equations and Gaussian closure approximations are introduced to obtain a closed form of moment equations. For wide-sense stationary stochastic processes, such as process variations and/or white noise, the autocorrelation, as a matrix to evaluate the robustness of the circuit design, is completely characterized by the second-moment probabilistic characteristics in time-domain. The method employed here is an enhanced version of the method described in Section 4.1 with an extension for transient analysis. Additionally, numerical method is given for the efficient solution of stochastic differentials for noise analysis of integrated circuits.

4.5.1   Stochastic MNA for Process Variability Analysis

The fundamental notion for the study of spatial statistics is that of stochastic (random) process defined as a collection of random variables on a set of temporal or spatial locations. Generally, the second-order stationary (wide sense stationary WSS) process model is employed, but other more strict criteria of stationarity are possible. This model implies that the mean is constant and the covariance only depends on the separation between any two points. In the second-order stationary process only the first and second moments of the process remain invariant. The covariance and correlation functions capture how the co-dependence of random variables at different locations changes with the separation distance. These functions are unambiguously defined only for stationary processes. For example, the random process describing the behavior of the transistor length L is stationary only if there is nonsystematic spatial variation of the mean L. If the process is not stationary, the correlation function is not a reliable measure of codependence and correlation. Once the systematic wafer-level and field-level dependencies are removed, thereby making the process stationary, the true correlation was found to be negligibly small. From a statistical modeling perspective, systematic variations affect all transistors in a given circuit equally. Thus, systematic parametric variations can be represented by a deviation in the parameter mean of every transistor in the circuit.

In general, for time-domain analysis, modified nodal analysis (MNA) leads to a nonlinear ordinary differential equation (ODE) or differential algebraic equations (DAE) system that, in most cases, is transformed in a nonlinear algebraic system by means of linear multistep integration methods [353, 354] and, at each integration step, a Newton-like method is used to solve this nonlinear algebraic system. Therefore, from a numerical point of view, the equations modeling a dynamic circuit are transformed to equivalent linear equations at each iteration of the Newton method and at each time instant of the time-domain analysis. Thus, we can say that the time-domain analysis of a nonlinear dynamic circuit consists of the successive solutions of many linear circuits approximating the original (nonlinear and dynamic) circuit at specific operating points. The MNA system can be written in the compact form as

$$F(\mathbf{p}', \mathbf{p}, t) + B(\mathbf{p}, t) \cdot \eta = 0 \tag{4.33}$$

where $\mathbf{p}$ is the vector of stochastic processes that represents the state variables (e.g. node voltages) of the circuit and $\boldsymbol{\eta}$ is a vector of wide-sense stationary processes as described in Section 4.1.2. $B(\mathbf{p},t)$ is an $N \times B_c$ current-controlled branches matrix, the entries of which are functions of the state $\mathbf{p}$ and possibly t. Every column of $B(\mathbf{p},t)$ corresponds to $\boldsymbol{\eta}$, and has normally either one or two nonzero entries. The rows correspond to either a node equation or a branch equation of an inductor or a voltage source. Equation (4.33) represents a system of nonlinear stochastic differential equations, which formulate a system of stochastic algebraic and differential equations that describe the dynamics of the nonlinear circuit that lead to the MNA equations when the random sources $\boldsymbol{\eta}$ are set to zero. Solving (4.33) means to determine the probability density function P of the random vector $\mathbf{p}(t)$ at each time instant t. Formally the probability density of the random variable $\mathbf{p}$ is given as

$$P(\mathbf{p}) = |\Gamma(\mathbf{p})| N(h^{-1}(\mathbf{p}) | \mathbf{m}, \Sigma) \tag{4.34}$$

where $|\Gamma(\mathbf{p})|$ is the determinant of the Jacobian matrix of the inverse transform $h^{-1}(p)$ with h a nonlinear function of $\boldsymbol{\eta}$. However, generally it is not possible to handle this distribution directly since it is non-Gaussian for all but linear h. To obtain a closed form of moment equations, Gaussian closure approximations are introduced to truncate the infinite hierarchy as clarified in Section 4.2. It is worth noticing that the main difficulty in solving (4.33) is related to the nonlinearity, thus it may be convenient to look for an approximation that can be found after partitioning the space of the stochastic source variables $\boldsymbol{\eta}$ in a given number of subdomains, and then solving the equation in each subdomain by means of a linear truncated Taylor approximation as explained in Section 4.2.

Let $x_0 = x(\eta_0, t)$ be the generic point around which to linearize, and with the change of variable $\boldsymbol{\xi} = \mathbf{x} - x_0 = [(\mathbf{p} - p_0)^T, (\boldsymbol{\eta} - \eta_0)^T]^T$, the first-order Taylor linearization of (4.33) in x_0 yields

$$C(x_0)\boldsymbol{\xi}' + (G(x_0) + C'(x_0))\boldsymbol{\xi} = 0 \tag{4.35}$$

where $G(x) = B'(x)$ and $C(x) = F'(x)$. Transient analysis requires only the solution of the deterministic version of (4.33), e.g. by means of a conventional circuit simulator, and of (4.35) with a method capable of dealing with linear stochastic differential equations with stochasticity that enters only through the initial conditions. Since (4.35) is a linear homogeneous equation in $\boldsymbol{\xi}$, its solution, will always be proportional to $\boldsymbol{\eta} - \eta_0$. Equation (4.35) can be written as

$$\boldsymbol{\xi}'(x_0) = E(x_0)\xi_0 + F(x_0)\eta_0 \tag{4.36}$$

Equation (4.36) is a system of stochastic differential equations which is linear in the narrow sense (right-hand side is linear in ξ and the coefficient matrix for the vector of variation sources is independent of ξ) [355]. Since these stochastic processes have regular properties, they can be considered as a family of classical problems for the individual sample paths and be treated with the classical methods of the theory of linear stochastic differential equations. By expanding every element of $\boldsymbol{\xi}(t)$ with

$$\boldsymbol{\xi}_i(t) = \Gamma(t)(\boldsymbol{\eta} - \eta_0) = \sum_{j=1}^{m} \alpha_{ij}(t) \cdot \eta_j \tag{4.37}$$

for m elements of a vector $\boldsymbol{\eta}$. As long as $\alpha_j(t)$ is obtained, the expression for $\boldsymbol{\xi}(t)$ is known, so that the covariance matrix of the solution can be written as

$$\Sigma_{\xi\xi} = \Gamma \Sigma_{\eta\eta} \Gamma^T \tag{4.38}$$

Defining $\alpha_j(t) = (\alpha_{1j}, \alpha_{2j}, \ldots, \alpha_{nj})^T$ and $F_j(t) = (F_{1j}, F_{2j}, \ldots, F_{nj})^T$, the requirement for $\alpha(t)$ is

$$\alpha_j'(t) = E(t)\alpha_j + F(t) \tag{4.39}$$

Equation (4.39) is an ordinary differential equation, which can be solved by a fast numerical method.

4.5.2 Stochastic MNA for Noise Analysis

The most important types of electrical noise sources (thermal, shot, and flicker noise) in passive elements and integrated circuit devices have been investigated extensively, and appropriate models have been derived [349] as stationary and in [350] as nonstationary noise sources. In this section the model description as defined in [350] is adapted, where thermal and shot noise are expressed as delta-correlated noise processes having independent values at every time point, modeled as modulated white noise processes. These noise processes correspond to the current noise sources which are included in the models of the integrated-circuit devices. Since these nonstationary models differ from a wide-sense stationary stochastic process models for process parameter variations, they cannot be treated directly as in Section 4.4.1.

The MNA formulation of the stochastic process that describe random influences that fluctuate rapidly and irregularly (i.e. white noise χ) can be written as

$$F(\mathbf{r}',\mathbf{r},t) + B(\mathbf{r},t) \cdot \chi = 0 \qquad (4.40)$$

where $\mathbf{r}$ is the vector of stochastic processes that represents the state variables (e.g. node voltages) of the circuit, χ is a vector of wide-sense white Gaussian processes and $B(\mathbf{r},t)$ is state and time dependent modulation for the vector of noise sources. Equations such as (4.40) cannot be treated as an ordinary differential equation using classical differential calculus due to the characteristics of white noise process χ. Since the magnitude of the noise content in a signal is much smaller in comparison to the magnitude of the signal itself in any functional circuit, a system of nonlinear stochastic differential equations described in (4.40) can be linearized under similar assumptions as noted in the previous section. Now, including the noise content description, (4.36) can be expressed in general form as

$$\lambda'(t) = E(t)\lambda + F(t)\chi \qquad (4.41)$$

where $\lambda = [(\mathbf{r} - r_0)^\mathrm{T}, (\chi - \chi_0)^\mathrm{T}]^\mathrm{T}$. Equation (4.41) is interpreted as an Ito system of stochastic differential equations. Now rewriting (4.41) in the more natural differential form

$$d\lambda(t) = E(t)\lambda dt + F(t)dw \qquad (4.42)$$

where $dw(t) = \chi(t)dt$ is substituted with a vector of Wiener process w. If the functions $E(t)$ and $F(t)$ are measurable and bounded on the time interval of interest, there exists a unique solution for every initial value $\lambda(t_0)$ [355].

If λ is a Gaussian stochastic process, then it is completely characterized by its mean and correlation function. From Ito's theorem on stochastic differentials

$$
\begin{aligned}
d(\lambda(t)\lambda^T(t))/dt &= \lambda(t) \cdot d(\lambda^T(t))/dt \\
&+ d(\lambda(t))/dt \cdot \lambda^T(t) + F(t) \cdot F^T(t)dt
\end{aligned} \qquad (4.43)
$$

and expanding (4.43) with (4.42), noting that λ and dw are uncorrelated, variance-covariance matrix $\mathbf{K}(t)$ of $\lambda(t)$ with the initial value $K(0) = E[\lambda\ \lambda^T]$ can be expressed in differential Lyapunov matrix equation form as [355]

$$d\mathbf{K}(t)/dt = \mathbf{E}(t)\mathbf{K}(t) + \mathbf{K}(t)\mathbf{E}^T(t) + \mathbf{F}(t)\mathbf{F}^T(t) \qquad (4.44)$$

Note that the mean of the noise variables is always zero for most integrated circuits. In view of the symmetry of $K(t)$, (4.44) represents a system of linear ordinary differential equations with time-varying coefficients. To obtain a numerical solution, (4.44) has to be discretized in time using a suitable scheme, such as any linear multi-step method, or a Runge-Kutta method. For circuit simulation, implicit linear multi-step methods, and especially the trapezoidal method and the backward differentiation formula were found to be most suitable [356]. If backward Euler is applied to (4.44), the differential Lyapunov matrix equation can be written in a special form referred to as the continuous-time algebraic Lyapunov matrix equation

$$P_r K(t_r) + K(t_r) P_r^T + Q_r = 0 \qquad (4.45)$$

$K(t)$ at time point t_r is calculated by solving the system of linear equations in (4.45). Such continuous time Lyapunov equations have a unique solution $K(t)$, which is symmetric and positive semidefinite. Several iterative techniques have been proposed for the solution of the algebraic Lyapunov matrix equation (4.45) arising in some specific problems where the matrix P_r is large and sparse [357]. The Bartels-Stewart method [358] has been the method of choice for solving small to medium scale Lyapunov equations such as in analog circuits, and Hammarling's method remains the one and only reference for directly computing the Cholesky factor of the solution $K(t_r)$ of (4.45) for small to medium systems. For the backward stability analysis of Bartels-Stewart algorithm, see [359]. Extensions of these methods to generalized Lyapunov equations are described in [360]. The Bartels-Stewart algorithm is now standard and is presented in textbooks [361, 362]. In this method, first P_r is reduced to upper Hessenberg form by means of Householder transformations, and then the QR-algorithm is applied to the Hessenberg form to calculate the real Schur decomposition [361] to transform (19) to a triangular system which can be solved efficiently by forward or backward substitutions of the matrix P_r

$$S = U^T P_r U \qquad (4.46)$$

where the real Schur form S is upper quasi-triangular and U is orthonormal. The formulation in this analysis utilizes a similar scheme. The transformation matrices are accumulated at each step to form U [358]. Now, setting

$$\bar{K} = U^T K(t_r) U$$
$$\bar{Q} = U^T Q_r U \qquad (4.47)$$

then (4.45) becomes

$$S\bar{K} + \bar{K}S^T = -\bar{Q} \qquad (4.48)$$

To find unique solution, (4.48) is partitioned as

$$\mathbf{S} = \begin{bmatrix} \mathbf{S}_1 & \mathbf{s} \\ \mathbf{0} & \upsilon_n \end{bmatrix} \quad \mathbf{\bar{K}} = \begin{bmatrix} \mathbf{K}_1 & \mathbf{k} \\ \mathbf{k}^T & k_{nn} \end{bmatrix} \quad \mathbf{Q} = \begin{bmatrix} \mathbf{Q}_1 & \mathbf{q} \\ \mathbf{q}^T & q_{nn} \end{bmatrix}$$

where $\mathbf{S}_1, \mathbf{K}_1, \mathbf{Q}_1 \in C^{(n-1) \times (n-1)}$; $s, k, q \in C^{(n-1)}$. The system in (4.48) then gives three equations

$$(\upsilon_n + \bar{\upsilon}_n)k_{nn} + q_{nn} = 0 \tag{4.49}$$

$$(\mathbf{S}_1 + \bar{\upsilon}_n\mathbf{I})\mathbf{k} + \mathbf{q} + k_{nn}\mathbf{s} = \mathbf{0} \tag{4.50}$$

$$\mathbf{S}_1\mathbf{K}_1 + \mathbf{K}_1\mathbf{S}_1^T + \mathbf{Q}_1 + \mathbf{sk}^T + \mathbf{ks}^T = \mathbf{0} \tag{4.51}$$

k_{nn} can be obtained from (4.49) and set in (4.50) to solve for k. Once k is known, (4.51) becomes a Lyapunov equation which has the same structure as (4.48) but of order $(n-1)$, as

$$\mathbf{S}_1\mathbf{K}_1 + \mathbf{K}_1\mathbf{S}_1^T = -\mathbf{Q}_1 - \mathbf{sk}^T - \mathbf{ks}^T \tag{4.52}$$

The same process can be applied to (4.52) until $\mathbf{S}_1$ is of the order -1. Note under the condition that $i = 1, \ldots, n$ at the k-th step ($k = 1, 2, \ldots, n$) of this process, a unique solution vector of length $(n + 1-k)$ and a reduced triangular matrix equation of order $(n-k)$ can be obtained. Since $\mathbf{U}$ is orthonormal, once (4.45) is solved for $\mathbf{\bar{K}}$, then $\mathbf{K}$ (t_r) can be computed using

$$\mathbf{K}(t_r) = \mathbf{U}\mathbf{\bar{K}}\mathbf{U}^T \tag{4.53}$$

Large dense Lyapunov equations such as those in large scale digital circuits can be solved by sign function based techniques [363], which perform well on parallel computers. Krylov subspace methods, which are related to matrix polynomials have been proposed [364] as well. Relatively large sparse Lyapunov equations can be solved by (standard) iterative approaches, e.g., [365]. In this analysis, a low rank version of the iterative method [366], which is related to rational matrix functions is applied. The postulated iteration [365] for the Lyapunov equation (4.45) is given by $\mathbf{K}(0) = 0$ and

$$\begin{aligned} (P_r + \gamma_i I_n)K_{i-1/2} &= -Q_r - K_{i-1}(P_r^T - \gamma_i I_n) \\ (P_r + \bar{\gamma}_i I_n)K_i^T &= -Q_r - K_{i-1/2}^T(P_r^T - \bar{\gamma}_i I_n) \end{aligned} \tag{4.54}$$

for $i = 1, 2, \ldots$ This method generates a sequence of matrices $\mathbf{K}_i$ which often converges very fast towards the solution, provided that the iteration shift parameters

γ_i are chosen (sub)optimally. For a more efficient implementation of the method, iterates are in this analysis replaced by their Cholesky factors, i.e., $K_i = L_i L_i^H$ and reformulated in terms of the factors L_i. The low rank Cholesky factors L_i are not uniquely determined. Different ways to generate them exist [366]. Note that the number of iteration steps i_{max} needs not be fixed a priori. However, if the Lyapunov equation should be solved as accurate as possible, correct results are usually achieved for low values of stopping criteria that are slightly larger than the machine precision.

4.5.3   Application Example

The effectiveness of the approaches was evaluated on several circuits exhibiting different distinctive features in a variety of applications. As a representative example of the results that can be obtained, the statistical simulation is applied to the characterization of the G_m-C-OTA biquad filter shown in Section 4.3. The calculated frequency and transient response of the filter are illustrated in Fig. 4.35. In comparison with Monte Carlo analysis (it can be shown that 1,500 iterations are necessary to accurately represent performance function) difference is less than 1% and 3% for mean and variance, respectively, while achieving significant gain on the cpu-time (12.2 vs 845.3 s). For noise simulations only the shot and thermal noise sources are included as including the flicker noise sources increases the simulation time due to the large time constants introduced by the networks for flicker noise source synbook. It is assumed that the time series r are composed of a smoothly varying function plus additive Gaussian white noise χ (Fig. 4.36a), and that at any point r can be represented by a low order polynomial (a truncated local Taylor series approximation). Noise estimation is robust to a few arbitrary spikes or discontinuities in the function or its derivatives (Fig. 4.36b). This is achieved by trimming off the tails of the distributions and then using percentiles to reverse the desired variance. However, this process increases simulation time and introduces bias in the results. Inadvertently, this bias is a function of the series length and as such predictable, so the last steps in noise estimation are to filter out that predicted bias from the estimated variance. The results of the estimation of the noise variance are illustrated in Fig. 4.37a. In Fig. 4.37b the filtered piecewise-linear and smoothed estimates of the probabilities of the model in each time step is plotted. It can be seen that it takes some time for the piecewise-linear estimates to respond to model transitions. As expected, smoothing reduces this lag as well as giving substantially better overall performance.

   The quality criterion adopted for estimating parameter variations is the mean-squared error criterion, mainly because it represents the energy in the error signal, is easy to differentiate and provides the possibilities to assign the weights (Fig. 4.38a). The Bartels-Stewart algorithm and Hammarling's method carried out explicitly (as done in Matlab) can exploit the advantages provided by modern high

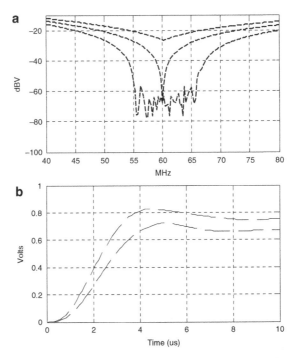

Fig. 4.35 (a) Gm-C-OTA biquad filter frequency response. Middle line designates the nominal behavior, (b) transient response of Gm-C biquad filter, minimum and maximum values are shown

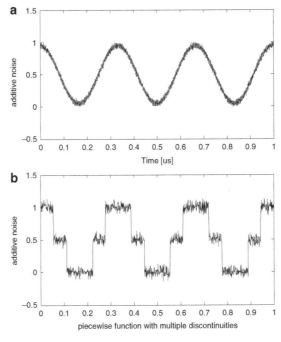

Fig. 4.36 (a) Time with additive Gaussian noise, (b) noise estimation for functions with multiple discontinuities

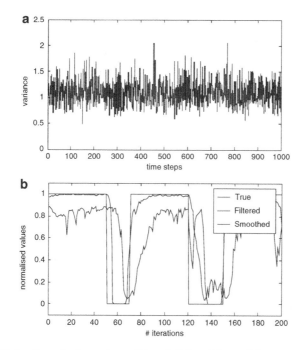

Fig. 4.37 (a) Estimation of noise variance, (b) probability of the model

performance computer hardware, which contains several levels of cache memories. For the recursive algorithms presented here it is observed that a faster lowest level kernel solver (with suitable block size) leads to an efficient solver of triangular matrix equations. For models with large dimension of the current N_c and node N_v branches, usually the matrix $\mathbf{P}_r$ has a banded or a sparse structure and applying the Bartels-Stewart type algorithm becomes impractical due to the Schur decompositions (or Hessenberg-Schur), which cost expensive $O(N^3)$ flops. In comparison with the standard Matlab function *lyap.m*, the cpu-time shows that computing the Cholesky factor directly is faster by approximately N flops. Similarly, when the original matrix equation is real, using real arithmetic is faster than using complex arithmetic. Hence, in this analysis, it is resorted to iterative projection methods when N_c and N_v are large, and the Bartels-Stewart type algorithms including the ones presented become suitable for the reduced small to medium matrix equations.

The approximate solution of Lyapunov equation is given by the low rank Cholesky factor L, for which $LL^H \sim K$. L has typically fewer columns than rows. In general, L can be a complex matrix, but the product LL^H is real. More precisely, the complex low rank Cholesky factor delivered by the iteration is transformed into a real low rank Cholesky factor of the same size, such that both low rank Cholesky factor products are identical. However, doing this requires additional computation. The iteration is stopped after a priori defined iteration steps

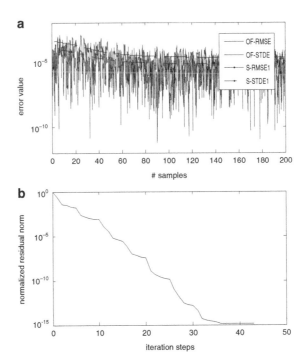

Fig. 4.38 (**a**) RMSE of estimating parameter variation with optimal piece-wise approximation and smoothing algorithm. (**b**) Stopping criterion: maximal number of iteration steps

(Fig. 4.38b) as in [367]. In comparison with 1,500 Monte Carlo iterations difference is less than 1% and 4% for mean and variance, respectively, with considerable cpu-time reduction (1241.7 vs 18.6 s).

4.6 Conclusion

The increasing costs associated with testing and debugging of complex mixed-signal electronic systems has motivated various research efforts to explore efficient testing methodologies. Structural, fault-model based testing has been recognized as a promising alternative or addition to expensive specification testing. The presented inexpensive structural methodology intends to facilitate test pattern generation at wafer level test, thereby providing a quantitative estimate of the effectiveness and completeness of the testing process. In the fault-based, structural testing, the fault model cannot only describe the fault effect more clearly and offer clues to derive the test stimuli, but also makes it more feasible to modify the input stimuli and to estimate its fault coverage. This approach is well established for digital circuits for which there are numerous test pattern generators and fault simulators based on a pass/fail decision, as is the case of the single-stuck-at fault models. Recent studies have revealed, however, that faults which shift the operating point of a transistor level analog circuit can be detected by inexpensive quasi-static testing or power supply current monitoring. With that in mind, the fault model utilizes the sensitivity of the circuit's quasi-static node voltages to process variations and consequently the current deviance so as to differentiate the faulty behavior. With the Karhunen-Loève expansion method, the parameters of the transistors are modeled as stochastic processes over the spatial domain of a die, thus making parameters of any two devices on the die, two different correlated random variables.

In fault-based approaches to analog testing the response from the device under test during manufacture is compared with the pre-compiled fault responses and a decision of rejection of the device is based on this comparison. It has been recognized that failures in analog circuits may assume a continuum of values, and a band-fault approach has been proposed for linear analog circuits where the signature for each fault assumed the form of a band and the decision of pass/fail made on the basis of comparison of bands. Since normally the fault list is quite large in analog fault simulation, it is very important to generate the tolerance window fast and efficiently. Therefore, the bounds on the circuit response are calculated by performing a mathematical evaluation of the performance function attributes against physical tolerance limits defined by device mismatch yielding the significant CPU-time savings. The Bayes risk is computed for all stimuli and for each fault in the fault list. The stimuli for which the Bayes risk is minimal, is taken as the test vector for the fault under consideration. The Neyman-Person statistical detector, which is a special case of the Bayes test, provides a workable solution when the a priori probabilities are unknown, or the Bayes risk is difficult to evaluate or set objectively. The tests are generated and evaluated taking into account the potential fault masking effects of process spread on the faulty circuit responses. The test generator technique also allows the test procedure to test only for the most likely group of faults induced by a manufacturing process.

To overcome system-test limitations of the structural current-based testing, the device-under-test is partitioned into smaller blocks with only limited additional hardware by means of the power-scan DfT technique. The variety and number of cores in modern Systems-on-Chip (SoC) and their nature type, e.g. analog, complicate the testing phase of individual blocks, of combinations of blocks and ultimately of the whole system. The problem in the analog domain is that it is much more difficult to scan signals over long distances in a chip and across its boundary to the outside world, since rapid signal degradation is very likely to occur. The DfT intends to facilitate structural, signature-based testing, by providing the means to observe the current (or voltage) signatures of individual cores (or parts of) instead of observing the current (or voltage) signature of the whole analog SoC. Additionally, as in case of multi-step A/D converters, such a DfT lessen the impact of overlap between the conversion ranges implemented to obtain high linearity, which can either mask faults or give an incorrect fault interpretation. As the results of test pattern generator indicate, most quasi-static failures in various blocks of the 12-bit two-step/multi-step A/D converter, depending on the degree of partitioning, are detectable through power-supply current structural test offering more than 20-fold reduction in test time in comparison to more traditional, functional histogram-based static or FFT-based dynamic ADC test. Only limitations of the quasi-static approach are due to the dynamic nature of the response such as faults in the inactive parts of the comparator's decision stage and storage latch. Furthermore, the fault coverage obtained shows that the resistor-based D/A converter is not suitable for current signature-based testing without additional, application specific, adjustments.

Besides DfT, BIST approaches are similarly an efficient way to help decrease the test development and debugging costs. The analog circuits are usually tested using functional approaches, often requiring a large data volume processing, high accuracy and high speed ATEs. In addition, these analog cores are normally very sensitive to noise and loading effects, which limit the external monitoring and make their test a difficult task. BIST schemes consist on moving part of the required test resources (test stimuli generation, response evaluation, test control circuitry, etc.) from the ATE to the chip. On-chip evaluation and generation of periodic signals are of undoubted interest from this point of view as most of the analog systems can be characterized and tested (frequency domain specifications, linearity, etc.) using this kind of stimuli. The method for sine-wave signal generation shown in this book relay on the programmable integrator, which allows discrete-time (for high-resolution, but lower-speed) or continuous-time (for high-speed, although with lower-resolution) periodic analog signal generation. The method has the attributes of digital programming and control capability, robustness and reduced area overhead, which make it suitable for BIST applications. To determine code transition levels to estimate the linearity of A/D converters the histogram procedure offers reduced sample number and test duration in comparison to the other statistical algorithms. However, to achieve a uniform stimulus condition over the A/D converter range, linearity of the input signal has to be at least three-bit more linear then device-under-test; goal, which is very difficult to achieve for the test of the very high-resolution A/D converters. Continuous research efforts explore methods to relax this requirement on the source linearity. For this purpose, survey and analysis of the several on-going methods is presented. Additionally, some remarks on the loopback method, promising configuration which allows all-digital test are given.

Statistical simulation is one of the foremost steps in the evaluation of successful high-performance IC designs due to process variations and circuit noise that strongly affect devices behavior in today's deep submicron technologies. In this book, rather than estimating statistical behavior of the circuit by a population of realizations, integrated circuits are described as a set of stochastic differential equations. For wide-sense stationary stochastic processes, such as process variations and/or white noise, Gaussian closure approximations are introduced to obtain a closed form of moment equations. The effectiveness of the approaches was evaluated on several circuits with continuous-time biquad filter as a representative example. As the results indicate, the suggested numerical methods provide accurate and efficient solutions of stochastic differentials for both, process variation and noise analysis of various scales of integrated circuits.

Chapter 5
Multi-Step Analog to Digital Converter Debugging

5.1 Concept of Sensor Networks

CMOS technologies move steadily toward finer geometries, which provide higher digital capacity, lower dynamic power consumption and smaller area resulting in integration of whole systems, or large parts of systems, on the same chip. However, due to technology scaling ICs are becoming more susceptible to variations in process parameters and noise effects like power supply noise, cross-talk reduced supply voltage and threshold voltage operation. Likewise, imperfection at the manufacturing stage, with a raw factory yield between 50% and 95%, depending on the maturity of the process technology, silicon area, and extending the use of 193 nm lithography for sub-65 nm CMOS technology, where Resolution Enhancement Techniques are no longer sufficient for accurate device definition, significantly impact circuit performance. With increased system complexity and reduced access to internal nodes, the task of accurately testing these devices is becoming a major bottleneck. The large number of parameters required to fully specify the performance of mixed-signal circuits and the presence of both analog and digital signals in these circuits make the testing expensive and a time consuming task. Particularly for nanometer CMOS ICs, the large number of metal layers with increasing metal densities, prevents physical probing of the signals for debug purposes. Since parameter variations depend on unforeseen operational conditions, chips may fail despite they pass standard test procedures.

Traditional test methods for analog circuits rely on specification testing, in which some or all response parameters are checked for conformity to the design specifications. However, specification testing is time consuming and hence, also expensive. Although several attempts [368–377] have been made to alleviate increasing test difficulties of A/D converter testing and debugging, none of these methods provides the possibilities for early identification of excessive process parameter variations. In [368], DSP techniques for data analysis are utilized. However, the technique requires intensive computation and on-chip availability of both, A/D and D/A converter. In [371], processing core circuits are incorporated into a VXI

A. Zjajo and J. Pineda de Gyvez, *Low-Power High-Resolution Analog to Digital Converters*, Analog Circuits and Signal Processing, DOI 10.1007/978-90-481-9725-5_5, © Springer Science+Business Media B.V. 2011

bus-based system, which performs both static and dynamic tests. A similar system with external instruments is developed in [372]. A large amount of sampled data must be collected to support both methods. The approach in [373] relies on analog circuitry and reference voltages for measurements and allows testing of only D/A converter-based A/D converters. In the oscillation-test method in [374], the impact of the control logic delay and the imperfect analog BIST circuitry on the test accuracy is not assessed. In [375], the linearity of the A/D converters is tested by monitoring the LSB externally. In [376] an efficient polynomial-fitting algorithm for D/A converter and A/D converter BIST is proposed. The drawback is again the need of both, on-chip A/D converter and D/A converter. The viability of a histogram-based BIST approach in case of a sine-wave input test signal is investigated in [377]. Applying the sequential decomposition to the test procedure, although reducing the additional circuitry implies that a high number of input test patterns are required to complete the test.

Functional faults in each of the analog component in the multi-step A/D converters affects the transfer function differently [378] and analyzing this property forms the basis of the approach described in this chapter [379]. To enhance observation of important design and technology parameters, such as temperature, threshold voltage, etc., and provide valuable information, which can be used to guide the test and allow the estimation of selected performance figures, dedicated sensors are embedded within the functional cores [380, 381]. Additionally, by monitoring on-chip process deviation the method intends to facilitate fast identification of excessive process parameter variations and to provide a reliable and complementary method to quickly discard faulty circuits in wafer and final tests without testing the complete device. Such a test method reduces the cost associated to production tests, since this early detection of the faulty circuits avoids running an important fraction of traditional tests. Detecting the faulty devices at the wafer level has additional advantage that packaging costs (which commonly represent 25% of the total system cost) can be avoided. Economic considerations are only one of the advantages of providing die-level process variation observability. Other advantages include increased fault coverage and improved process control, diagnostic capabilities, reduced IC performance characterization time-cycle, simplified test program development and easier system-level diagnostics.

5.1.1 Observation Strategy

From a circuit design perspective parametric process variations can be divided into inter-die and intra-die variations. Inter-die variations such as the process temperature, equipments properties, wafer polishing, wafer placement, etc. affect all transistors in a given circuit equally. For the purposes of circuit design, it is usually assumed that each component or contribution in inter-die variation is due to different physical and independent sources; therefore, the variation component can be represented by a deviation in the parameter mean of the circuit. Intra-die variations are deviations occurring within a die. These variations may have a variety of sources that depend on the physics of the manufacturing steps (optical proximity effect, dopant fluctuation, line edge roughness, etc.) and the effect of these non-idealities (noise, mismatch) may limit the minimal signal that can be processed and the accuracy of the circuit behavior. For linear systems, the non-linearities of the devices generate distortion components of the signals limiting the maximal signal that can be processed correctly. Although, certain circuit techniques such as using small modulation index for the bias current to reduce the effect of distortion non-idealities, large device sizes to lower mismatch and utilizing low-impedance level to limit the thermal noise signals, these measures have, however, important consequences on the power consumption and operation speed of the system.

In general, the design margins for mixed signal designs depend significantly on process parameters and their distributions across the wafer, within a wafer lot and between wafer lots, which is especially relevant for mismatch. Measurement of these fluctuations is paramount for stable control of transistor properties and statistical monitoring and the evaluation of these effects enables the efficient development of the test patterns and test and debugging methods, as well as ensures good yields. IC manufacturing facilities try to realize constant quality by applying various methods to analyze and control their process. Some of the quality control tools include, e.g. histograms, check sheets, pareto charts, cause and effect diagrams, defect concentration diagrams, scatter diagrams, control charts, time series models and statistical quality control tools, e.g. process capability indices, and time series. Process control monitoring (PCM) data (electrical parameters, e.g. MOS transistor threshold voltage, gate width, capacitor Q-value, contact chain resistance, thin-film resistor properties, etc. measured from all the test dice on each wafer) is required to be able to utilize these quality control tools. Making decisions about if the product or process is acceptable, is by no means an easy task, e.g. if the process/product is in control and acceptable, or in control but unacceptable, or out of control but acceptable.

When uncertain, additional tests for the process and/or the product may be required for making the decision. Masks for wafers are generally designed so that a wafer after being fully processed through the IC manufacturing process will contain several test dice. The area consumed by a test die is usually quite large, i.e. sometimes comparable to several ordinary production dice. Measuring the electrical properties from the test dice gives an estimate of the quality of the lot processing, and device requirement to fulfill a priori specifications, e.g. temperature range, speed. Finally the IC devices are tested for functionality at the block level in the wafer probing stage, and the yield of each wafer is appended to the data. The tester creates suitable test patterns and connects signal generators online. It digitizes the measurement signals and finally determines, according to the test limits, if the device performs acceptably or not. Then, the wafer is diced, and the working dice are assembled into packages. The components are then re-tested, usually in elevated temperature to make sure they are within specification.

Silicon wafers produced in a semiconductor fabrication facility routinely go through electrical and optical measurements to determine how well the electrical parameters fit within the allowed limits. The yield is determined by the outcome of the wafer probing (electrical testing), carried out before dicing. The simplest form

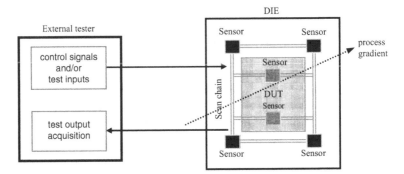

Fig. 5.1 Observation strategy

of yield information is the aggregate pass/fail statistics of the device, where the yield is usually expressed as a percentage of good dice per all dice on the wafer. In principle, yield loss can be caused by several factors, e.g. wafer defects and contamination, IC manufacturing process defects and contamination, process variations, packaging problems, and design errors or inconsiderate design implementations or methods. Constant testing in various stages is of utmost importance for minimizing costs and improving quality. Figure 5.1 depicts the observation strategy block diagram for dice wafer probing. A family of built-in process variation sensing circuits is placed (at least) at each corner of the device-under-test as it maximizes the sensing capability of process variations due to process gradients or are embedded in the device-under-test itself. Depending on the size of DUT, additional sensors can be placed in and around device-under-test to form the additional statistical mass. Figure 5.2 depicts the test strategy block diagram applied to a two-step/multi-step A/D converter.

5.1.2 Integrated Sensor

The complete test scheme including the die-level process monitor circuit, detector, reference ladder and the switch matrix to select the reference levels for decision window, the interface to the external world, control blocks to sequence events during test, the scan chain to transport the pass/fail decision, and the external tester is illustrated in Fig. 5.3. The analog decision is converted into pass/fail (digital) signals through the data decision circuit. The interface circuitry, allows the external controllability of the test, and also feeds out the decision of the detector to a scan chain. The test control block (TCB) selects through a test multiplexer (TMX) the individual die-level process monitor circuit measurement. Select, reference and calibration signals are offered to the detector through this interface circuitry. Digital control logic can be inserted on the chip or done externally.

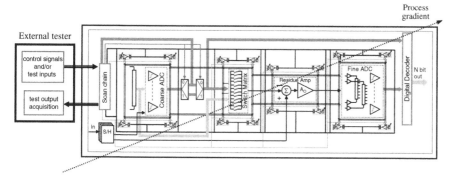

Fig. 5.2 Concept illustrated on multi-step A/D converter

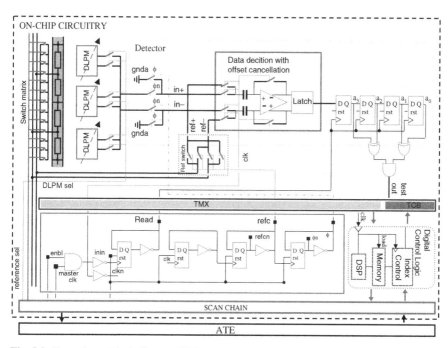

Fig. 5.3 Test scheme block diagram [379]

The die-level process monitor (DLPM) circuits can be extracted from selected structures in the device-under-test. This selection, although easily generalized, relies on the knowledge and analysis of the device-under-test itself, so the resulting DLPM circuits and reference voltages are related to the device-under-test specifications and performance figures under study. Alternatively, die-level process monitor circuits can be designed as an array of transistor pairs as well, each of the different size, where one pair of the n-channel and p-channel transistors is selected through the internal decoding/selection circuitry. Loading for each transistor pair can be extracted from the device-under-test or set independently. The data detector compares the output of the die level process monitor against a comparison reference window, whose voltage values (corresponding to the required LSB values) are selected from the reference ladder or set externally. The reference voltages defining the decision windows are related to the device-under-test specifications and performance figures under study. By sweeping the reference voltage until a change in the decision occurs, it is possible to detect the tolerance of the die-level process monitor circuit under test, which in turn is a mirror of the actual circuit component in the device-under-test. This information can be used to assess whether the whole device-under-test is likely to be faulty, or to adjust the test limits in the ATE to test the device-under-test.

Die-level process monitor circuit testing is based on a pass/fail condition of a window rather than on a single threshold. In contrast to single threshold decisions, testing against a decision window requires differential measurements. Due to the differential nature of the measurements, two runs with interchanged detector references are needed in each test to ensure a proper pass/fail decision. This double-measurement protocol allows the definition of a pass/fail window, instead of a single pass/fail level. Since the result of each run is a digital one-bit signal, the computation of the test result can be done either on-chip adding some simple logic to the detector, or off-chip using resources located in the tester itself. Two runs m_{t1} and m_{t2} are needed with interchanged data decision circuit references, consisting of two thresholds $m_{t1,2l}$ and $m_{t1,2r}$. If a test is successful, the measurement point plus uncertainty due to noise, $m_{t1,2} + \varsigma$, will lie within the range given by $(m_{t1,2l}, m_{t1,2r})$, where ς is the uncertainty due to noise. As a result, the following inequalities hold,

$$m_{t1,2l} \leqslant m_{t1,2} + \varsigma \leqslant m_{t1,2r}$$
$$m_{t1,2l} - \max(\varsigma) \leqslant m_{t1,2} \leqslant m_{t1,2r} - \min(\varsigma) \tag{5.1}$$

Assuming noise ς falls in the range of $(-\Delta, \Delta)$, $m_{t1,2}$ satisfies the following inequality detection thresholds in the presence of measurement noise is

$$m_{t1,2l} - \Delta \leqslant m_{t1,2} \leqslant m_{t1,2r} + \Delta \tag{5.2}$$

The reference voltages defining the decision windows are related to the device-under-test specifications and performance figures under study. By sweeping the reference voltage until a change in the decision occurs, information about the process variations can be extracted. The performance of the detector in terms of resolution and robustness against process variations is a major concern for the intended application. The robustness against process variations is provided by an auto-zeroing scheme. If a better resolution is required, the efficiency of this auto-zeroing can be improved, at the expenses of area overhead, by increasing the value of the input capacitors and/or the preamplifier gain. However, the auto-zeroing scheme does not assure the functionality of the comparator. For instance, a stuck-at fault affecting the output memory element will not be corrected and it will result in a faulty detector. For this reason, a previous test stage to test the detector functionality has to be added to the test protocol. Figure 5.4 illustrates the timing

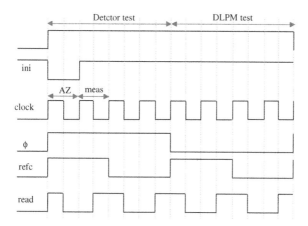

Fig. 5.4 Clocking scheme

diagram of the required control signals (external and on-chip generated ones). When the *Enable* signal is high, the system enters in test mode. In the test mode, two main phases can be distinguished according to the state of signal ϕ. If ϕ is high, the inputs of the detector are shorted to analog ground to perform a test of the detector itself, whereas if ϕ is low the particular die-level process monitor circuit is connected to the detector and tested. Each of these phases takes four master clock periods, two with the reference signal set to the upper limit of the comparison window and the other two with the reference set to the lower limit. During the detector auto-test, the change of the reference should cause the output to change state, since the input is set to zero.

The detection of this change is a quick and easy proof of the functionality of the detector. During the die-level process monitor circuit test, the output of the DLPM is sequentially compared with the references to determine whether the measurement is inside the expected window or not. In both cases, a simple shift register triggered by the signal labeled *Read* acquires the detector output. The rising edges of the *Read* signal are located at the hold state of the detector. The test output will be a 4-bit signal, labeled $a_0a_1a_2a_3$, which codifies the four different states. The overall result of the test is given by $T = (a_0 \oplus a_1)\&(a_2 \oplus a_3)$. This test result can be computed either on-chip in a DSP unit, as depicted in Fig. 5.3, or off-chip. Once the result is available (either the test result itself or the 4-bit number $a_0a_1a_2a_3$ without processing) it can be fed to a scan chain scheme for its later extraction.

Notice that the control signals related to the scan chain are not shown in the timing diagram. In addition, it is important to remark that the system features an additional test mode to test all the flip-flops used in the test scheme (not represented in the figure for simplicity). When this test mode is activated, the flip-flops are isolated from the rest of the circuitry and connected together as a shift register. Additional test input/output for this purpose are also available. Different

Fig. 5.5 (**a**) Test result as a function of flip-flops outputs. (**b**) Comparator switching

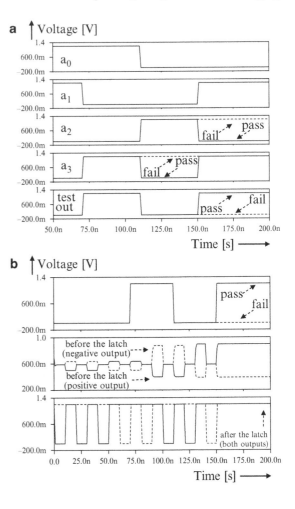

simulations have been carried out as illustrated in Fig. 5.5 to verify the functionality. A 'pass' die-level process monitor circuit test event denotes the measurement inside the comparison window, while a 'fail' DLPM test event is obtained with a slightly narrower comparison window. At the end of the evaluation time the test output is a go/no-go digital signal, which combines the result of the detector test and the die-level process monitor circuit test. Note that the implementation of the clock generation circuitry needs a control signal to set the initial conditions in the D-flip-flops to a known value. This signal can be externally or internally generated, for instance it can be triggered by the rising edge of the *enable* signal. All the flip-flops used are scannable and there is a flip-flop test enable signal for that purpose. For a proper definition of the comparison window, the digital correction and the offset cancellation implemented in the actual design have to be taken into account. Detailed description on how to define the comparison window is explained in following section.

5.1.3   Decision Window and Application Limits

In multi-step A/D converters, high linearity is obtained by extensive usage of correction and calibration procedures. Providing structural DfT and BIST capabilities to this kind of A/D converters is difficult since the effects of the correction mechanism must be taken into account. Overlap between the conversion ranges of two stages has to be considered, otherwise, there may exist conflicting operational situations that can either mask faults or give an incorrect fault interpretation. Although a multi-step A/D converter makes use of considerable amount of digital logic, most of its signal-processing functions are executed in the analog domain. The conversion process therefore is susceptible to analog circuit and device impairments. The primary error sources present in each stage in a multi-step A/D converter are systematic decision stage offset errors λ, stage gain errors η, and errors in the internal reference voltages γ. The offset errors include offset caused by component mismatch, self-heating effects, comparator hysteresis or noise. The gain error group includes all the errors in the amplifying circuit, including technology variations and finite gain and offset of the operational amplifier. The reference voltage errors are caused by resistor ladder variations and noise, as well as to errors in the switch matrix, which are mainly due to charge injection in the transmission gate. The input-referred error e_{in} that is equivalent to the contributions of all the individual error sources can be expressed as

$$e_{in} = e_1 + \sum_{i=1}^{k-1} \frac{e_{i+1}}{G^i} \text{ with } e_i \leqslant \frac{V_{FS}}{2^{N+1}} G^{i-1} \tag{5.3}$$

which is the limit of the A/D converter error arising from each error source to less than ½ LSB, where k is the number of the stages i, V_{FS} is full scale input signal and G is the gain of the stage. Decision stage offset moves A/D converter decision levels. If the correction range is not exceeded by the combination of all errors that shift the first-stage A/D converter decision levels, the effect of the first-stage A/D converter decision stage offset is eliminated by the digital correction, leaving the input-referred offset as the only effect of a sub-range offset. An offset on the residue amplifier gives a *dc* shift of the next-stage A/D converter reference with respect to the preceding stage A/D converter and sub-D/A converter range. The effect of coarse A/D converter offset is studied by examining plots of the ideal residue versus the input illustrated in Fig. 5.6a; note that the fault provokes over-range and level shifting errors. Processing these data with the rest of the A/D converter, including the correction logic is shown in Fig. 5.6b. Digital correction does not mask all errors, and hence the circuit is faulty: on the other hand, since the window comparator threshold has been exceeded the fault is also a detected fault.

D/A converter offset can be replaced by an input-referred stage offset and an offset in series with the coarse A/D converter as shown in Fig. 5.7. The non-compensated remaining offset at the input of each A/D converter comparator due

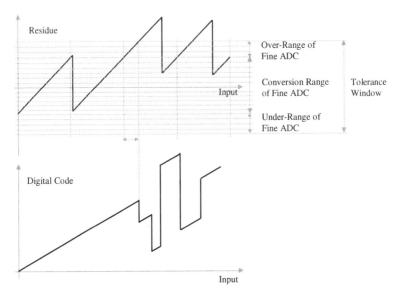

Fig. 5.6 (**a**) Coarse ADC transfer characteristics in the presence of offset error. (**b**) Faulty digitally corrected ADC transfer characteristic

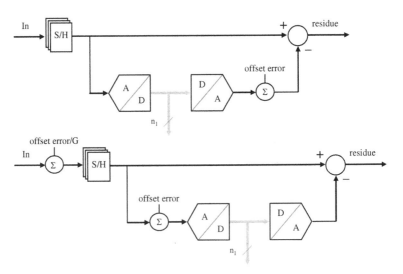

Fig. 5.7 (**a**) Block diagram of a first stage A/D converter with offset error in D/A converter. (**b**) Equivalent diagram with rearranged offset errors

to the decision stage offset is given by $V_{off/NC}^{D} = V_{off}^{D}/G^{i-1}$, where $V_{off/NC}^{D}$ is the input referred non-compensated offset, V_{off}^{D} is the decision stage offset, and G^{i-1} is the gain of the preceding stage. Imposing a $\pm\frac{1}{2}$ LSB maximum deviation leads to the definition of the comparison window:

$$\Delta V = G^i V^C_{off} \Rightarrow -\frac{V_{FS}}{2^{N+1}} G^i G^{i-1} \leqslant \Delta V \leqslant \frac{V_{FS}}{2^{N+1}} G^i G^{i-1} \tag{5.4}$$

where G^i is the gain of the decision stage-based die-level process monitor circuit.

The gain requirements are straightforward. An error in per stage gain causes non-linearity in the transfer characteristic from input to output of the multi-step A/D converter. A gain error in the residue amplifier scales the total range of residue signal and causes an error in the analog input to the next stage when applied to any nonzero residue, which will result in residue signal not fitting in the fine A/D converter range. If the error in the analog input to the next stage is more than one part in 2^r (where r is the resolution remaining after the inter-stage gain error), it will result in a conversion error that is not removed by digital correction. Moreover, if the inter-stage gain is smaller than the ideal value, a fixed number of missing codes at every MSB transition can occur [55] (i.e., constant DNL errors or constant jumps in INL at every transition of the bits resolved by the first stage). Dual-residue signal processing [61] spreads the errors of the residue amplifiers over the whole fine range, which results in an improved linearity. An equivalent block diagram in which the D/A converter gain error Δ_i has been replaced by three gain errors: one in series with the stage input, one in series with the coarse A/D converter, and one in series with the stage output is illustrated in Fig. 5.8. If the correction range is not exceeded by the combination of all errors that shift the coarse A/D converter decision levels, the effect of the gain error in series with the coarse A/D converter is eliminated by the digital correction. The two remaining gain errors contribute inter-stage gain errors, which have the same effect on A/D converter linearity as the residue amplifier gain errors.

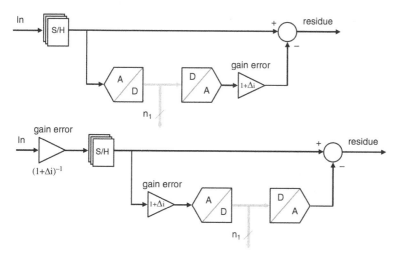

Fig. 5.8 (a) Block diagram of a first stage of examined A/D converter with gain error in DAC. (b) Equivalent diagram with rearranged gain errors

An error in the range of the fine A/D converter results in an error similar to a residue amplifier gain error. The gain of the sub-tractor and amplifier should therefore be lined with the fine A/D converter range. To limit resulting nonlinearity to $\pm\frac{1}{2}$LSB, $|G\sigma_i V_{res}|$ $\leq G\times(V_{FS}/2^{N+1})$. The error in the residue amplifier is proportional to $G\times V_{res}$, thus, the effect of the gain error is largest when $G\times V_{res}$ is maximum. The references of the D/A converter and the subtraction of the input signal and the D/A converter output determine the achievable accuracy of the total A/D converter. The residue signal V_{res} is incorrect exactly by the amount of the D/A converter nonlinearity

$$V_{res} = GV_{in} - DAC_{out} - \delta_l \tag{5.5}$$

where DAC_{out} is the ideal output of the D/A converter, G is the gain and δ_l is D/A converter nonlinearity error. D/A converter errors result in each linear segment of the residue transfer curve being shifted up or down by different static random values. Hence, D/A converter errors result in non-constant missing codes at every MSB transition. To limit resulting D/A converter nonlinearity to less than $\frac{1}{2}$ LSB, $|\delta_l|_{min} \leq V_{FS}/2^{N+1}$. The comparison window for internal reference voltages deviations is, thus, given by:

$$\left.\begin{array}{l} \Delta V_{|max} = V_{FS}\frac{\Delta R}{\sum\limits_{j=1}^{N} R_j} \\[2em] \Delta V = I_{ref}\Delta R \end{array}\right\} \Rightarrow -\frac{I_{ref}\sum\limits_{j=1}^{N} R_j}{2^{N+1}} \leqslant \Delta V \leqslant \frac{I_{ref}\sum\limits_{j=1}^{N} R_j}{2^{N+1}} \tag{5.6}$$

where I_{ref} is the reference current in the resistor ladder die-level process monitor circuit, V_{FS} is the full scale of the converter, R_j is the value of each resistor in the resistor ladder, and N is the total number of resistor in the ladder.

5.1.4 Die-Level Process Monitors Circuit Design

To illustrate the concept, consider only a simple 4 bit flash stage as shown in Fig. 5.9, consisting of a reference ladder and 16 comparators. From the previous analysis it can be concluded that the gain, decision and reference ladder are crucial to the proper converter performance. To mimic the device-under-test behavior, the gain-based and decision stage-based die-level process monitor are extracted (replicated) from the device-under-test as illustrated in Figs. 5.10 and 5.11, where the stage gain-based and the decision stage-based die-level process monitor matches the actual flash converter comparator.

The type of latch employed is determined by the resolution of the stage. For a low-resolution quantization per stage, a dynamic latch is more customary since it dissipates less power than the static latch. While the latch circuits regenerate the difference signals, the large voltage variations on regeneration nodes will introduce the instantaneous large currents. Through parasitic gate-source and gate-drain

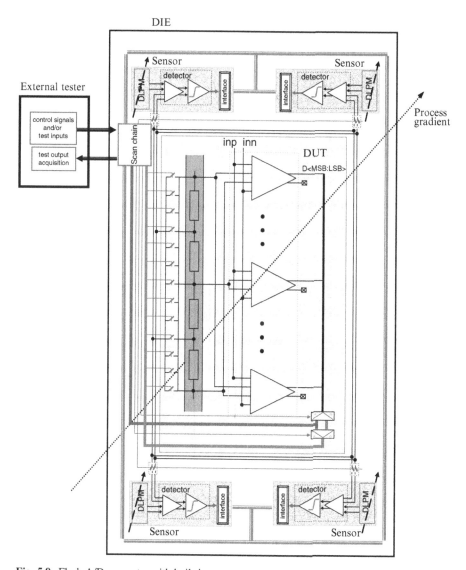

Fig. 5.9 Flash A/D converter with built-in sensors

capacitances of transistors, the instantaneous currents are coupled to the inputs of the comparators, making the disturbances unacceptable. In flash A/D converters where a large number of comparators are switched on or off at the same time, the summation of variations came from regeneration nodes may become unexpectedly large and directly results in false quantization code output [172]. It is for this reason that the static latch is preferable for higher-resolution implementations. The key requirement which determines the power dissipation during the comparison process

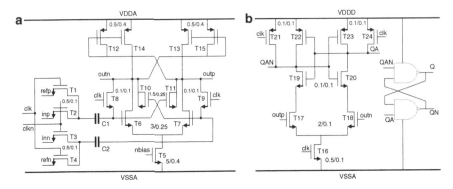

Fig. 5.10 Data decision with offset calibration: (**a**) Preamplifier and offset calibration circuit and (**b**) data decision stages

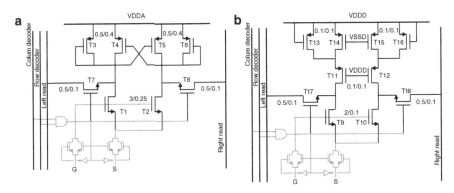

Fig. 5.11 Die-level process monitors: (**a**) stage gain-based and (**b**) decision stage-based

is the accuracy, i.e. how accurately the comparator can make a decision in a given time period. As typical cross-coupled latch comparators exhibit large offset voltage, several pre-amplifiers are placed before the regenerative latch to amplify the signal for accurate comparison. The power dissipation in the regenerative latch is relatively small compared to the preamp power, as only dynamic power is dissipated in the regenerative latch and low offset pre-amp stages usually require *dc* bias currents. Therefore, the power dissipation is directly related to how many preamp stages are required, and the number of stages is determined by the required amplification factor before a reliable comparison can be made by the regenerative comparator. If high gain is required from a single stage preamp, the large value of the load resistor must be used, which in turn slows down the amplification process with an increased *RC*-constant at the output. In situations like this, the gain is distributed among several cascaded low gain stages to speed up the process. During this process care must be also taken to design a low noise pre-amp stage since its own circuit noise is amplified through its gain. For instance, if the input signal is held constant close to the comparator threshold, the thermal noise from both circuits and input sampling switches is also amplified through the preamp gain. The preamp stage is usually implemented with a source-coupled pair, and its power-to-thermal-noise relationship is similar to that of the S/H circuit case where the key block is the high gain op amp, except that the preamp is usually in the open-loop configuration. Also, $1/f$ noise must be considered since it appears like a slowly varying offset of the comparator for high speed operation. Periodic offset cancellation at a rate much higher than the $1/f$ noise corner frequency, usually every clock period, can reduce this effect. The analysis for noise is omitted here since it is relatively straightforward compared to that of the amplifier in the feedback as in S/H circuits.

Another major factor which affects the accuracy of the comparator is the offset voltage caused by the mismatches from process variations. This includes charge injection mismatches from input switches, threshold and transistor-dimensions mismatches between cross-coupled devices. To lessen the impact of mismatch, several schemes, such as inserting a preamplifier [1] in front of the latch, adding a chopper amplifier [2] and auto-zero scheme to sample an offset in the capacitor in front of the latch or digital background calibration [18] have been developed. In the auto-zero scheme, during the offset sampling period, the output of the first stage caused by its offset voltage is sampled on the sampling capacitor of the second stage. In the next clock phase, when the actual comparison is to be made, the stored voltage on the second stage sampling capacitor effectively cancels out the offset of the first amplifier, and a very accurate comparison can be made. For this cancellation technique, notice that the gain of the first stage must be chosen relatively low so that the output voltage due to its offset does not rail out of the range (or supply). One observation is that the offset voltage of the dynamic comparator circuit cannot be cancelled by this technique because the positive feedback amplifies even a small offset voltage to the supply rails and therefore no information on the offset voltage can be obtained at the output of the comparator. As a result, this technique requires a preamp with a dc bias current and therefore static power to reduce offset voltage. If an input signal is sampled on a capacitor before comparison, the capacitance value must be carefully chosen to reduce various non-idealities in addition to the kT/C noise. In some multi-step A/D converters realizations, the comparator often has its own input sampling capacitor to eliminate the dedicated input S/H circuit [62].

Figure 5.10 shows a sensor together with an auto-zeroing scheme to cancel a possible sensor offset. The switched-capacitor comparator operates on a two phase non-overlapping clock. The differencing network samples V_{ref} during phase clk onto capacitor C, while the input is shorted giving differential zero. During phase $clkn$, the input signal V_{in} is applied at the inputs of both capacitors, causing an input differential voltage to appear at the input of the comparator preamp. At the end of $clkn$ the regenerative flip-flop is latched to make the comparison and produce digital levels at the output. The charge injection from T_8 will cause an offset voltage $\Delta V = \Delta Q/(C + C_p)$. Requirement on the input bandwidth sets the magnitude of ΔV, and the higher the sampling bandwidth is, larger the ΔV. Circuit technique employed to limit charge injection include bottom plate sampling, use of dummy switch or reducing the gate channel length (Section 3.3). Another important consideration for choosing C is the signal attenuation due to C_p. At the input of the amplifier, the input capacitance of the amplifier and the parasitic capacitance from the switch attenuate the input signal by $C/(C + C_p)$ and effectively reduces the amplification. Based on matching and common-mode charge injection errors, C was chosen to be near minimum size, approximately $5fF$.

The comparison references needed to define the sensor decision windows are controlled through the *dc* signals labeled *refp* and *refn* in the figure. The die-level process monitor measurements are directly related to asymmetries between the branches composing the circuit, giving an estimation of the offset when both DLPM inputs are grounded or set at predefined common-mode voltage. As shown in Fig. 5.12a, the gain-based monitor consists of the circuitry replicated from the observed A/D converter gain stage, which consists of a differential input pair (transistors T_1 and T_2) with active loading (T_3 and T_4) and some additional gain (transistors T_5 and T_6) to increase the monitor's resolution and transistors T_7 and T_8 to connect to read lines (lines leading to a programmable data decision circuit). The

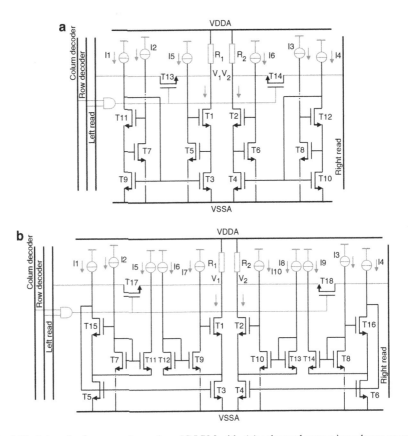

Fig. 5.12 Internal reference voltages-based DLPM with, (a) enhanced output impedance current mirror, (b) modified wide-swing current mirror with enhanced output impedance

different device arrangements in the matrix include device orientation and the nested device environment. The matrix is placed several times on the chip to obtain information from different chip locations and distance behavior. Similarly, as shown in Fig. 5.12b, in the decision stage monitor circuit the latch (transistors T_{12} to T_{17}) has been broken to allow a *dc* current flow through the device needed for the intended set of measurements. In addition to these two, internal reference voltages monitoring circuits as shown in Fig. 5.12 sense the mismatching between two of the unit resistors used in the actual resistor ladder design. The current that flow through the resistors (whose values are extracted from the ladder itself) is fixed using a current mirror. Since the current is fixed, the voltage drop between the nodes labeled V_1 and V_2 is a measurement of the mismatching between the resistors. The feedback amplifier is realized by the common-source amplifier consisting of T_5 and its current source I_5. The amplifier keeps the drain-source voltage across T_3 as stable as possible, irrespective of the output voltage. The addition of this amplifier ideal increases the output impedance by a factor equal to one plus the loop gain over that which would occur for a classical cascode current mirror. Assuming the output impedance of current source I_5 is roughly equal to r_{ds5} the loop gain can be approximated by $(g_{m5}\, r_{ds5})/2$. The circuit consisting of T_7, T_9, T_{11}, I_1 and I_2 operates almost identically to a diode-connected transistor; however it is employed instead to guarantee that all transistor bias voltages are accurately matched to those of the output circuitry consisting of T_1, T_3, T_5 and I_5. As a consequence I_{R1} will very accurately match I_1 [382].

As transistors T_3 and T_9 are biased to have drain-source voltages ($V_{DS3} = V_{DS9} = V_{eff5} + V_T$) larger than the minimum required, V_{eff3}, this can pose a limitation in very low power supply technologies. Alternative realization [383], illustrated in Fig. 5.12b combines the wide-swing current mirror with the enhanced output-impedance circuit. Here, diode-connected transistors used as level shifters have been added in front of the common-source enhancement amplifier. At the output side, the level shifter is the diode-connected transistor T_7, biased with current I_2. The circuitry at the input acts as diode-connected transistor while ensuring that all bias voltages are matched to the output circuitry to the I_{R1} accurately matches I_1. In the case in which $I_7 = I_1/7$ all transistors are biased with nearly the same current density, except T_7 and T_9. As a result, all transistors have the same effective gate-source voltages, except for T_7 and T_9, which have twice effective gate-source voltage as they are biased the four times the current density. Thus, the gate voltage of T_9 equals $V_{G9} = 2V_{eff} + V_T$ and the drain-source voltage T_3 is given by $V_{DS3} = V_{G9} - V_{GS12} = V_{eff}$. Therefore, T_3 is biased on the edge of the triode region. Although the power dissipation of the circuit is almost doubled over that of a classical cascode current mirror, by biasing the enhancement circuitry at lower densities, albeit at the expense of speed, e.g. additional poles introduced by the enhancement circuitry are at lower frequencies, sufficient saving on power dissipation can be made.

Thus, three generalized strategies can be extracted from the presented die-level process monitor circuits: gain stage-, decision stage- and resistor ladder-DLPMs. Gain stages, such as the residue amplifiers in the A/D converter, can be tested using the same strategy developed for testing the preamplifier in the A/D converter. The methodology can be directly translated to any gain stage, allowing the detection of mismatch issues through the measurement of output offset. Decision stages in the A/D converter can be tested via adapting the decision stage die-level process monitor strategy to each particular design. This strategy is based on breaking the regeneration feedback in the latch, and then sensing process mismatches through the measurement of output offset.

Internal reference voltages can be tested adapting the same scheme proposed for the resistor ladder DLPM, which gives a measurement of resistor mismatching through the measurement of a voltage drop. To increase the monitor resolution, some additional gain can be inserted into the DLPMs between input differential pair and the loading. By sweeping the reference voltage until a change in the decision occurs, information about the process variation can be extracted as shown in Fig. 5.13a.

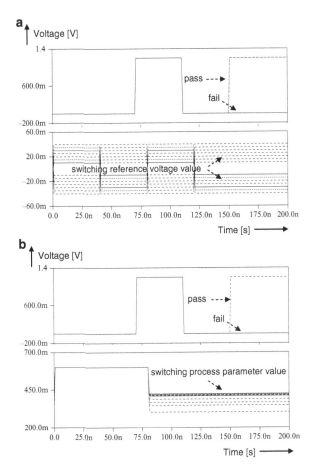

Fig. 5.13 (a) Sweeping the reference voltages to extract the DLPMs offset voltage values, illustrated on the decision DLPM, (b) testing the device-under-test by setting the discrimination window to ½ LSB level and sweeping the process parameters values

Similarly, by setting the discrimination window to ½ LSB level, variations due to the mismatch can be detected (Fig. 5.13b). Discrimination window for various die-level process monitors have been defined according the rules of the multi-step A/D converter error model from Section 5.2. By extracting the die-level process monitor circuit from the device-under-test itself, the monitoring circuit accomplish some desirable properties: (i) it is designed to maximize the sensitivity of the circuit to the target parameter to be measured, (ii) it matches the physical layout of the extracted device under test, (iii) it is small and stand alone, and consumes no power while in off state, and (iv) the design of die-level process monitor is flexible enough to be applied in several ways depending on the system-on-chip to which it is added.

5.1.5   Temperature Sensor

To convert temperature to a digital value, both a well-defined temperature-dependent signal and a temperature-independent reference signal are required. Both can be derived utilizing exponential characteristics of bipolar devices for both negative- and positive temperature coefficient quantities in the form of the thermal voltage and the silicon bandgap voltage. For constant collector current, base-emitter voltage has negative temperature dependence around room temperature. This negative temperature dependence is cancelled by a proportional-to-absolute temperature (PTAT) dependence of the amplified difference of two base-emitter junctions biased at fixed but at unequal current densities resulting in the relation directly proportional to the absolute temperature. This proportionality is quite accurate and holds even when the collector currents are temperature dependent, as long as their ratio remains fixed. In a n-well CMOS process, both lateral npn and pnp transistors and vertical or substrate pnp transistors are used for this purpose. As the lateral transistors have low current gains and their exponential current voltage characteristic is limited to a narrow range of currents, the substrate transistors are preferred. In the vertical bipolar transistors, a p^+ region inside an n-well serves as the emitter and the n-well itself as the base of the bipolar transistors. The p- type substrate acts as the collector and as a consequence, all their collectors are connected together, implying that they cannot be employed in a circuit unless the collector is connected to the ground. These transistors have reasonable current gains and high output resistance, but their main limitation is the series base resistance, which can be high due to the large lateral dimensions between the base contact and the effective emitter region. To minimize errors due to this base resistance, the maximum collector currents through the transistors are constrained to be less than 0.1 mA.

The slope of the base-emitter voltage depends on process parameters and the absolute value of the collector current. Its extrapolated value at 0 K, however, is insensitive to process spread and current level. The base-emitter voltage is also sensitive to stress. Fortunately, substrate pnp transistors are much less stress-sensitive than other bipolar transistors [384]. In contrast with the base-emitter voltage V_{be}, ΔV_{be} is independent of process parameters and the absolute value of the collector currents. Often a multiplicative factor is included in the equation for ΔV_{be} to model the influence of the reverse Early effect and other nonidealities [385]. If V_{be} and ΔV_{be} are generated using transistors biased at approximately the same current density, an equal multiplicative factor will appear in the base-emitter voltage. ΔV_{be} is insensitive to stress [386]. Its temperature coefficient is, however, typically an order of magnitude smaller than that of (depending on the collector current ratio).

The nominally zero temperature coefficient is usually exploited in a temperature-independent reference generation circuits such as bandgap–reference illustrated in Fig. 5.14a. In general, accurate measure of the on-chip temperature is acquired or through generated proportional-to-absolute temperature current or the generated proportional-to-absolute temperature voltage. In the previous case, the reference voltage is converted into current by utilizing an opamp and a resistor. The absolute accuracy of the output current will depend on both the absolute accuracies of the voltage reference and of the resistor. Most of the uncertainty will depend on this resistor and its temperature coefficient. In a bandgap voltage reference, an amplified version of ΔV_{be} is added to V_{be} to yield a temperature-independent reference voltage V_{ref}. The negative voltage-temperature gradient of the base-emitter junction of the transistor Q_1 is compensated by a proportional-to-absolute temperature voltage across the resistor R_1, thereby creating an almost constant reference voltage V_{ref} [387]. The amplifier A senses voltages at its inputs, driving the top terminals of R_1 and R_2 ($R_1 = R_2$) such that these voltages settle to approximately equal voltages. The reference voltage is obtained at the output of the amplifier (rather than at its input). Due to the asymmetries, the inaccuracy of the circuit is mainly determined by the offset of the opamp, which directly adds to ΔV_{be}. To lower the effect of offset, the opamp incorporates large devices. Similarly, the collector currents of bipolar transistors Q_1 and Q_2 are rationed by a pre-defined factor, e.g. transistors are multiple parallel connections of unit devices.

The conceptual view of the temperature sensor is illustrated in Fig. 5.14b. The left half of this circuit is the temperature sense-circuit, which is similar to a

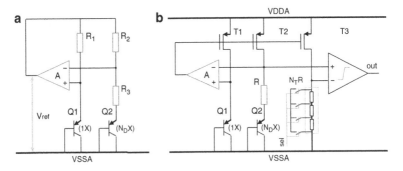

Fig. 5.14 (a) Typical implementation of bandgap reference circuit. (b) Temperature sensor – conceptual view

conventional bandgap reference circuit. The right part, comprising a voltage comparator, creates the output signal of the temperature switch. To enable a certain temperature detection, two signals are required, both with a well defined but different temperature dependence; an increasing proportional-to-absolute temperature voltage across the resistor network $N_T R$ and decreasing voltage at the comparator positive input generates temperature decisions. As the bias currents of the bipolar transistors are in fact proportional to absolute temperature, proportional-to-absolute temperature current I_{D3} can be generated by a topology comprising bipolar transistors Q_1 and Q_2, resistor R, an amplifier and CMOS transistors $T_1 - T_3$.

The temperature sensor as shown in Fig. 5.15 is based on a modified temperature sensor [388]. The amplifier consists of a non-cascoded OTA with positive feedback to increase the loop-gain. The amplifier output voltage is relatively independent of the supply voltage so long as its open-loop gain is sufficiently high. If input voltages are equal to zero, the input differential pair of the amplifier may turn off. To prevent this, a start-up circuit consisting of transistors $T_5 - T_6$ is added, which drives the circuit out of the degenerate bias point when the supply is tuned on. The diode-connected device T_5 provides a current path from the supply through T_6 to ground upon start-up. The scan chain delivers a four-bit value for the setting of the resistor value $N_T R$. The input of the comparator consists of a differential source-couple stage, followed by two amplifying stages and one digital inverter. To provide stable bandgap reference voltage < 1 V as illustrated in Fig. 5.16a, zero-temperature coefficient voltage ≈ 1.2 V is firstly converted into current through transistor T_{22} and then summed up to lower reference voltage through R_2 and $N_R R$. The opamp has sufficient gain to equalize its input voltages. Since these nodes are the same, the currents from these nodes to ground must be the same as well. The current through R_1 is therefore proportional-to-absolute temperature. This current is also flowing through the output transistor T_{22}. The curvature of V_{be} will also be present in the reference voltage. For a current, which is independent of temperature, the curvature correction is in the same order of magnitude of mismatch. By allowing a proportional-to-absolute temperature current to have a small positive temperature coefficient, the second-order component of the curvature is eliminated. Such a

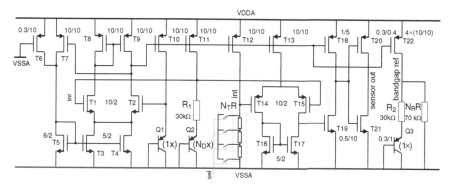

Fig. 5.15 Temperature sensor – schematic view

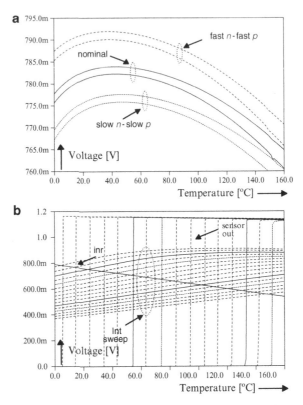

Fig. 5.16 (a) Bandgap reference voltage; nominal, fast-fast and slow-slow process corners. (b) Sixteen selection levels in the temperature sensor

temperature-dependent reference voltage cancels the second-order nonlinearity originating from V_{be}. For this purpose, a transistor Q_3 is added. In essence, a term based on the ratio of the resistors $N_R R$ and R_2 is generated which subtracts the V_{be} of a junction with a constant current for the V_{be} of a junction with the proportional-to-absolute temperature current. To accurately define this ratio, all resistors are constructed of identical unit resistors. The voltage across R_2 is curvature-compensated by adjusting the base-emitter voltage V_{be} of a substrate *pnp* transistor with adjustable resistors $N_R R$. The proportional-to-absolute temperature voltage across this resistor compensates for the proportional-to-absolute temperature-type spread on V_{be} of a transistor Q_3.

In the test silicon, four bits for a sixteen selection levels are chosen for the temperature settings, resulting in a temperature range from 0°C to 150°C with a resolution of 9°C as illustrated in Fig. 5.16b. If a higher accuracy is required, a selection D/A converter with higher resolution is necessary. For the robustness, the circuit is completely balanced and matched both in the layout and in the bias conditions of devices, cancelling all disturbances and non-idealities to the first order.

5.2 Estimation of Die-Level Process Variations

5.2.1 Expectation-Maximization Algorithm

In general, a circuit design is optimized for parametric yield so that the majority of manufactured circuits meet the performance specifications. The complexity of yield estimation, coupled with the iterative nature of the design process, makes yield maximization computationally prohibitive. As a result, circuit designs are verified using models corresponding to a set of worst-case conditions of the process parameters. Worst-case analysis refers to the process of determining the values of the process parameters in these worst-case conditions and the corresponding worst-case circuit performance values. Worst-case analysis is very efficient in terms of designer effort, and thus has become the most widely practiced technique for statistical verification. However, the worst-case performance values obtained are extremely pessimistic and as a result lead to unnecessary large and power hungry designs in order to reach the desired specifications. Thus, it would be advantageous to choose a more relaxed design condition. Statistical data extracted through the sensor measurements allow not only possibilities to enhance observation of important design and technology parameters, but to characterize current process variability conditions (process corners) of certain parameters of interest, enabling optimized design environment as well. As the number of on-chip sensors is finite due to area limitations, additional informations are obtained through statistical techniques. Although, in statistics several methods, such as listwise [389] and pairwise [390] deletion and structural equation modelling [391] would provide estimates of the selected performance figures from the incomplete data, imputation method (e.g. substitution of some plausible value for a missing datapoint) and its special case, multiple imputations based on expectation-maximization (EM) algorithm [392, 393] offers maximum likelihood estimates. The work in this section [394] utilizes therefore the maximum-likelihood method and adjusted support vector machine (SVM) classifier [395, 396] as a very efficient method for test guidance based on the information obtained through monitoring process variations, since it simultaneously minimizes the empirical classification error and maximizes the geometric margin.

Given the observation vector of the sensor's observations $x_i \in X$, the estimation of an unknown parameter vector $\theta \in \Theta$ designating true values of die-level process parameter variation would be an easy task if the source signal vector $y_i \in Y$, assumed to be realizations of the random variables which are independent and identically distributed according to the probability $p_{XY/\Theta}(x,y/\theta)$, were also available. The maximum likelihood (ML) estimation involves estimation of θ for which the observed data is the most likely, e.g. marginal probability $p_{X/\Theta}(x/\theta)$ is a maximum. The parameters θ involve parameters (μ_y, Σ_y), $y \in Y$ of Gaussian components and the values of the discrete distribution $p_{Y/\Theta}(y/\theta)$, $y \in Y$. The $p_{X/\Theta}(x/\theta)$ is the Gaussian mixture model given by the weighted sum of the Gaussian distributions. The logarithm of the probability $p(T_X/\theta)$ is referred to as the log-likelihood $L(\theta/T_X)$ of θ

with respect to T_X. The input set T_X is given by $T_X = \{(x_1, \ldots, x_l)\}$, which contains only vectors of observations x_i. The log-likelihood can be factorized as

$$L(\theta|T_X) = \log p(T_X|\theta) = \sum_{i=1}^{l} \sum_{y \in Y} p_{X|Y\Theta}(x_i|y_i, \theta) p_{Y|\Theta}(y_i|\theta) \qquad (5.7)$$

The problem of maximum likelihood estimation from the incomplete data T_X can be defined as

$$\theta^* = \arg\max_{\theta \in \Theta} L(\theta|T_X) = \arg\max_{\theta \in \Theta} \sum_{i=1}^{l} \sum_{y \in Y} p_{X|Y\Theta}(x_i|y_i, \theta) p_{Y|\Theta}(y_i|\theta) \qquad (5.8)$$

Obtaining optimum estimates through ML method thus involves two steps: computing the likelihood function and maximization over the set of all admissible sequences. To evaluate the contribution of the random parameter θ, analysis of the likelihood function requires computing an expectation over the joint statistics of the random parameter vector, a task that is analytically intractable. Even if the likelihood function can be obtained analytically off line, however, it is invariably a nonlinear function of θ, which makes the maximization step (which must be performed in real time) computationally infeasible. In such cases, the expectation-maximization algorithm [392, 393, 397–400] may provide a solution, albeit iterative, to the ML estimation problem.

The EM algorithm is most useful in cases where it is possible to derive the maximum likelihood estimates for complete data analytically. If some observations are missing, the algorithm allows obtaining the maximum likelihood estimates of the unknown parameters by a computational procedure which iterates, until convergence, between two steps [392, 400]. As the main statistical concern is parameter estimation and in most cases this is best achieved by the use of maximum likelihood theory, the EM algorithm substitutes the missing data in the log likelihood function, not in the incomplete data set; the missing values are substituted by the conditional expectations of their functions, as they appear in the log-likelihood function. If the assumed model is a member of Gaussian population, then the replaced values corresponds to the expected values of the sufficient statistics for the unknown parameters. In particular, for such densities, the incomplete-data set is the set of observations, whereas each element of the complete-data set can be defined to be a two-component vector consisting of an observation and an indicator specifying which component of the mixture occurred during that observation.

Instead of using the traditional incomplete-data density in the estimation process, the EM algorithm uses the properties of the complete-data density. In doing so, it can often make the estimation problem more tractable and also yield good estimates of the parameters for small sample sizes [397]. Thus, with regard to implementation, the EM algorithm holds a significant advantage over traditional steepest-descent methods acting on the incomplete-data likelihood equation.

Moreover, the EM algorithm provides the values of the log-likelihood function corresponding to the maximum likelihood estimates based uniquely on the observed data. Similarly, with traditional implementations, the determination of the efficient root of the incomplete-data likelihood equation requires the determination first of all roots of the likelihood equation. The EM algorithm can be viewed as an alternative to maximizing the likelihood function over the complete-data set. In particular, since likelihood function is unknown, maximize its expectation given the available pertinent information, namely, the observed data and the current estimate of the parameters is maximized instead. The EM algorithm builds a sequence of parameter estimates $\theta^{(0)}$, $\theta^{(1)}$,...,$\theta^{(t)}$, such that the log-likelihood $L(\theta^{(t)}/T_X)$ monotonically increases, i.e., $L(\theta^{(0)}/T_X) < L(\theta^{(1)}/T_X) <... < L(\theta^{(t)}/T_X)$ until a stationary point $L(\theta^{(t-1)}/T_X) = L(\theta^{(t)}/T_X)$ is achieved. Using Bayes rule and writing explicitly only the unknown parameter θ, the log likelihood of x_i can be written as

$$\log p(T_X|\theta) = \log p(X, Y|\theta), \theta(t) - \log p_{X,Y|X}(X, Y|X), \theta(t) \tag{5.9}$$

Taking expectations on both sides of the above equation given x_i and θ, where $\theta^{(t)}$ is an available estimate of θ,

$$\log p(T_X|\theta) = E_{\theta^{(t)}}\{\log p(X, Y|\theta)|X, \theta(t)\} - E_{\theta^{(t)}}\{\log p_{X,Y|X}(X, Y|X)|X, \theta(t)\} =$$
$$= Q_n(\theta|\theta^{(t)}) - P(\theta|\theta^{(t)}) \tag{5.10}$$

By Jensen's inequality, the relation holds that

$$P(\theta|\theta^{(t)}) \leqslant P(\theta^{(t)}|\theta^{(t)}) \tag{5.11}$$

Therefore, a new estimate θ in the next iteration step that makes $Q(\theta/\theta^{(t)}) \geq Q(\theta/\theta^{(t)})$ leads also to

$$\log p(T_X|\theta) \geqslant \log p(T_X|\theta^{(t)}) \tag{5.12}$$

In each iteration, two steps, called E-step and M-step, are involved. In the E-step, the EM algorithm form the auxiliary function $Q(\theta/\theta^{(t)})$, which determines the expectation of log-likelihood of the complete data based on the incomplete data and the current parameter

$$Q(\theta|\theta^{(t)}) = E(\log p(X, Y|\theta)|X, \theta(t)) \tag{5.13}$$

In the M-step, the algorithm determines a new parameter maximizing Q

$$\theta^{(t+1)} = \arg\max_{\theta} Q(\theta|\theta^{(t)}) \tag{5.14}$$

At each step of the EM iteration, the likelihood function can be shown to be non-decreasing [400]; if it is also bounded (which is mostly the case in practice), then the algorithm converges. In [392] is proved that an iterative maximization of $Q(\theta/\theta^{(t)})$ will lead to a maximum likelihood estimation of θ. For a broad class of probability density functions, including Gaussian mixture densities, at each iteration the new parameter estimate θ can be explicitly solved as the stationary point corresponding to the unique maximum of $Q(\theta/\theta^{(t)})$ [397]. As in direct maximum likelihood estimation, the EM algorithm only leads to locally optimal estimates of model parameters, with each resulting estimate depending upon the initial parameters chosen to start the iterative estimation.

EM algorithm

Initialization
− Initialize the data set $T = \{(x_1, \ldots, x_l)\}$
− Initialize the parameter $\theta^{(0)}$

Data collection
− Collect N samples from the DLPMs

Update parameter estimate
1. Calculate $Q(\theta/\theta^{(t)}) = E(log\ p(X,Y/\ \theta)/X,\theta^{(t)})$ – E step
2. Re-estimate θ by maximizing the θ-function
   $\theta^{(t+1)} = argmax_\theta\ Q(\theta/\theta^{(t)})$, estimate mean and variance – M step
3. Increase the iteration index, t
4. Stop when a stationary point $L(\theta^{(t-1)}/T_X) = L(\theta^{(t)}/T_X)$ is found.

5.2.2 Support Vector Machine Limits Estimator

When an optimum estimate of parameter distribution is obtained, next step is to update the test limit values utilizing adjusted support vector machine classifier. Assuming that the input vectors belong to a priori and a posteriori classes, e.g. test limits, the goal is to decide which class a new measurement data will be in. Each new measurement is viewed as an r-dimensional vector (a list of r numbers) and the SVM classifier maps (separates) the input vectors into r-1-dimensional hyperplane in feature space Z where a linear decision surface is constructed through certain non-linear mapping. Although several classifiers are available, such as quadratic, boosting, neural networks, Bayesian networks, adjusted support vector machine (SVM) classifier is especially resourceful, since it simultaneously minimize the empirical classification error and maximize the geometric margin.

Let $D = \{x_i,c_i)/x_i \in R^r,\ c_i \in \{-1,1\}\}^n_{i=1}$ be the input vectors belonging to a priori and a posteriori classes, where the c_i is either 1 or -1, indicating the class to which the data x_i from the input vector belongs. Similarly, let

$$w \cdot x + b = 0 \qquad (5.15)$$

be the optimal hyperplane in feature space. The vector w is a normal vector perpendicular to the hyperplane. The parameter $b///w//$ determines offset of the hyperplane from the origin along the normal vector w. To maximize the margin, or distance between the parallel hyperplanes that are as far apart as possible while still separating the data, w and b have to be chosen as such that they minimize $\|w\|$ subject to the optimization problem described by

$$c_i(w \cdot x_i + b) \geq 1 \tag{5.16}$$

for all $1 \leq i \leq n$. The optimization problem is difficult to solve because it depends on $\|w\|$, which involves a square root. If the input vectors belonging to a priori and a posteriori classes cannot be separated by a hyperplane, the margin between patterns of the two classes becomes arbitrary small, resulting in the value of the functional vector of parameters turning arbitrary large. Maximizing the quadratic programming optimization therefore either reaches a maximum (in this case one has constructed the hyperplane with the maximal margin), or the maximum is found that exceeds some given (large) constant. By altering the equation by substituting $\|w\|$ with $\frac{1}{2}\|w\|^2$ without changing the solution (the minimum of the original and the modified equation have the same w and b), the optimization problem can now be solved by standard quadratic programming optimization [395]. The input vectors belonging to a priori and a posteriori classes are divided into a number of sub-sets. The quadratic programming problem is solved incrementally, covering all the sub-sets of classes constructing the optimal separating hyperplane for the full data set. Note that during this process the value of the functional vector of parameters is monotonically increasing, since more and more training vectors are considered in the optimization leading to a smaller and smaller separation between the two classes. Writing the classification rule in its unconstrained dual form reveals that the maximum margin hyperplane and therefore the classification task is only a function of the support vectors, e.g. the training data that lie on the margin.

$$\max \sum_{i=1}^{n} \alpha_i - \frac{1}{2} \sum_{i,j} \alpha_i \alpha_j c_i c_j x_i^T x_j \tag{5.17}$$

subject to $\alpha_i \geq 0$ and $\sum_{i=1}^{n} \alpha_i c_i = 0$,

$$w = \sum_i \alpha_i c_i x_i \tag{5.18}$$

where the α terms constitute the weight vector in terms of the training set. Additionally, a modified maximum margin technique in [395] allows for mislabelled examples. If there exists no hyperplane that can divide the a priori and a posteriori classes, the modified maximum margin technique finds a hyperplane that separate the training set with a minimal number of errors. The method introduces non-negative variables, ξ_i, which measure the degree of misclassification of the data x_i

$$c_i(w \cdot x_i + b) \geq 1 - \xi_i \tag{5.19}$$

for all $1 \leq i \leq n$ and $\sigma > 0$. The objective function is then increased by a function which penalizes non-zero ξ_i, and the optimization becomes a tradeoff between a large margin, and a small error penalty. If the penalty function is linear, the optimization problem now transforms to

$$\min \frac{1}{2} \|w\|^2 + C \sum_i \xi_i^\sigma \qquad (5.20)$$

such that (5.16) holds for all $1 \leq i \leq n$. For sufficiently large C and sufficiently small σ, the vector w and constant b that minimize the functional (5.20) under constraints in (5.16), determine the hyperplane that minimizes the number of errors on the training set and separate the rest of the elements with maximal margin. This constraint in (5.16) along with the objective of minimizing $\|w\|$ can be solved using Lagrange multipliers. The key advantage of a linear penalty function is that the variables ξ_i vanish from the dual problem, with the constant C appearing only as an additional constraint on the Lagrange multipliers. Similarly, non-linear penalty functions can be employed, particularly to reduce the effect of outliers on the classifier; however, the problem can become non-convex and thus, finding a global solution can become considerably more difficult task.

In general, the discriminant function of an arbitrary classifier does not have a meaning of probability e.g. support vector machine classifier. However, the probabilistic output of the classifier can help in post-processing, for instance in combining more classifiers together. Fitting a sigmoid function to the classifier output is a one way to solve this problem. Let $T_{XY} = \{(x_1, y_1), \ldots, (x_l, y_l)\}$ is the training set composed of the vectors $x_i \in X$ and the corresponding binary hidden states $y_i \in Y$. It is assumed that the training set T_{XY} to be identically and independently distributed from underlying distribution. Let $f{:}X \subseteq R^n \to R$ be a discriminant function trained from the data T_{XY}. The parameters θ of a posteriori distribution $p_{Y|F\Theta}(y|f(x), \theta)$ of the hidden state y given the value of the discriminant function $f(x)$ are estimated by the maximum-likelihood method

$$\theta = \arg \max_{\theta'} \sum_{i=1}^{l} \log p_{Y|F\Theta}(y_i|f(x_i)|\theta') \qquad (5.21)$$

where the distribution $p_{Y|F\Theta}(y|f(x),\theta)$ is modeled by a sigmoid function determined by parameters θ.

5.3 Debugging of Multi-Step A/D Converter Stages

The fundamental observation, upon which the debugging of A/D converter stages is motivated, is that practical analog-to-digital converters are prone to exhibit errors. To be more precise, a practical converter is likely to exhibit deviations from the ideal operation of sample, hold and quantization. The goal of debugging is to evaluate if the true output from the converter is within the tolerable deviation of the ideal output. The debugging methods considered are applied after the converter stages, thus operating on the digital signal provided from the output. One of the fundamental constraints is therefore that the internal signals and states of the analog-to-digital converter stages under consideration are unavailable.

Nowadays, A/D converters are widely used in various applications. A/D conversion in radio receivers, for instance, imposes special demands on the converter, and a trend in receiver design has been to move the digitization closer to the receiving antenna. Flexibility in configuration and lower cost are two reasons for striving in this direction. Meanwhile, the carrier frequency, as well as the bandwidth, is increasing, calling for higher sampling rates and increasing analog input bandwidth. The linearity of the A/D converter is also a key characteristic, and the specifications of the system in which the A/D converter is a part, impose requirements on linearity of the converter. To meet these stringent performance requirements, technology and design techniques are pushed to the limit, making them prone to errors. A similar case arises when an A/D converter is integrated on the same chip as a digital signal processor (DSP). In this case there is often a tradeoff between optimum-design point for the performance of the DSP and for the A/D converter. The DSP would typically be manufactured using a chip process with smaller geometry an lower supply voltage than what is beneficial for the A/D converter, mainly in order to keep down power consumption and facilitate higher computational power. The A/D converter would then, again, suffer from manufacturing parameters that are less suited for high-precision analog design.

5.3.1 Quality Criterion

Before proceeding with debugging of an A/D converter stages, firstly a measure of error, e.g. an estimator of the loss function, is introduced. In other words, a mechanism is necessary to distinguish, whether the A/D converter performance is acceptable or not. A quality criterion is generally speaking a function that given the input and output to a system calculates the deviation inflicted by the system. Most common quality criterion measures are based on the distance between the output and the input, and are therefore denoted distance measures. That is, the deviation is a function of the absolute difference between output and input, and not of the input or output themselves. In the multi-dimensional case this corresponds to the deviation being a function of the norm (length) of the difference vector. Two commonly used distance measures are the absolute error and the squared error. The quality criterion usually adopted for an estimator of the loss function is the mean-squared error criterion, mainly because it represents the energy in the error signal, is easy to differentiate and provides the possibilities to assign the weights.

As shown in Section 5.1.3, besides timing errors, the primary error sources present in a multi-step A/D converter are systematic decision stage offset errors λ, stage gain errors η, and errors in the internal reference voltages γ. To find a parameter vector $W^T = [\eta, \gamma, \lambda]$ that gives the best-fit line, the least-squares method attempts to locate a function which closely approximate the transition points by minimizing the sum of the squares of the ordinate differences (called residuals) between points generated by the function and corresponding points in the data. Suppose that $D_{out,i}$ represents the output code which results from an input voltage of V_i, $f(V_i) = aV_i + b$. If the sum of the squares of the residuals, $\Gamma(W)$

$$\Gamma_{MSE}(W) = \sum_{i=0}^{2^n-1} (D_{out,i} - f(V_i, W))^2 \tag{5.22}$$

is minimized, then the function $f(V_i)$ have the property that $f(V_i) \approx D_{out,i}$. $f(V_i)$ is the best-fit line, whose slope is given by a, and the point which the line intercepts the y-axis in given by the b. The design of A/D converter incorporates both defining the quantization regions $f(V_i)$ and assigning suitable values to represent each level with the reconstruction points. In most cases the $D_{out,i}$ is a function of the input $f(V_i)$, so that the expected value is taken with respect to $f(V_i)$ only. Optimal reconstruction points for minimizing the mean-square error have been derived in [401].

Although the mean-squared error criterion is very commonly used, especially from a signal processing point of view, other criteria can be considered. From an A/D converter characterization point of view one, the reconstruction levels might be considered to be an inherent parameter of the A/D converter under test and not of the input signal as was the case in the mean-squared error criterion. The midpoint strategy is based on the assumption that the A/D converter acts as a staircase quantizer. It is based on the assumption that the reconstruction value associated with a specific quantization region should be the midpoint of that region. If the quantization regions should deviate from the ideal ones, then the output values should be changed accordingly. The midpoint approach in fact is consistent with mean-squared error criterion approach if each quantization region is symmetric. Two such signals are the uniform noise and the deterministic ramp, which provide symmetric PDFs within each quantization region, save the regions at the extremes of the signal range where the signal may occupy only part of the region. In the minimum harmonic estimation method [402, 403], on the other hand, estimation values are selected in such a way that the harmonic distortion generated by the A/D converter is minimized. The method uses single sinewaves, and the estimation tables are built using error basis functions, usually two-dimensional Gaussian basis functions in a phase-plane indexing scheme. The basis function coefficients are selected using minimization of the power in the selected number of first harmonics to the test frequency.

The estimation values dependent not only on the characteristics of the A/D converter under test, but also on the test signal itself (through the PDF of the signal). It is therefore of vital importance that the estimation routine is carefully designed as it can yield an estimation system that is heavily biased towards a specific signal, since the estimation values were trained using that signal type. On the other hand, if prior knowledge says that the A/D converter will be used to convert signals of a specific class, it is straightforward to evaluate the system using the same class of signals. Using estimation signals with a uniform PDF can be considered to lead to unbiased calibration results. In this case, both mean-squared and midpoint strategy, coincide. Although, there are many specific measures for describing the performance of an A/D converter – signal-to-noise-and-distortion ratio, spurious-free dynamic range, effective number of bits, total harmonic distortion, … to mention a few, which assess the precision and quality of A/D converters, most of the specialized measures result in fairly complicated expressions that do not provide results of practical use. Exceptions are signal-to-noise-and-distortion ratio and effective number of bits which are both closely related to the mean-squared error criterion; therefore, most results expressed as mean-squared error can be transferred to results on signal-to-noise-and-distortion ratio and effective number of bits as shown in Appendix B.

5.3.2 Estimation Method

Even though extensive research [404–407] has been done to estimate the various errors in different A/D converter architectures, use of DfT and dedicated sensors for analysis of multi-step ADC to update parameter estimates have been negligible. The influence of the architecture on analog-to-digital converter modeling is investigated in [404], and in [405] with use of some additional sensor circuitry, pipeline A/D converter are evaluated in terms of their response to substrate noises globally existing in a chip. In [406], the differential nonlinearity test data is employed for fault location and identification of the analog components in the flash converter and in [407] is shown how a given calibration data set may be used to extract estimates of specific error performance. Functional fault in each of the analog component in the multi-step ADC affects the transfer function differently [378] and analyzing this property form the basis of approach in [408]. The A/D converter is here seen as a static function that maps the analog input signal to a digital output signal. The static parameters are determined by the analog errors in various A/D converter components and therefore, a major challenge in A/D converter test and debugging is to estimate the contribution of those individual errors to the overall A/D converter linearity parameters. The A/D converter characteristics may also change when it is used, e.g., due to temperature change. This means that the A/D converter has to be reevaluated at regular intervals through temperature sensors to examine its performance. Each stage of the A/D converter under test is evaluated experimentally, i.e., a signal is fed to the input of each stage of the A/D converter and the transfer characteristics of each stage of the A/D converter are determined from the outcome. Most of the evaluation methods require that a reference signal is available in the digital domain, this being the signal that the actual stage output of the A/D converter is compared with. This reference signal is in the ideal case a perfect, infinite resolution, sampled version of the signal applied to the A/D converter under test. In a practical situation, the reference signal must be estimated in some way. This can be accomplished by incorporating auxiliary devices such as a reference A/D converter, sampling the same signal as the A/D converter under test [409], or a D/A converter feeding a digitally generated signal to the A/D converter under test

[403, 410]. Another alternative is to estimate the reference signal by applying signal processing methods to the output of the A/D converter under test. Special case of this exists; in [411], sinewave reference signals in conjunction with optimal filtering techniques extract an estimate of the reference signal. Some of the methods do not rely on any digital reference signal. In [409], a method is proposed that estimates the integral nonlinearity (INL) from the output code histogram. In [411], a hybrid system utilizing the minimum mean-squared approach followed by a lowpass filter is proposed. The filtering is possible since the system is aimed at over-sampling applications, so that the signal of interest only can reside in the lower part of the spectrum. The method is aided with the sinewave histogram method and Bayesian estimation.

The overall examined multi-step A/D converter consists primarily of non-critical low-power components, such as low-resolution quantizers, switches and open-loop amplifiers. In $m + n$ multi-step A/D converter the m most significant bits are found from the first resistance ladder and the n least significant bits are generated from the second resistance ladder. Usually, the full range of the second resistance ladder is longer than one step in the first ladder, as explained in Section 3.2. With this over-range compensation in the second ladder, the static errors can be corrected since the signal still lies in the range of the second ladder. This means that the output of the ADC is redundant and it is not possible, from the digital output, to find the values from each subranging step without employing dedicated DfT as clarified in Section 5.4. Since the separate A/D converter stages have to be verified, it is necessary to fix the circuit during testing in such a way that every stage is tested for their full input range. To set the inputs of the separate A/D converter stage at the wanted values, a chain is available in the switch-ladder circuit. So, for mid-range A/D converter measurements, it is necessary to fix the coarse A/D converter values since they determine mid-range A/D converter references, and for testing of the fine A/D converter both coarse and mid ADCs decisions have to be predetermined. The response of the individual A/D converter stage is then routed to the test bus. The sub-D/A converter settings are controlled by serial shift of data through a scan chain that connects all sub-D/A converter registers in serial. It is possible to freeze the contents of the sub-D/A converter registers in normal mode and shift out the data via the scan chain to capture the current sub-D/A converter settings. A test control bit is required per sub-D/A converter to adjust (increase) the reference current to obtain an optimal fit of sub-D/A converter output range to the A/D converter input range. The references of the sub-D/A converter and the subtraction of the input signal and the sub-D/A converter output determine the achievable accuracy of the total A/D converter. The residue signal V_{res} is incorrect exactly by the amount of the sub-D/A converter nonlinearity caused by errors in the internal reference voltages γ

$$V_{res} = \eta V_{in} - (s - 1)\gamma V_{ref} - \lambda V_{offset} \tag{5.23}$$

where s is the observed stage. To obtain a digital representation of (5.23) each term is divided with V_{ref}

$$D_{out} = \eta D_{in} - (s - 1)\gamma - \lambda D_{os} \tag{5.24}$$

where $D_{out} = V_{res}/V_{ref}, D_{in} = V_{in}/V_{ref},$ and $D_{os} = V_{offset}/V_{ref}.$ By denoting the kth stage input voltage as $D_{in,k} = V_{in,k}/V_{ref},$ the kth stage output voltage as $D_{out} = V_{res,k}/V_{ref},$ and the kth stage decision D_k, a recursive relationship when (5.24) is applied to each stage in sequence becomes

$$
\begin{aligned}
D_{out} = D_{out,3} &= \{[\ldots]\eta_2 - (D_2 - 1)\gamma_2 - \lambda_2 D_{os,2}\}\eta_3 \\
&\quad - (D_3 - 1)\gamma_3 - \lambda_3 D_{os,3} = \\
&= D_{in,N}\eta_N \ldots \eta_1 - (D_3 - 1)\gamma_3 - \lambda_3 D_{os,3}
\end{aligned}
\tag{5.25}
$$

Such a model is useful to economically generate an adaptive filtering algorithm look-up table for error estimation and fault isolation. Although in estimation theory several methods are available to estimate the desired response $D_{out}(t)$, the steepest-descent method (SDM) algorithm offers the smallest number of operations per iteration, gives unbiased estimates and require less memory than comparable algorithms based on Hessian matrix, such as for instance the Gauss-Newton, Levenberg-Marquardt and BFS methods. By feeding the different input values D_{in} to each stage at iteration time t, as shown in Fig. 5.17, unknown filter output the desired response $D_{out}(t)$ becomes

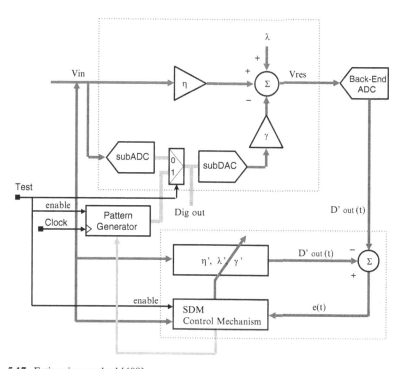

Fig. 5.17 Estimation method [408]

$$D_{out}(t) = D_{in}(t) \times W \tag{5.26}$$

where weights $W^T = [\eta, \gamma, \lambda]$ are used to describe the behavior of the filter.

Statistical data extracted through the die-level process monitor measurements and on-chip temperature information and their maximum-likelihood calculated through EM algorithm provide the estimates $(W')^T = [\eta', \gamma', \lambda']$ with an initial value. By forcing the input signals to the predefined values in each stage, the input to a filter is controlled and residue voltages are obtained. The desired output is digitized; D_{out} is collected from the back-end A/D converter and subtracted from the corresponding nominal value. This desired response is then supplied to filter for processing.

Essentially, based on the predefined inputs and current error estimates $W^T = [\eta, \gamma, \lambda]$, the steepest-descent algorithm involves creation of an estimation error, e, by comparing the estimated output $D'_{out}(t)$ to a desired response $D_{out}(t)$ and the automatic adjustment of the input weights $(W')^T = [\eta', \gamma', \lambda']$ in accordance with the estimation error e

$$W'(t+1) = W'(t) - \mu \times D_{in}(t) \times e(t) \tag{5.27}$$

where the scaling factor used to update $W'(t + 1)$ is the step-size parameter, denoted by μ. $D_{out}(t)$ and $D_{in}(t)$ are matrices with 2^{n-1} rows and three columns, where n is the resolution of the stage. The step size, μ, decrease in each iteration until the input weights decrease, i.e., until $W'(t + 1) < W'(t)$. The estimation error, e, is the difference between the desired response and the actual steepest-descent filter output

$$e(t) = D'_{out}(t) - D_{out}(t) \tag{5.28}$$

and based on the current estimate of the weight vector, W', $D'(t)$ is

$$D'_{out}(t) = D_{in}(t) \times W'(t) \tag{5.29}$$

At each iteration, the algorithm requires knowledge of the most recent values, $D_{in}(t)$, $D_{out}(t)$ and $W'(t)$. During the course of adaptation, the algorithm recurs numerous times to effectively average the estimate and finally find the best estimate of weight W. The temporarily residue voltage in input D_{in} need to be updated after each iteration time to improve the accuracy, which can be done by using the current error estimate W'.

It is important to note that because of the imbalanced utilization and diversity of circuitry at different sections of an integrated circuit, temperature can vary significantly from one die area to another and that these fluctuations in the die temperature influence the device characteristics thereby altering the performance of integrated circuits. Furthermore, the increase in the doping concentration and the enhanced electric fields with technology scaling tend to affect the rate of change of the device parameter variations when the temperature fluctuates. The device parameters that are affected by temperature fluctuations are the carrier mobility, the saturation velocity, the parasitic drain/source resistance, and the threshold voltage. The absolute values of threshold voltage, carrier mobility, and the saturation velocity degrade as the temperature is increased. The degradation in carrier mobility tends to lower the drain current produced by a MOSFET. Although both saturation velocity and mobility have a negative temperature dependence, saturation velocity displays a relatively weaker dependence since electric field at which the carrier drift velocity saturates increases with the temperature. Additionally, as the transistor currents become higher while the supply voltages shrink, the drain/source series resistance becomes increasingly effective on the I–V characteristics of devices in scaled CMOS technologies. The drain/source resistance increases approximately linearly with the temperature. The increase in the drain/source resistance with temperature reduces the drain current. Threshold voltage degradation with temperature, however, tends to enhance the drain current because of the increase in gate overdrive. The effective variation of transistor current is determined by the variation of the dominant device parameter when the temperature fluctuates. On average the variation of the threshold voltage due to the temperature change is between -4 and -2 mV/°C depending on doping level. For a change of 10°C this results in significant variation from the 500 mV design parameter commonly used for the 90 nm technology node. In the implemented system, the temperature sensors register any on-chip temperature changes, and the estimation algorithm update the W' with a forgetting factor, ζ [413]. The estimate at time $t + 1$ is

$$W'(t+1) = \zeta W'(t) + (1 - \zeta)W^0(t+1)$$
$$0 < \zeta \leqslant 1$$

(5.30)

where $W^0(t + 1)$ is an estimate prior to the registered temperature change.

Algorithm

Initialization
– Initialize the input vector $D_{in}(0)$
– Force the inputs and collect the desired output $D_{out}(0)$
– Measure and set the initial value of the weights $W'(0)$
– Initialize the steepest descent update step $\mu = 1$
– Initialize the forgetting factor ζ

Data collection
– Collect N samples from the DLPM and temperature sensors
– Collect N samples from the AD converter

Update parameter estimate

1. Update the input vector $D_{in}(t + 1)$ based on current available $W(t)$
2. Calculate the error estimate $W'(t)$
3. Generate the output estimate $D'_{out}(t) = D_{in}(t) \times W'(t)$
4. Calculate the estimation error $e(t) = D'_{out}(t) - D_{out}(t)$
5. Calculate the error estimate $W'(t + 1) = W'(t) - \mu \times D_{in}(t) \times e(t)$
6. If $W'(t + 1) > W'(t)$ decrease step size μ and repeat step 5
7. Increase the iteration index, t and repeat steps 1–6 for best estimate

(continued)

Algorithm
8. Denote the final value of W' by W_l'
9. If temperature changes update W' with new estimate W_l'

5.4   DfT for Full Accessability of Multi-Step Converters

Modern systems-on-chip (SoC) integrate digital, analog and mixed-mode modules, e.g. mixed-signal, or RF analog and digital, etc., on the same chip. This level of integration is further complicated by the use of third-party cores obtained from virtual library descriptions of the final IC block. Furthermore, the variety and number of cores and their nature type, e.g. analog, complicate the testing phase of individual blocks, of combinations of blocks and ultimately of the whole system. Additionally, the large number of parameters required to fully specify the performance of mixed-signal circuits such as analog to digital converters and the presence of both analog and digital signals in these circuits make the testing expensive and time consuming task. In multi-step A/D converter, high linearity is obtained by extensive usage of correction and calibration procedures. Providing structural DfT and BIST capabilities to this kind of A/D converters is complex since the effects of correction mechanism must be taken into account. Overlap between the conversion ranges of two stages have to be considered, otherwise, there may exists conflicting operational situations that can either mask faults or give an incorrect fault interpretation. Additionally, DfT have to permit multi-step A/D converter re-configuration in such a way that all sub-blocks are tested for their full input range allowing full functional observability and controllability [415, 416]. A system has to be partitioned into sub-blocks to have access to internal nodes such that each isolated sub-block receives the proper stimuli for testing. Similarly, to allow analog structural testing [263] and be able to observe the current (or voltage) signatures of individual cores instead of observing the current (or voltage) signature of the whole analog SoC as well, individual cores have to be able to turn on and off such that the core(s) *dc*, *ac*, and/or transient characteristics can be tested in isolation or together with other cores of an analog SoC.

   IEEE 1149.1 standard boundary scan has been successfully incorporated into digital designs and has considerably simplified the test and debugging of advanced electronic devices. Similarly, IEEE 1149.4 [253] is regarded as a mixed signal counterpart of the 1149.1 boundary scan for digital circuits, where new pins and analog switches to support analog signals are added. The 1149.4 standard specifies the same signal pins that are associated with the 1149.1 standard and it is compliant with the digital test access port (TAP) and boundary architecture. The IEEE 1141.4 standard employs two analog test pins; one to apply test signals and the other to route the response waveforms to the measurement equipment. The external analog-test bus, accesses the internal bus under the control of the test bus interface circuit. The test bus interface circuit allows the internal test-bus lines to connect to either or both analog test pins, isolates the internal test bus when it is not in use to eliminate unwanted noise interference, or connects the bus to one of two control voltages. The 1149.4 standard primary target is to perform a simple chip-to-chip interconnect testing similar to that used in traditional digital boundary scan. However, it can also be utilized to perform internal analog circuit testing [417].

Consider the case of three-step/multi-step A/D converter illustrated in Fig. 5.18. The input signal is sampled by a three-times interleaved sample-and-hold, eliminating the need for re-sampling of the signal after each quantization stage. The S/H splits and buffers the analog delay line sampled signal that is then fed to three A/D

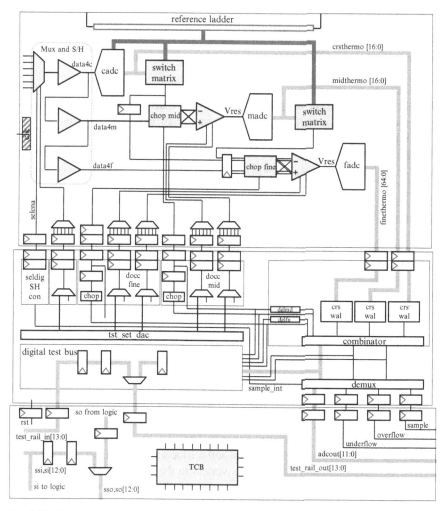

Fig. 5.18 Block diagram of the Multi-step A/D converter and test control circuitry [416]

converters, namely, the coarse (4 bits), the mid (4 bits) and the fine (6 bits). The quantization result of the coarse A/D converter is used to select the references for the mid quantization in the next clock phase. The selected references are combined with the held input signal in two dual-residue amplifiers, which are offset calibrated. The mid A/D converter quantizes the output signals of these mid-residue amplifiers. The outputs from both coarse and mid A/D converters are combined in order to select proper references for the fine quantization. These references are combined with the sampled input signal in two, also offset calibrated, dual-residue amplifiers. The amplified residue signals are applied to a fine A/D converter. Faults in coarse A/D converter, which produces the c most significant bits, affect the step sizes in the transfer function of the c bits in each step. Faults in mid A/D converter affect the step sizes in the transfer function of the m mid bits, and repeat themselves in all the c bits. Similarly, faults in fine A/D converter affect the step sizes in the transfer function of the f fine bits, and repeat themselves in all the $c + m$ bits. On the other hand, faults in the first sub-D/A converter affect the transfer function of $c + m$, and faults in the second sub-D/A converter affect the transfer function of $c + m + f$ bits in a periodic manner.

The key to the debugging of the multi-step A/D converter is thus to select and separate the information on the faults from the transfer function. For at-speed testing and debugging of the analog performance of the A/D converter it is not only imperative to have all 12 digital outputs and the two out of range signals available at the device pins, but to be able to perform debugging of each stage the output signals of the coarse, mid and fine A/D converters need to be observable too. Debugging of each stage is performed sequentially starting from the first stage. Each stage is tested separately - at a lower speed – enabling the use of standard industrial analog waveform generators. To allow coherent testing, the clock signal of the A/D converter has to be fully controllable by the tester at all times. Adding all these requests together leads to an output test bus that needs to be 14 bits wide. The connections of the test bus are not only restricted to the test of the analog part. For digital testing the test bus is also used to carry digital data from scan chains. The test-shell contains all functional control logic, the digital test-bus, a test control block (TCB) and a CTAG isolation chain for digital input/output to and from other IP/cores as illustrated in Fig. 5.18. Further, logic necessary for creating certain control signals for the analog circuit parts, and for the scan-chains a bypass mechanism, controlled by the test control block, is available as well.

In the coarse A/D converter, faults in the analog components internal to the converter cause deviation, from ideal, of the transfer function of the coarse A/D converter by changing the step sizes in the transfer function. The fault cases which include resistor value, comparator's offset and comparator's bias current out of specification fault result in different patterns. The number of peaks and the location of the peak data identify the type of fault and the location of the fault. Since there is no feedback from mid and fine A/D converters to the coarse result value, it is not necessary to set these two A/D converters at the fixed value to test coarse A/D converter. Calibration D/A converter settings do not show in coarse A/D converter results; the calibration system however should remain operative. Random calibration cycles are not allowed to prevent interference with test results. The response of the mid A/D converter cannot directly be tested using the normal A/D converter output data due to an overlap in the A/D converter ranges. Nevertheless, by setting the coarse exor output signals using the scan chain through this block, known values are assigned to the mid switch. The residue signals are now used to verify the mid A/D converter separately by observing the mid A/D converter output bits via the test bus. Faults in mid A/D converter affect the step sizes in the transfer function of the m mid bits, and repeat themselves in all the c bits.

For the mid A/D converter test the chopper signals required for calibration need to be operative. After completing the mid A/D converter test, the chopper signal have to be verified by setting the chopper input to the two predefined conditions and analyze the mid A/D converter data to verify offsets. Since calibration D/A converter settings do show in mid A/D converter results, the D/A converter is set to a known value to prevent interference with mid ADC test results.

Similarly to the mid A/D converter, the fine A/D converter cannot be monitored directly due to the overlap in the A/D converter ranges. Through the available scan chains in the coarse exor and the switch-ladder, control signals are applied to the both mid and fine switch. The predefined input signals are extracted when the A/D converter work in normal application mode with a normal input signal. At a certain moment the scan chains are set to a hold mode to acquire the requested value. Now, the residue signals derived through the predefined input signals evaluate the fine A/D converter performance. For the fine A/D converter test the chopper signals need to be active. To verify offsets, similar procedure as in mid A/D converter is followed. The calibration D/A converter settings have to be known and set to a known value to prevent interference with test results. The digital control block for all three tests operates normally; provides clock pulses, chopper signals and sets the calibration D/A converters in a known condition.

5.4.1 Test Control Block

To control testability within the A/D converter, CTAG compliant test control blocks as shown in Fig. 5.19a are implemented. All test control blocks sections consist of a shift register and a shadow (or update) register. The shift register is implemented with standard D-type flip-flops, while the shadow register includes an asynchronous reset function which will be, by default, active during the device functional mode. The control reset must be directly from an IC pin, and as such a dedicated pin is required. Several control signals are available: (i) *tcb_enable*, which is the asynchronous reset for the test control block and directly controls the shadow register, (ii) *tcb_tdi*, the input to the shift-register portion of the test control block and is used to load the required test mode data, (iii) *tcb_tck*, the clock of the shift-register, (iv) *tcb_update*, which is connected to the clock ports on the update register and is used to transfer the contents of the shift-register to the update register, (v) *tcb_hold*, which is used to disable the shift function of the shift-register,

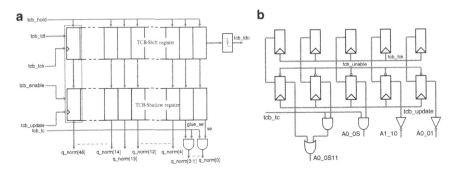

Fig. 5.19 (a) TCB implementation. (b) Contents of TCB slices determining logic

(vi) *tcb_tc*, which toggles the *se* to control normal and shift mode, (vii) *tcb_tdo*, the data output of the test control block and can be connected to the *tcb_tdi* of a preceding TCB block, or directly to an IC pin, and (viii) *q_norm [xx]*, which are the output signals of the test control block. It is possible to switch from any (test) mode to any other (test) mode without having to go through the test control block reset state.

Each TCB-Shift register has one accompanied a TCB-Shadow register, which is parallel loaded at signal *tcb_update*. All TCB-Shift register flip-flops change state on the positive edge of *tcb_tck*. To provide an output for the TCB-block, an anti-skew flip-flop on the negative edge of the *tcb_tck* is added to the last TCB-Shift register. All static test control signals are asynchronously resetable by *tcb_enable* (active low). The TCB-Shift mode is controlled by the *tcb_hold* signal, which switches between shift mode and hold mode. When *tcb_hold* is set to digital one, the registers hold their current state. When *tcb_hold* is set to 0, the register is in shift mode, with data clocking on the positive edge of *tcb_clk*. The A/D converter-test control block contains a total of 215 bits, divided over 211 normal slices and three slices which are gated with the *tcb_tc*. The *se* slice is a special one with a transport mode. The logic after the shadow flipflop(s) determines the slice type. Parts of test control block slices determining logic is shown in Fig. 5.19b.

5.4.2 Analog Test Control Block

Although hierarchically not placed in the test-shell, the two sub test control blocks in the analog top block will be considered as part of the CTAG TCB. Since the separate cores of A/D converter have to be verified, it is necessary to fix the circuit during testing in such a way that all cores are tested for their full input range. In the analog part, there are seven scan chains of which three need special attention. To set the inputs of the separate cores at the wanted values, a chain is available in the switch-ladder circuit. So, for mid-range A/D converter measurements, it is neces-sary to fix the coarse A/D converter values since they determine mid-range A/D converter references, and for testing of the fine A/D converter both coarse and mid A/D converters decisions have to be predetermined. The response of the individual cores is then routed to the test bus. The sub-D/A converter settings are controlled by serial shift of data through a scan chain that connects all sub-D/A converter registers in serial as shown in Fig. 5.20. In order to create extra margin in the timing, signals are shifted half a period, which translates as running on a negative clock in application mode. To avoid the problems, which can arise by integrating

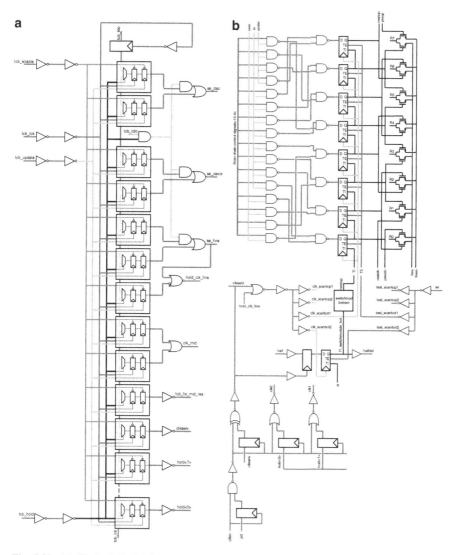

Fig. 5.20 (a) Clock shell. (b) Control logic for the switch ladder

this chain in a much longer chain in the larger system, the polarity of the clock control signal *clkswld* can be switched at will. This is implemented through the *hold_clk_fine* signal, which is in fact a *clk_fine_off* signal (results in a '0' signal) as shown in Fig. 5.20b.

Note that *clkswld* cannot be set off in the clock block. In the clock gates is an override by the scan-enable signal present, making it possible to shift new values in or through the chain, without the need for reprogramming the test control block even if the fix signal has been set accidentally. The needed inversion for application mode is done locally in the switch ladder logic followed by a local clock-tree, needed for the 128 flip-flops and one anti-skew element. It is possible to freeze the contents of the sub-D/A converter registers in normal mode and shift out the data via the scan chain to capture the current sub-D/A converter settings. A Test Control Bit is required per sub-D/A converter to adjust (increase) the reference current to obtain an optimal fit of the sub-D/A converter output range to the A/D converter input range. Beside the chain in the control logic for the switch ladder, the second chain available is in the coarse and mid exor block, which consists of test-points implemented in an analog way with tri-state inverters as shown in Fig. 5.21: for two coarse bits. In normal application these test-points do not have a function, and the clock for those flip-flops is switched off. During scan-tests, when the clock for the test-points is running, the application values are sampled.

The third scan chain available facilitates supply current readings of individual cores [263] as shown in Section 4.2 by turning on/off the biasing network of the cores under consideration in an individual manner. By placing the switches at the ground nodes of the core's biasing circuit and not at the ground nodes of the analog core itself, an impact on the core's bias point due to voltage drop caused by switch on-resistance is limited. To ensure that a core is totally off, e.g. that it does not have floating nodes, a local power-down and local clock-down signals are made available.

Because faults in the output registers mask other faults, the A/D converter is first analyzed for digital faults. Fault cases where all or some faults are digital faults are

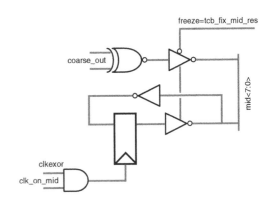

Fig. 5.21 Test-point structure of coarse exor

considered digital faults because fault masking in the digital output circuit restricts the ability to debug faults in the analog components. The debugging of the faults is performed separately for faults in the digital and in the analog components. If the digital circuit is fault free, the debugging techniques for faults in analog components as described in Section 5.3 are applied to the A/D converter to obtain the fault location and error value of the faulty analog components.

The digital circuitry in the multi-step A/D converter can be divided into two sections. The first is the purely digital circuit which consists of the output registers and the control circuits. To facilitate the application of the test vectors to the digital circuit, the scan path design for test technique is used for the registers. To isolate digital test-shell as a separate core for testing, the digital I/O's of the core are provided with hold circuits (controlled by the local test control block bits *hold_inp* and *hold_out*) according to the CTAG protocol. The hold flip-flops are contained in the surround scan chain *ssi/sso* (Fig. 5.22). The other digital section in the A/D converter is the encoder which is embedded in the sub-A/D converter stages. In the case where the encoder design is a simple encoder consisting of a latch or flip-flop circuitry, a level detection circuitry, and a NOR logic encoder, functional test data from the corresponding stage can be used to verify the

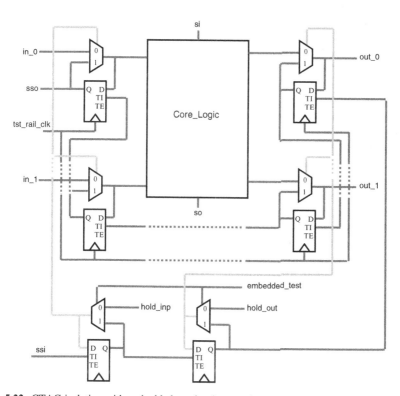

Fig. 5.22 CTAG isolation with embedded test implementation

encoder. It can be deduced that if the linearity of the sub-A/D converter stage which contains a simple encoder do not show any missing code errors, the encoder is fault free. For A/D converters with non-simple encoders, the flip-flops at the output of the comparators in the sub-A/D converter stages are converted into a scan chain. Digital test and debugging vectors are then applied serially through this scan chain to the encoder.

5.5 Debugging of Time-Interleaved Systems

System-on-chip (SoC) realizations require an A/D converter to be embedded on a large digital IC. To achieve the lowest cost, the system-on-chip has to be implemented in state-of-the-art CMOS technologies and must be area and power efficient and avoid the need for trimming to achieve the required accuracy. The rapidly decreasing feature size and power supply voltage of deep-submicron CMOS technology increases the pressure on converter requirements. The multi-step A/D converter architecture as shown in Chapter 3, allows the design of a high-speed, power efficient converter with less latency and less digital logic then a pipelined A/D converter. In such a system, the differential input signal is sampled with a time-interleaved sample-and-hold (S/H) circuit. In a time interleaved system, the sampling process is divided between N S/H units, which should be identical. The input signal is connected to all the units and the N different S/H units sequentially sample the input signal. This means that each unit has the time NT_s to complete a sampling process, while the overall sampling interval is T_s. The output signals from the S/H units are further quantized in the conversion stages before they are multiplexed together to form one output digital signal, which has a sampling interval of T_s. To achieve the sequential sampling, the clock signal is delayed with iT_s to the ith S/H unit. However, the sample-and-hold circuit must still be fast enough to track the high frequency input signal. This is one limit to the number of S/H units that can be used in the time interleaved system. As extensively elaborated in Section 3.3.2, although time-interleaved principle provides high-speed and low power solution, its implementation, due to the manufacturing process, introduces several static and dynamic mismatch errors between interleaved units (offset, gain, bandwidth and time [133–136]), which limit the system performance. Offset mismatch causes fixed pattern noise in the sample-and-hold. It arises from op-amp offset mismatches and charge injection mismatches across the sample and hold units and results in constant amplitude offset in each S/H unit. For a *dc* input, each sample and hold unit may produce a different output and the period of this error signal is N/f_s. Due to the time mismatch caused by clock skew, clock jitter, phase noise or clock cross-coupling with input signal, delay times of the clock between the different S/H units are not equal. As a result the input signal will be periodically but non-uniformly sampled. The time mismatch errors cause frequency dependant noise in the system, which is the largest at the zero-crossings with a period of N/fs and is modulated by the input frequency f_{in}. Similarly, if the gains of each S/H unit are different, the basic error occurs with a period of N/fs but the magnitude of the error is modulated by the input frequency f_{in}. In both cases, noise spectrum peaks at $fs/N \pm f_{in}$. The random variation in S/H unit's internal capacitance, transconductance, hold capacitance as well as input capacitance and kickback noise of the subsequent stages, as seen from one S/H unit, degrade output settling behavior and circuit gain-bandwidth product differently. This error occurs with a period of N/fs, but the magnitude of the error, which is frequency and amplitude dependant, is modulated by the input frequency f_{in} with noise spectrum spurious at $fs/N \pm f_{in}$.

Various methods for estimating static and dynamic mismatch effects in time-interleaved systems have been proposed [95, 96, 418–423]. In [418], the knowledge of multitone signals in DSL modems is used to estimate offset and time errors. In [95, 96, 419], a pseudo random signal is employed for background calibration of offset and gain errors. Distortion in time interleaved systems, with an algorithm for estimation and interpolation of the time errors with a sinusoidal input signal is discussed in [133, 135]. In [420], an efficient interpolation algorithm is presented to estimate time error with a ramp input signal. In [421], a blind time error estimation method and in [422] blind equalization of the time, gain, and offset mismatch errors was presented, assuming only that the input signal is band limited to the Nyquist frequency. In [423] blind estimation method for a rather new problem, bandwidth mismatch in an interleaved sampling system is introduced and the explicit formulas for its effects are derived.

Following notation in [422], the nominal sampling interval, without time errors, is denoted T_s. The sampling frequency is denoted $f_s = 1/T_s$ and the angular sampling frequency, $\omega_s = 2\pi fs$. N denotes the number of S/H units in the time interleaved array, which means that the sampling interval for each S/H unit is NT_s. The time, offset, bandwidth and gain errors are denoted Δ^0_{ti}, Δ^0_{oi}, Δ^0_{bi}, and Δ^0_{gi}; $i = 0, \ldots, N-1$ respectively. The vector notation $\Delta^0_t = [\Delta^0_{t0} \ldots \Delta^0_{tN-1}]$ is used to denote all the time errors. The offset, bandwidth and gain errors are denoted similarly. It is assumed that there are no other errors in the S/H units except the four types of mismatch errors. Similarly, $x(t)$ is the analog input signal and $y_i[k]$ are the output subsequences from the N S/H units, sampled with time errors

$$y_i[k] = (1 + \Delta^0_{g_i})(1 + \Delta^0_{b_i})x((kN + i)T_s + \Delta^0_{t_i}) + \Delta^0_{o_i} \qquad (5.31)$$

Uniform sampling of $x(t)$ at the Nyquist rate results in samples $x(t = nT_s)$ that contain all the information about $x(t)$. If the input signal is band limited to the Nyquist frequency and the error parameters are known, the input signal can be perfectly reconstructed from the irregular samples [424]. The amplitude offset errors are removed by subtracting the offset error parameters from the respective subsequences. Similarly, the gain errors can be removed, after the offset errors are removed, by dividing the subsequences by the respective S/H gain. At the end the bandwidth errors are removed. Reconstructured output signal z_i can be then expressed as

$$z_i^{(\Delta_{o_i}^0)}[k] = y_i[k] - \Delta_{o_i}^0 = (1 + \Delta_{g_i}^0)(1 + \Delta_{b_i}^0)x((kN + i)T_s + \Delta_{t_i}^0)$$

$$z_i^{(\Delta_{o_i}^0, \Delta_{g_i}^0)}[k] = z_i^{(\Delta_{o_i}^0)}[k]/(1 + \Delta_{g_i}^0) = (1 + \Delta_{b_i}^0)x((kN + i)T_s + \Delta_{t_i}^0) \qquad (5.32)$$

$$z_i^{(\Delta_{o_i}^0, \Delta_{g_i}^0, \Delta_{b_i}^0)}[k] = z_i^{(\Delta_{o_i}^0, \Delta_{g_i}^0)}[k]/(1 + \Delta_{b_i}^0) = x((kN + i)T_s + \Delta_{t_i}^0)$$

In [425] a minimal sampling rate for an arbitrary sampling method that allows perfect reconstruction is developed. In [426] it was shown that only special cases of bandpass signals can be perfectly reconstructed from their uniform samples at the minimal rate of $2 \times Bandwidth$ [samples/sec], while in [427] a reconstruction scheme that recovers any bandpass signal exactly is provided. In [428] and [429] a blind multi-coset sampling strategy that is called universal in [429] is suggested. Multi-coset sampling is a selection of certain samples from uniform sampling. The uniform grid is divided into blocks of L consecutive samples. A constant set C of length p describes the indices of p samples that are kept in each block while the rest are zeroed out. The set $C = \{c_i\}_{i=1}$ is referred to as the sampling pattern where $0 \le c_1 < c_2 < \ldots < c_p \le L - 1$. Define the ith sampling sequence for $1 \le i \le p$ as

$$x_{c_i}[n] = \begin{cases} x(t = nT_s) & n = mL + c_i \\ 0 & \text{otherwise} \end{cases} \qquad (5.33)$$

The sampling stage is implemented by p uniform sampling sequences with period $1/(LT_s)$, where the ith sampling sequence is shifted by $c_i T_s$ from the origin. Therefore, a multi-coset system is uniquely characterized by the parameters L, p and the sampling pattern C. Direct calculations show that [429]

$$Z_c^{(\Delta_b^0)}(e^{j2\pi f T_s}) = \frac{1}{LT} \sum_{r=0}^{L-1} (e^{j2\pi c_i r/L})X(f + \tfrac{r}{LT}) \qquad (5.34)$$
$$\forall f \in F_0 = (0, \tfrac{1}{LT}), 1 \le i \le p$$

where $Z_{ci}(e^{j2\pi f T})$ is the discrete-time Fourier transform (DTFT) *of $x_{ci}[n]$* band-limited to F. For our purposes it is convenient to express previous equation in a matrix form as

$$y(f) = Ax(f), \quad \forall f \in F_0 \qquad (5.35)$$

where $y(f)$ is a vector of length p whose ith element is $Z_{ci}(e^{j2\pi f T})$, and the vector $x(f)$ contains L unknowns for each f

$$x_i(f) = X(f + \tfrac{i}{LT}), \quad 0 \le i \le L - 1, \quad f \in F_0 \qquad (5.36)$$

(continued)
The matrix A depends on the parameters L, p and the set C but not on $x(t)$ and is defined by

$$A_{ik} = \frac{1}{LT_s} e^{j2\pi c_i k/L} \tag{5.37}$$

With the inverse time discrete Fourier transform (ITDFT) the time error reconstructed signal can be derived

$$z_c^{(\Delta_t^0)}[k] = ITDFT(Z^{(\Delta_t^0)}(e^{j2\pi f T_s})) \tag{5.38}$$

The quality criterion commonly adopted for an estimator of the loss function is the mean-squared error criterion, mainly because it represents the energy in the error signal, is easy to differentiate and provides the possibilities to assign the weights (Section 5.3)

$$W_t^{(M)}(\Delta_t) = \sum_{i=1}^{L-1} \sum_{j=0}^{i-1} \left(\frac{1}{L} \sum_{k=1}^{M} (z_i^{(\Delta_{t_i})}[k])^2 - (z_j^{(\Delta_{t_j})}[k])^2 \right)^2 \tag{5.39}$$

Similarly, to estimate the offset errors, it is assumed that the mean value of the output from each S/H unit corresponds to the respective offset errors. Additionally, it is assumed that there are no other errors in the S/H units except the offset errors. To estimate the gain errors, it is assumed that the gain errors are only errors present in the circuit.

Algorithm

Initialization
- Choose a batch size, M, for each iteration
- Initialize the steepest descent update $\mu = 1$
- Initialize the parameter estimates $\Delta_{oi}^{(0)} = 0$, $\Delta_{gi}^{(0)} = 0$, $\Delta_{bi}^{(0)} = 0$, $\Delta_{ti}^{(0)} = 0$, $i = 0, \ldots, N-1$
- Initialize the forgetting factor ζ

Data collection
- Collect M samples from DUT

Update parameter estimate
1. Calculate the reconstructed signal according to (5.35) and (5.41)
2. Calculate the gradients of the loss function $\nabla W_o(\Delta_o), \nabla W_g(\Delta_g), \nabla W_b(\Delta_b), \nabla W_t(\Delta_t)$
3. Update the parameter estimate $\Delta^{(i+1)} = \Delta^{(i)} - \mu \times rW(\Delta^{(i)})$
4. If $\Delta^{(i+1)} > \Delta^{(i)}$ decrease step size μ and repeat step 3
5. Increase the iteration index, i, and repeat steps 1–4 for best estimate
6. Denote the final value of $\Delta^{(i+1)}$ by $\Delta_F^{(i+1)}$
7. If condition change update $\Delta_F^{(i+1)}$ with new estimate $\Delta_{F'}^{(i+1)}$

The measure of the gain error is the variance of the output from each S/H unit corresponding to the respective gain of the S/H unit. Equally, to estimate the bandwidth errors, it is assumed that the bandwidth errors are only errors present in the circuit. The measure of the bandwidth error is the variance of the gain-bandwidth product from each S/H unit corresponding to the respective gain-bandwidth product of the S/H unit. Note that Δ_{bi} is a function of the input frequency as well as the bandwidth. Also, remark that phase variation due to the bandwidth mismatch is a nonlinear function of the input frequency while the phase mismatch due to the timing skew is its linear function. The offset, gain and bandwidth loss function can be then defined as

$$W_o^{(M)}(\Delta_o) = \sum_{i=1}^{L-1} \sum_{j=0}^{i-1} \left(\frac{1}{L} \sum_{k=1}^{M} (z_i^{(\Delta_{oi})}[k])^2 - (z_j^{(\Delta_{oj})}[k])^2 \right)^2$$

$$W_g^{(M)}(\Delta_g) = \sum_{i=1}^{L-1} \sum_{j=0}^{i-1} \left(\frac{1}{L} \sum_{k=1}^{M} (z_i^{(\Delta_{gi})}[k])^2 - (z_j^{(\Delta_{gj})}[k])^2 \right)^2 \qquad (5.40)$$

$$W_b^{(M)}(\Delta_b) = \sum_{i=1}^{L-1} \sum_{j=0}^{i-1} \left(\frac{1}{L} \sum_{k=1}^{M} (z_i^{(\Delta_{bi})}[k])^2 - (z_j^{(\Delta_{bj})}[k])^2 \right)^2$$

The minimizing arguments of the loss functions give the mismatch error estimates. However, the mismatch errors may change slowly with for instance temperature and aging, so the parameter estimates have to be adaptively updated with new data. Since the minimizing argument cannot be calculated analytically, a numerical minimization algorithm instead is employed.

Although in estimation theory several methods are available to estimate the desired response, a gradient search method [430] offers the smallest number of operations per iteration and does not require correlation function calculation. Essentially, the algorithm involves creation of an estimation error by comparing the estimated output to a desired response and the automatic adjustment of the input weights in accordance with the estimation error

$$\Delta^{(i+1)} = \Delta^{(i)} - \mu \nabla W(\Delta^{(i)}) \qquad (5.41)$$

where the scaling factor used to update $\Delta^{(i+1)}$ is the step-size parameter, denoted by μ. The step size, μ, decrease in each iteration until the input weights decrease, i.e., until $\Delta^{(i+1)} < \Delta^{(i)}$. If condition on-chip change, e.g. temperature, the estimation algorithm update the $\Delta^{(i+1)}$ with a forgetting factor, ζ [413]. The new estimate is given as

$$\Delta^{(i+1)} = \zeta \Delta^{(i)} + (1 - \zeta) \Delta_F^{(i+1)}$$
$$0 < \zeta \leqslant 1 \qquad (5.42)$$

where $\Delta_F^{(i+1)}$ is an estimate prior to the registered temperature change.

5.6   Foreground Calibration

As elaborated in Chapter 3, a practical converter is likely to exhibit deviations from the ideal operation of sample, hold and quantization. Debugging techniques, such as those described in previous sections, evaluate if the true output from the converter is within the tolerable deviation of the ideal output. The information obtained in this manner can than be re-used to supplement the circuit calibration. A wide variety of calibration techniques to minimize or correct the steps causing discontinuities in the A/D converter's stage transfer functions has been proposed [29, 95, 96, 117, 203–211]. The mismatch and error attached to each step can either be averaged out, or their magnitude can be measured and corrected. Analog calibration methods include in this context the techniques in which adjusting or compensation of component values is performed with analog circuitry, while the calculation and storing of the correction coefficient can be digital. However, digital methods have gained much more popularity, mainly because of the increased computational capacity, their good and well predefined accuracy, and flexibility.

Besides the classification to error-averaging, analog, and digital calibration methods, the techniques can be divided into foreground and background methods, depending on whether the normal operation is interrupted or not. As illustrated in Fig. 5.23a, foreground calibration requires the operation of an A/D converter to be interrupted so that a known input sequence can be applied to the A/D converter,

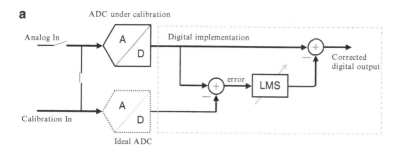

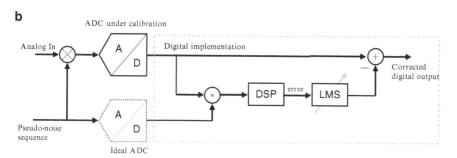

Fig. 5.23 Principle of (**a**) foreground calibration and (**b**) background calibration

where by comparing the output of the A/D converter to the expected A/D converter output under ideal conditions the impact of missing codes can be quantified and corrected. Ideal A/D converter, drawn with dashed lines in Fig. 5.23a is not physically implemented; digital output is already known since calibration input is known. In foreground schemes calibration can be achieved within a small number of clock cycles as the error signal is highly correlated with the error sources causing the missing codes. However, the A/D converter has to be taken offline every time calibration is performed, which in some applications may not be possible.

On the other hand, background calibration shown in Fig. 5.23b, such as [34] highlighted in Section 3.5.2 continuously measures and removes missing codes. Similar to foreground calibration, ideal A/D converter in background calibration technique is not physically implemented; digital output is already known since pseudo-noise sequence is known. Typically, the background calibration techniques are developed from the same algorithms as the foreground methods by adding hardware or software to perform the calibration coefficient measurements transparent to the normal operation, although, a number of algorithms are originally intended for background operation. In a statistical scheme the input of the stage under calibration is effectively modulated by a known pseudo-random sequence, where by correlating the digital output of the A/D converter with the known pseudo-random sequence the impact of missing codes can be determined. In essence, the input signal is multiplied with a pseudo binary random sequence of + 1 and − 1 before A/D conversion. The digital output signal is then multiplied with the same sequence, to reconstruct the signal. To avoid significantly altering the A/D converter output spectrum the pseudo-noise sequence is typically made very long to avoid correlations with the analog input, as well as small in amplitude so that the injected pseudo-random sequence which appears as an additional white noise source at the output only consumes a small portion of the dynamic range.

In general, the maximum sample rate and attainable resolution are equal for foreground and background calibration techniques, while the latter approach ends up with more complex realizations requiring redundant hardware or excessive digital signal processing. With background calibration schemes however, since the adjustment of the compensation values is with ± 1 at a time, a large number of clock cycles are required before the end value is reached. In A/D converters which use background-statistical techniques long calibration times can lead to excessive test times, limiting IC production throughput and thus reducing revenue. For example, with four million calibration cycles, even with a reasonably high sampling rate of 40 MS/s, 100 ms would be required at minimum to test the A/D converter. For higher resolution and/or lower speed A/D converters the test time can be much higher [431].

Unlike in the analog calibration technique where components are measured and then corrected until they match, in the foreground digital calibration technique component ratios are measured, quantized by following stages of the A/D converter and then stored digitally. The results of the measurements are later used with the digital output of the analog to digital converter to generate a corrected digital output with improved linearity. This technique has the disadvantage that it requires the use of digital adders not required by the analog technique. However, the technique greatly relaxes the design requirements for some of the analog components and results in a design that is very reliable and robust. In each stage of the multi-step, the comparators divide the range of possible signal values into a set of smaller ranges or bins. Associated with each bin is a digital code and D/A converter weight. Each stage of the A/D converter assigns the signal to a certain bin and generates a residue signal. When the conversion is complete, the signal has a bin assignment for each stage of the A/D converter. Thus, the signal is assigned a digital code and weight by each stage of the A/D converter. To reconstruct a linear digital representation of the input signal, the weights are all summed.

When the linear reconstruction of the inputs is done, the results for both signals should match each other. If w is used to denote the weights from the i stage of interest and D is used to denote the code resulting from the rest of the A/D converter, then $code_i = w_i + D_i$ and $code_{i+1} = w_{i+1} + D_{i+1}$. For codes to match, $w_{i+1} - w_i = D_i - D_{i+1} = G(V_{Ri} - V_{Ri+1})$. $D_i - D_{i+1}$ is the height of the discontinuity. Thus, the height of the discontinuity is the difference in the weights corresponding to the bins on either side of the discontinuity. Therefore, the weights are assigned by measuring each of the discontinuities: assume that a signal V_i close to the threshold for the discontinuity is applied to the input of the stage of interest. Two measurements are then made: in the first case, the D/A converter level immediately below the comparator threshold is subtracted from the input signal. The resulting residue for this case is then amplified and quantized by the rest of the multi-step A/D converter obtaining the digitized output D_i. Next, the input signal is kept the same and the D/A converter level immediately above the comparator threshold is subtracted from the input signal. The residue for this case is also amplified and quantized by the rest of the multi-step A/D converter. For this case, the residue output D_{i+1} is obtained. $D_i - D_{i+1}$ is then computed to obtain the result for the weight difference. This measurement is done at each threshold.

A digitally calibrated multi-step A/D converter operation is similar to the operation of an uncalibrated converter with the following exception. Instead of incorporating the A/D converter output bits from each stage directly into the digital output word, the bits from each stage are instead used as the address to a lookup table. A decoder translates the address into a signal that selects the appropriate D/A converter weight from the lookup table. This D/A converter weight is then added to the digital result from the previous stage. In the last stage, the final digital output word is the sum of each of the selected DAC weights, one from each stage of the converter. Recall that residue signal V_{i+1} of $i + 1$ stage can be described by

$$V_{i+1} = \eta_i V_i + (1 - D_i) \cdot \gamma_i V_{ref} + \lambda_i \qquad (5.43)$$

where D_i is the output from the sub-A/D converter of the stage. Dividing both sides with V_{ref}, the stage i input signal V_i can be found as

$$V_i/V_{ref} = 1/\eta_i(V_{i+1}/V_{ref} - (1 - D_i) \cdot \gamma_i - \lambda_i/V_{ref}) \qquad (5.44)$$

This equation forms the basis of calibration algorithm. Stage gain error η_i, internal reference voltage error γ_i and systematic offset error λ are estimated by using the algorithm described in Section 5.3. In order to recover the correct value of V_i, V_{i+1}/V_{ref} have to be recovered first. A digital code D_{out} is obtained by normalizing the analog input voltage V_{in} to reference voltage V_{ref}.

If the (N-i) bits back-end A/D converter is an ideal A/D converter, the digital output D_{BE} of input voltage V_{i+1} written in offset code format can be expressed as

$$D_{BE} = b_1 b_2 ... b_N = b_1 2^{N-1} + b_1 2^{N-2} + ... + b_{N-1} 2 + b_N \qquad (5.45)$$

Multiplying the offset-coding D_{BE} by V_{LSB} and removing the offset, the approximate value of V_{i+1} is then calculated as

$$V_{in}/V_{ref} \approx (D_{BE} - 2^{N-i-1})/2^{N-i-1} = b_{i+1} + b_{i+2} 2^{-1} + ... + b_{N-1} 2^{-(N-i-2)}$$
$$+ b_N 2^{-(N-i-1)} - 1 + 1/2^N$$

$$(5.46)$$

Assuming the error parameters for each stage which will be calibrated are known from the algorithm described in Section 5.3, the procedure to calibrate the entire multi-step A/D converter can be described as: (i) firstly, calibration is started from stage i to find the residue voltage V_{i+1} from back-end A/D converter output D_{BE}, (ii) next, the stage input V_i based on the algorithm output V_{i+1} is recovered, and (iii) at the end, first two steps to calibrate stage $i-1$ are repeated until the first stage is reached. V_1 is the calibrated value of input signal V_{in}. The digital output is obtained by quantizing V_1 to N bits.

The accuracy of the calibration is dependent on the accuracy of the A/D converter used to perform the calibration. Differential non-linearity in the A/D converter used to measure the weights corrupts the measurement; the new A/D converter that includes the measured stage will have DNL at the comparator thresholds, which can be a limiting factor if the multi-step A/D converter is employed to calibrate itself. Fortunately, the DNL caused by the A/D converter during this measurement is scaled down since the measurement occurs on the amplified residue. Therefore, if the DNL of the A/D converter without the calibrated stage is x, the DNL of the A/D converter with the calibrated stage is x/G where G is the gain of the residue amplifier. Thus, it is possible to calibrate a multi-step A/D converter starting from back-end A/D converter output and working forward towards the front-end. Similarly, some accumulation of errors does occur during the calibration process. However, this accumulation of errors are kept to a minimum by designing the multi-step A/D converter in such a way that small variations in the input signal only affect the code produced by the end stages.

The design allows the inclusion of the calibration logic on the same chip as the A/D converter or its realization with an FPGA, which is combined with the A/D converter chip on the circuit board level. The calibration state machine controls the reference voltage switching in the D/A converters of the first two stages during the calibration hold phases. The input signals for the calculation and coding logic are the RSD-corrected digital words from the converter chip, the raw bits of the first and the second stage, the state-indicating bits, and an external reset signal. The calibration coefficients per stage are calculated as an average of four measurements and stored in a memory. During normal operation calibration coding is simply the addition of two coefficients, which are selected according to the raw bits of the two first stages, to each uncalibrated A/D converter output word.

5.7 Experimental Results

The validity of the various concepts described in this Chapter is evaluated on the three-step/multi-step A/D converter (Fig. 5.24) with dedicated embedded sensors fabricated in standard single poly, six metal 0.09-μm CMOS. The converter input signal is sampled by a three-time interleaved sample-and-hold, eliminating the need for re-sampling of the signal after each quantization stage. The S/H splits and buffers the analog delay line sampled signal that is then fed to three A/D converters, namely, the coarse (4 bits), the mid (4 bits) and the fine (six bits). The quantization result of the coarse A/D converter is used to select the references for the mid quantization in the next clock phase. The selected references are combined with the held input signal in two dual-residue amplifiers, which are offset calibrated. The mid A/D converter quantizes the output signals of these mid-residue amplifiers. The outputs from both coarse and mid A/D converters are combined in order to select proper references for the fine quantization. These references are combined with the

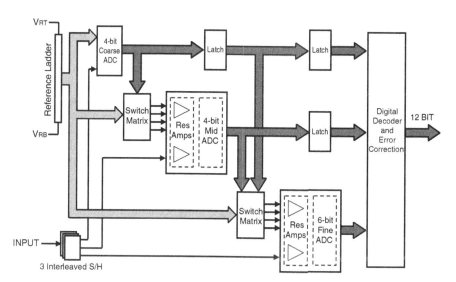

Fig. 5.24 Prototype three-step/multi-step A/D converter

sampled input signal in two, also offset calibrated, dual-residue amplifiers. The amplified residue signals are applied to a fine A/D converter.

The prototype was fabricated in standard single poly, six metal 0.09-μm CMOS with the core area of 0.6 mm^2 excluding bond pads. The A/D converter operates at 1.2 V supply voltage and dissipates 85 mW (without output buffers). As the CMOS technology scales to a smaller feature size, it is apparent that digital circuits will benefit from it in speed, power and area. These benefits cannot be applied to analog circuits directly simply because the figures of merit in analog circuits extend beyond speed, power and area. The digital scaling yields a combination of area reduction, bandwidth increase and reduction in power dissipation although at the cost of lower SNR and worse linearity. The focus of the scaling for the analog circuits can be shifted from the standard digital scaling to the performance increase analog scaling. In [369] is shown that the analog scaling can achieve the increase of the circuit bandwidth and the dynamic range at a fixed frequency, although under assumption that analog circuits occupy small part of a system on chip and therefore can be penalized by the increase of area and power. The techniques which contribute to the goals of the scaling can be categorized into two sections. On architectural level, the choice of per-stage resolution, coding and digital correction relaxes the requirements on some active circuit blocks which are conventionally known as power demanding blocks. On a circuit level, taking advantage of digital correction, dynamic comparators are used to eliminate static power consumption. A high speed, low voltage opamp with a pre-amplifier and cascode transconductance stage is employed to achieve fast settling and high dc gain. Lastly, the clock boosting circuit is utilized to reduce the on-resistance due to low supply voltage. The digital section in this converter design includes the clock generators, digital decoders and digital correction circuit.

The effect of technology scaling on source/drain parasitic capacitance is not obvious because it depends on doping density, junction depth, source/drain area scaling, etc. If it is assumed that parasitics are only affected by the scaled feature size, the ratio of the source/drain parasitic capacitance is roughly equal to the technology scaling constant s. Other sources of parasitics capacitance, such as wiring, overlapping, also change with technology. Although the line width of a metal line may decrease with technology, the coupling between two adjacent metal lines tends to increase because the minimum separation between metal lines is decreased. Other minor factors, such as metal resistance requirement, may determine the width for a long metal wire which will also affect the parasitic capacitance. In general, because the overall die size is about $(1/s)^2$ smaller, it can be assumed that the parasitic (if dominated by wiring and overlapping) is roughly $(1/s)^2$ as well. If all of the dimensions and voltages (and currents) are scaled following the suggestions from the analog scaling scenario, the transconductance increase by the factor $1/s^3$. The output impedance in saturation scales with s^3 so the intrinsic gain, $g_m r_o$ remains constant. With analog scaling, the maximum allowable voltage swings decrease by a factor s, lowering the dynamic range of the circuit. In order to restore the dynamic range, the transconductance of the transistors must increase by a factor $1/s^3$ because thermal noise voltages and currents scale with

$\sqrt{g_m}$. Since voltage scaling requires that $V_{GS} - V_T$ decrease by a factor of $1/s$, drain current I_D must increase by the same factor, increasing the power dissipation with $1/s$. Similarly, if C_{ox} is scaled up by $1/s$ and L and $(V_{GS} - V_T)$ are scaled down by s, then W must increase by $1/s^2$.

For the 12-bit resolution, the hold capacitors in the sample-and-hold circuit is determined by the noise limitation, therefore, the required capacitor size is the same in both technologies. In reality, since C_H is kept the same for noise reasons, f_T is higher in the scaled technology for a particular bias condition and a fixed g_m. This translates to an increase of speed with the same power dissipation. Furthermore, v_{DSAT} can also be lowered in the scaled technology to increase g_m. Also the parasitic capacitances for both the opamp input and source/drain output capacitances are smaller in the scaled technology than the 0.18-μm design improving the feedback factor as well as the settling time. If the C_L is determined by the parasitic capacitance, it is decreased by about a factor of s when the technology is scaled due to the reduced gate and source/drain capacitances. This means g_m can be scaled by a factor of s as well. If v_{DSAT} is kept constant, the current required can be reduced by roughly a factor of s, which translates to a power saving of $1/s$.

In Fig. 5.25 a micrograph of the test-chip is presented. Dedicated embedded sensors and the complete DfT occupy less than 10% of the overall area. Extra digital circuitry is added on the left, along with some dummy metal lines for process yield purposes. Additionally, the test-chip contains a temperature sensor (located between the coarse A/D converter and the fine residue amplifiers) and matrix of differential transistor pairs and ladder resistors divided into specific groups, which are placed in and around the partitioned multi-step A/D converter.

To allow characterization of current process variability conditions of certain parameters of interest, enabling the optimized design environment, each group of monitors target a specific error source as shown in Section 5.1.4. Repetitive single die-level process monitor measurements are performed to minimize noise errors.

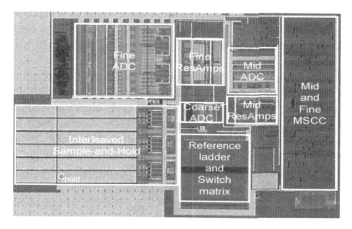

Fig. 5.25 Chip micrograph

Each die-level process monitor consists of multiple differential transistor pairs (with sources and drains connected in parallel within one path), which are spaced at different mutual distances to examine the effect of spatial dependence on offset. All sources are connected to a common point. All transistors in the same monitor have their gates connected to a common point. Special attention is paid in the layout to obtain a very low resistance in the gate path to eliminate systematic errors during the measurements; very wide source metal connections are used.

The clock lines are routed in the center of the active area where the appropriate phases are tapped off at the location of each stage in the circuit. Digital correction is at the lower right corner of the active area; and 12 bit output is produced at the pads on the bottom. Extra digital circuitry is added on the right, along with some dummy metal lines for process yield purposes. Similar to the converter described in Chapter 3, the differential circuit topology is used throughout the design and multiple substrate taps are placed close to the noise sensitive circuits to avoid the noise injection. For analog blocks, substrate taps are placed close to the n-channel transistors and connected to an analog ground nearby (for common-source configuration, substrate taps are connected to the source). For digital blocks, substrate taps are placed close to n-channel transistors and connected to a separate substrate pin to a dedicated output pad. This pad is then joined with ground on the evaluation board. An added advantage for placing substrate taps close or next to transistors is to minimize the body effect variation. For common-source devices, no body effect is found since the source and body are connected. For cascode devices, although the source potential may vary with respect to the body potential, the effect of V_T on the drain current is greatly reduced due to the source degeneration. No additional substrate taps are placed to avoid that they act as noise receptors to couple extra noise into the circuit. Separate V_{DD} and ground pins are used for each functional block not only to minimize the noise coupling between different circuit blocks, but also to reduce the overall impedance to ground. Multiple V_{DD} and ground pins are used throughout the chip. The digital V_{DD} and ground pins are separated from the analog ones. Within the analog section, V_{DD} and ground pins for different functional blocks are also separated to have more flexibility during the experiment. Each supply pin is connected to a Hewlett-Packard *HP3631A* voltage regulator and is also bypassed to ground with a 10 μF Tantalum capacitor and a 0.1 μF Ceramic chip capacitor.

The reference current sources are generated by *Keithley 224* external current source. For the experiment, the sinusoidal input signal is generated by an arbitrary waveform generator (*Tektronix AWG2021*). This signal comes on-board through a SMA connector and is applied to a transformer (Mini-Circuit PSCJ-2-1) which converts the single-ended signal to a balanced, differential signal. The outputs of the transformer are *dc* level-shifted to a common-mode input voltage and terminated with two 50 Ω matching resistors. The common-mode voltage of the test signal going into the A/D converter is set through matching resistors connected to a voltage reference and is nominally set at 0.6 V. The digital output of the A/D converter is buffered with an output buffer to the drive large parasitic capacitance of the lines on the board and probes from the logic analyzer. The digital outputs are captured by the logic analyzer (*Agilent 1682AD*). A clock signal is also provided to the logic analyzer to synchronize with the A/D converter. All the equipment is set by a LabView program and signal analysis is performed with MatLab.

5.7.1 Application of Results for A/D Test Window Generation/Update

By sweeping the reference voltage until a change in the decision occurs, information about the process variation effects is extracted. A discrimination window for various die-level process monitors is defined according to the rules of the multi-step A/D converter error model described in Section 5.1.3. The drain voltage of the different transistors in each die-level process monitor are accessed sequentially though a switch matrix which connects the drain of the transistor pairs under test to the voltage meter; the drains of the other transistors are left open. The switch matrix connects the gate of the transistor pairs under test to the gate voltage source and connects the gates of the other rows to ground. The analysis of critical dimensions shows a dependence of the poly-line width on the orientation. This cause performance differences between transistors with different orientations. For transistor pairs no systematic deviations are observed between different gate orientations. All transistors are biased in strong inversion by using gate voltages larger than V_T. Since the different transistors are measured sequentially the *dc* repeatability of the *dc* gate voltage source must be larger than the smallest gate-voltage offset to be measured. The repeatability of the source in measurement set-up was better than six digits, which is more than sufficient. The offset is estimated from the sample obtained by combining the results of the devices at minimum distance over all test-chips. The same statistical techniques are used as for the distance dependence.

In the coarse A/D converter illustrated in application example, some DNL errors are present as shown in Fig. 5.26a. Figures 5.26b and 5.27 illustrate histogram estimated from 140 samples extracted from a similar number of differential transistor pairs and ladder resistor measurements in the test matrix. As the number of on-chip die-level process monitors is finite due to area limitations, additional information is obtained through statistical techniques. The estimation of the parameters based on the EM-algorithm is illustrated in Fig. 5.28, corresponding to the maximum likelihood estimates based uniquely on the observed data. The EM algorithm allows obtaining the maximum likelihood estimates of the unknown parameters by a computational procedure which iterates, until convergence, between two steps. As the main statistical concern is parameter estimation and in most cases this is best achieved by the use of maximum likelihood theory, the EM algorithm substitutes the missing data in the log likelihood function, not in the incomplete data set; the missing values are substituted by the conditional expectations of their functions, as they appear in the log-likelihood function. To make the problem manageable, the model for the process parameter variation in this realization is assumed to follow Gaussian distribution. With that assumption, then the

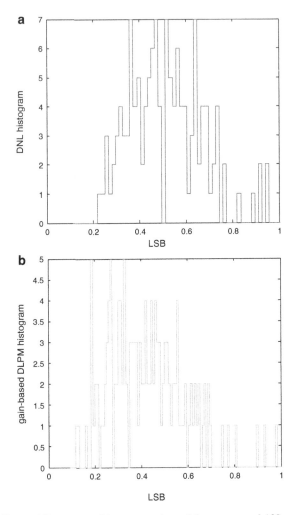

Fig. 5.26 (a) Coarse A/D converter histogram estimated from measured 100 samples, (b) gain-based DLPM histogram estimated from measured 140 samples (© IEEE 2008)

modeled values corresponds to the expected values of the sufficient statistics for the unknown parameters. For such densities, it can be said that the incomplete-data set is the set of observations, whereas each element of the complete-data set can be defined to be a two-component vector consisting of an observation and an indicator specifying which component of the mixture occurred during that observation. The mixtures of Gaussians are initialized by applying the EM equations to the observed mixtures of two univariate Gaussian components based on die-level process monitors and coarse A/D converter DNL measurements. The plot of the log-likelihood function $L(\theta^{(t)}/T_{XY})$ with respect to the number of iterations is

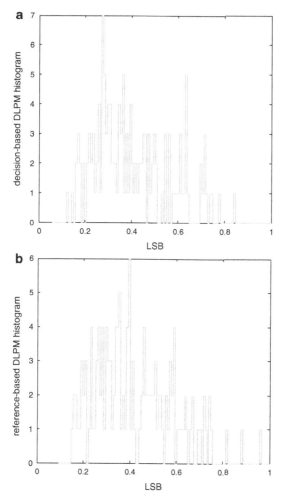

Fig. 5.27 DLPM histogram estimated from measured 140 of (a) decision stage-based DLPM and (b) reference-based DLPM (© IEEE 2008)

visualized in Fig. 5.29a. Each iteration is guaranteed to increase the likelihood, and finally the algorithm converges to a local maximum of the likelihood function in 12 iterations.

The mean μ and the variance σ of decision stage offset errors λ, stage gain errors η, and errors in the internal reference voltages γ is estimated. The convergence properties of the EM algorithm are discussed in detail in [393]. Recall that $\theta^{(t+1)}$ is the estimate for θ which maximizes the difference $\Delta(\theta/\theta^{(t)})$. Starting with the current estimate for θ, that is $\theta^{(t)}$, $\Delta(\theta/\theta^{(t)}) = 0$. Since $\theta^{(t+1)}$ is chosen to maximize $\Delta(\theta/\theta^{(t)})$, $\Delta(\theta^{(t+1)}/\theta^{(t)}) \geq \Delta(\theta^{(t)}/\theta^{(t)}) = 0$, so for each iteration the likelihood $L(\theta)$ is nondecreasing.

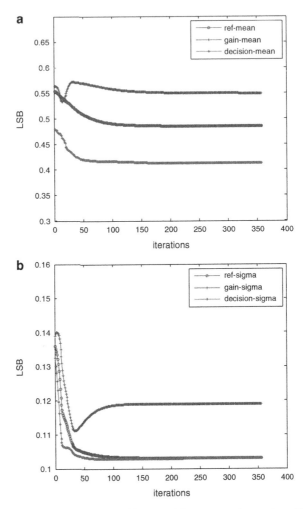

Fig. 5.28 (a) Estimating mean μ values of λ, η, γ with respect to the number of iterations of the EM. (b) Estimating variance σ values of λ, η, γ with respect to the number of iterations of the EM (© IEEE 2008)

When the algorithm reaches a fixed point for some $\theta^{(t)}$ the value $\theta^{(t)}$ maximizes Q (θ). Since L and Q are equal at $\theta^{(t)}$ if L and Q are differentiable at $\theta^{(t)}$, then $\theta^{(t)}$ must be a stationary point of L. The stationary point needs not, however, be a local maximum. In [393] it is shown that it is possible for the algorithm to converge to local minima.

When a measured parameter distribution is derived, the next step is to update the high and low limit values by adjusting the support vector machine (SVM) classifier (Fig. 5.29b) in the corresponding functional test specs of the device

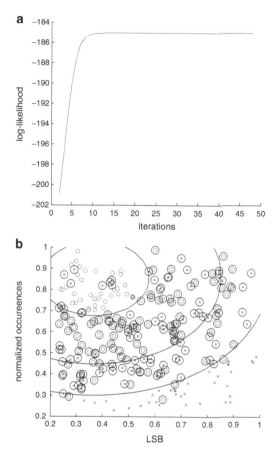

Fig. 5.29 (a) Log-likelihood with respect to the number of iterations of the EM. (b) Fitting a posteriori probability to the SVM output based on multiple runs of DLPM and DUT measurements (© IEEE 2008)

under test. This process related information allows design re-centering based upon the most failing die-level process monitors (test event, which is obtained with a slightly narrower comparison window), e.g. on the fly test limit setting. Through standard quadratic programming optimization, the input vectors belonging to a priori and a posteriori classes are divided into a number of sub-sets. The quadratic programming problem is solved incrementally, covering all the sub-sets of classes constructing the optimal separating hyperplane for the full data set. Note that during this process the value of the functional vector of parameters is monotonically increasing, since more and more training vectors are considered in the optimization leading to a smaller and smaller separation between the two classes. As illustrated in Fig. 5.29b, the high limit value can be updated in the corresponding functional test specs of the device under test with 0.35 LSB, which will lead to the increased yield.

5.7.2   Application of Results for A/D Converter Debugging and Calibration

The key to the debugging of the multi-step A/D converter is to select and separate the information on the faults from the transfer function. For at-speed testing and debugging of the analog performance of the A/D converter it is not only imperative to have all 12 digital outputs and the two out of range signals available at the device pins, but to be able to perform debugging of each stage the output signals of the coarse, mid and fine A/D converters need to be observable too. Debugging of each stage is performed sequentially starting from the first stage. Each stage is tested separately - at a lower speed – enabling the use of standard industrial analog waveform generators. To allow coherent testing, the clock signal of the A/D converter has to be fully controllable by the tester at all times. Adding all these requests together leads to an output test bus that needs to be 14 bits wide. The connections of the test bus are not only restricted to the test of the analog part. For digital testing the test bus is also used to carry digital data from scan chains. The test-shell contains all functional control logic, the digital test-bus, a test control block (TCB) and a CTAG isolation chain for digital input/output to and from other IP/cores. Further, logic necessary for creating certain control signals for the analog circuit parts, and for the scan-chains a bypass mechanism, controlled by the test control block, is available as well.

In the coarse A/D converter, faults in the analog components internal to the converter cause deviation, from ideal, of the transfer function of the coarse A/D converter by changing the step sizes in the transfer function. The fault cases which include resistor value, comparator's offset and comparator's bias current out of specification fault result in different patterns. The number of peaks and the location of the peak data identify the type of fault and the location of the fault. Since there is no feedback from mid and fine A/D converters to the coarse result value, it is not necessary to set these two A/D converters at the fixed value to test coarse A/D converter. Calibration D/A converter settings do not show in coarse A/D converter results; the calibration system however should remain operative. Random calibration cycles are not allowed to prevent interference with test results. The response of the mid A/D converter cannot directly be tested using the normal A/D converter output data due to an overlap in the A/D converter ranges. Nevertheless, by setting the coarse exor output signals using the scan chain through this block, known values are assigned to the mid switch. The residue signals are now used to verify the mid A/D converter separately by observing the mid A/D converter output bits via the test bus.

Faults in mid A/D converter affect the step sizes in the transfer function of the m mid bits, and repeat themselves in all coarse c bits. For the mid A/D converter test the chopper signals required for calibration need to be operative. After completing the mid A/D converter test, the chopper signal have to be verified by setting the chopper input to the two predefined conditions and analyze the mid A/D converter data to verify offsets. Since calibration D/A converter settings do show in mid A/D converter results, the D/A converter is set to a known value to prevent interference with mid ADC test results.

Similarly to the mid A/D converter, the fine A/D converter cannot be monitored directly due to the overlap in the A/D converter ranges. Through the available scan chains in the coarse exor and the switch-ladder, control signals are applied to the both mid and fine switch. The predefined input signals are extracted when the A/D converter works in a normal application mode with a normal input signal. At a certain moment the scan chains are set to a hold mode to acquire the requested value. Now, the residue signals derived through the predefined input signals evaluate the fine A/D converter performance. For the fine A/D converter test the chopper signals need to be active. To verify offsets, a similar procedure as in the mid A/D converter is followed. The calibration D/A converter settings have to be known and set to a known value to prevent interference with test results. The digital control block for all three tests operates normally; provides clock pulses, chopper signals and sets the calibration D/A converters in a known condition.

To evaluate the algorithms given in Sections 5.3 and 5.4 consider the test results shown in Tables 5.1 and 5.2. For each of the three-steps of the A/D converter in Table 5.1, different λ, η, and γ are generated randomly, so that the relative errors are

Table 5.1 Estimation results

	Actual value	Estimated value
γ_1	0.0229362	0.0259427
η_1	0.0121342	0.0116849
λ_1	0.0017936	0.0011347
γ_2	0.0328464	0.0342953
η_2	0.0154584	0.0142748
λ_2	0.0054635	0.0052347
γ_3	0.0417635	0.0424573
η_3	0.0173216	0.0165324

Table 5.2 Example of the calculated and estimated values

f_{in} (MHz)	21			43		
σ_o (%)	0.015	0.046	0.156	0.015	0.046	0.156
Cal. (dB)	76.77	67.23	56.78	76.77	67.23	56.78
Est. (dB)	77.50	67.96	57.50	77.50	67.96	57.50
σ_g (%)	0.01	0.05	0.10	0.01	0.05	0.10
Cal. (dB)	89.54	75.56	69.54	89.54	75.56	69.54
Est. (dB)	89.55	75.59	69.58	89.55	75.59	69.58
σ_b (%)	0.286	1.42	2.86	0.286	1.42	2.86
Cal. (dB)	84.06	70.14	64.06	78.62	64.70	58.62
Est. (dB)	83.60	70.14	64.29	78.59	64.85	58.96
σ_t (%)	0.01	0.05	0.10	0.01	0.05	0.10
Cal. (dB)	87.46	73.42	67.41	81.11	67.13	61.11
Est. (dB)	86.39	73.23	67.33	80.09	66.99	61.02

uniformly distributed in the interval $[-0.1, 0.1]$. At first, μ was set to 1/4 to speed up the algorithm, and then μ equal to 1/64 after 1,000 iteration times to improve the accuracy. Steady state is attained within $\sim 10^4$ clock cycles or effectively 0.22 ms (with $\mu = 1/64$).

In Table 5.2 spurious-free dynamic range is employed as a performance matrix. The circuit under test is the three time-interleaved S/H before prototyping. Validations have been performed for the entire S/H usable signal bandwidth and most probable limitation mechanisms, the time σ_t, offset σ_o, bandwidth σ_b, and gain σ_g variations. The two input frequencies are shown; for both of them, three values for any of the observed error mechanisms are chosen for the evaluation.

The advantage of the method is that it gives unbiased estimates, so that the estimation accuracy can be made arbitrarily good by increasing the amount of estimation data. Although the accuracy increase quite slowly with the amount of data, observed A/D converter, however, use very high sample rates (above 50 MS/s) so some million samples are collected in less than second ensuring the very fast converter evaluation.

The most significant A/D converter output bits have a strong correlation to the analog input signal, which is utilized to investigate the signal feedthrough from the output to the input by adding the possibility of scrambling the outgoing digital words with a pseudo-random bit-stream. The scrambling is realized by putting XOR gates before each output buffer and applying the random bit to their other input. For unscrambling, the random bits are taken out through an extra package pin.

The calibration technique was verified in all stages with full scale inputs. If the analog input to the calibrated A/D converter is such that the code transition is i, then the code transition of the ideal A/D converter is either i or $i + 1$. The offset between the digital outputs of these two converters for the range of analog inputs is denoted Δ_{i1} and Δ_{i2}, respectively. If a calibrated A/D converter has no errors in the internal reference voltages γ, and the stage gain errors η, the difference between the calibrated and the ideal A/D converter outputs is constant regardless of the analog input, thus $\Delta_{i1} = \Delta_{i2}$. If errors in the internal reference voltages γ and stage gain errors η are included, the calibrated A/D converter incurs unique missing codes. The difference between Δ_{i1} and Δ_{i2} precisely gives the error due to missing codes that occurs when the ideal A/D converter changes from i to $i + 1$. In a similar manner the unique error due to missing codes at all other transitions can be measured for the calibrated A/D converter.

With errors from the missing codes at each measured transition, the calibrated A/D converter stage is corrected by shifting the converter's digital output as a function of the transition points such that the overall transfer function of the calibrated A/D converter is free from missing codes. As long as the input is sufficiently rapid to generate a sufficient number of estimates of Δ_{i1}, Δ_{i2}, for all i, there is no constraint on the shape of the input signal to the A/D converter.

Constant offset between calibrated and ideal A/D converter appears as a common-mode shift in both Δ_{i1} and Δ_{i2}. Since the number of missing codes at each code transition is measured by subtracting Δ_{i2} from Δ_{i1}, the common mode is eliminated and thus input-referred offsets of calibrated A/D converter have no impact in the calibration scheme (under the practical assumption that the offsets are not large enough to saturate the output of the converter stages). To account for an overall internal reference voltages γ, stage gain errors η and systematic offset λ, the algorithm provides the estimates with the final values $(W')^T = [\gamma', \eta', \lambda']$. As ideal A/D converter offers an ideal reference for calibrated A/D converter, the error signal used for the algorithm adaptation (which is formed by the difference of the

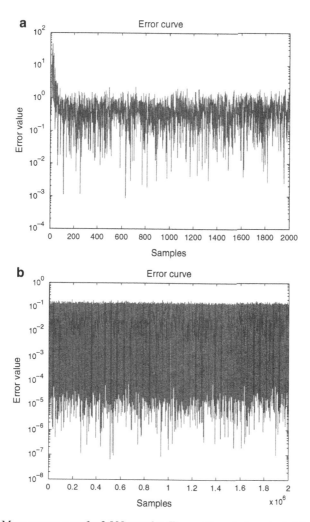

Fig. 5.30 (a) Mean-square error for 2,000 samples. The quality criterion adopted for an estimator is the mean-squared error criterion, mainly because it represents the energy in the error signal, is easy to differentiate and provides the possibilities to assign the weights. (b) Mean-square error for two million samples (© IEEE 2008)

two A/D converter outputs) is highly correlated with the error between them, thus steady state convergence of occurs within a relatively short time interval.

In Fig. 5.30 the correction parameters are shown. The largest correction values significantly decrease with the amount of samples. Calibration results of the mid A/D converter are shown in Figs. 5.31 and 5.32. The most of the errors that change quickly between adjacent levels are eliminated; some of the slow varying errors are however still left. This is caused by errors in the estimation of the amplitude distribution; slow variations in the errors cannot be distinguished from variations in the true amplitude distribution since only smoothness is assumed. For a sinusoidal signal the amplitude distribution looks like a bathtub. Because of the bathtub shape with high peaks near the edges the histogram is very sensitive to amplitude

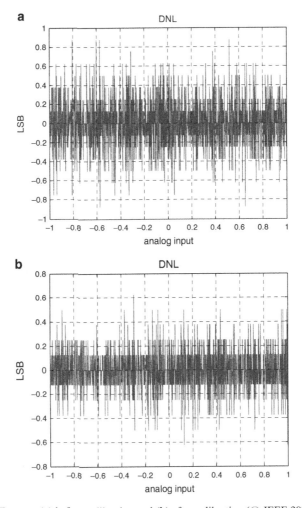

Fig. 5.31 DNL curve (**a**) before calibration and (**b**) after calibration (© IEEE 2008)

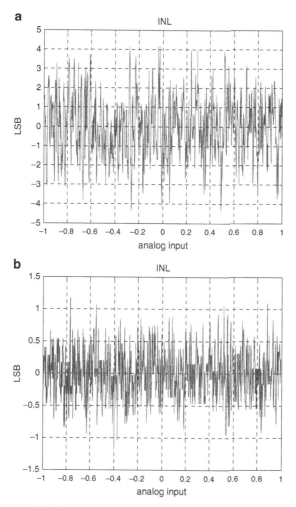

Fig. 5.32 INL curve (**a**) before calibration and (**b**) INL curve after calibration (© IEEE 2008)

changes in the input signal. The estimation is the most accurate for the middle codes. The errors near the edges of the excitation are, however, not completely eliminated due to the amplitude distribution, which has an abrupt change near the edges. Since only the static errors are handled in the algorithm, the errors can be assumed to approximately have a repetitive structure. This can be used to estimate the errors by extrapolation near the edges where the excitation is too low even to estimate the mismatch errors. However, the quality improvement is limited by the extrapolation that does not give perfect result, since the errors are not exactly periodical.

The peak improvement is about ±0.2 LSB for DNL measurement and ±2.9 LSB for INL. It is noted that the residual INL errors after calibration shown in Fig. 5.32

are due primarily to distortion from the fine A/D converter, as well as distortion from the front-end sample and hold, which sets the best achievable linearity for A/D converter.

5.8 Conclusion

The design margins for mixed signal designs depend significantly on process parameters and their distributions across the wafer, within a wafer lot and between wafer lots, which is especially relevant for mismatch. Measurement of these fluctuations is paramount for stable control of transistor properties and statistical monitoring and the evaluation of these effects enables the efficient development of the test patterns and test and debugging methods, as well as ensures good yields. With the use of dedicated sensors, which exploit knowledge of the circuit structure and the specific defect mechanisms, the method described in this book facilitates early and fast identification of excessive process parameter variation effects at the cost of at maximum 10% area overhead. The sensors allow the readout of local (within the core) performance parameters as well as the global distribution of these parameters. The flexibility of the concept allows the system to be easily extended with a variety of other performance sensors. The feasibility of the method for on-line and off-line debugging has been verified by experimental measurements from the silicon prototype fabricated in standard single poly, six metal 0.09-μm CMOS and evaluated on 12-bit multi-step A/D converter.

As the number of on-chip sensors is finite due to area limitations, additional informations are obtained through statistical techniques. Instead of using the traditional incomplete-data density in the estimation process, the implemented maximum-likelihood algorithm uses the properties of the complete-data density. In doing so, it can often make the estimation problem more tractable and also yield good estimates of the parameters for small sample sizes.

To allow the test guidance with the information obtained through monitoring process variations several classifiers are considered, such as quadratic, boosting, neural networks, or Bayesian networks. However, implemented adjusted support vector machine classifier is chosen, since it simultaneously minimize the empirical classification error and maximize the geometric margin.

Debugging method is applied after the converter stages and operates on the digital signal provided from the output. The quality criterion adopted for an estimator of the loss function is the mean-squared error criterion, mainly because it represents the energy in the error signal, is easy to differentiate and provides the possibilities to assign the weights. Additionally, employed adaptive filtering algorithm for error estimation offers the small number of operations per iteration and does not require correlation function calculation nor matrix inversions. To control the input to a filter, DfT is necessary, which allows forcing the input signal to the predefined values in each stage. The implemented design-for-test approach permits circuit re-configuration in such a way that all sub-blocks are tested for their full input range allowing full observability and controllability of the device under test. Adding testing capability does not degrade the converter performance and has low impact on area and power consumption.

Moreover, the debugging approach is extended to allow foreground calibration of the multi-step A/D converter. The presented calibration does not need any dedicated test signal and does not require a part of the conversion time. It works continuously and with every signal applied to the A/D converter.

Chapter 6
Conclusions and Recommendations

6.1 Summary of Results

With the fast advancement of CMOS fabrication technology, more and more signal-processing functions are implemented in the digital domain for a lower cost, lower power consumption, higher yield, and higher re-configurability. The trend of increasing integration level for integrated circuits has forced the A/D converter interface to reside on the same silicon in complex mixed-signal ICs containing mostly digital blocks for DSP and control. However, specifications of the converters in various applications emphasize high dynamic range and low spurious spectral performance. It is nontrivial to achieve this level of linearity in a monolithic environment where post-fabrication component trimming or calibration is cumbersome to implement for certain applications or/and for cost and manufacturability reasons. Additionally, as CMOS integrated circuits are accomplishing unprecedented integration levels, potential problems associated with device scaling – the short-channel effects – are also looming large as technology strides into the deep-submicron regime. The A/D conversion process involves sampling the applied analog input signal and quantizing it to its digital representation by comparing it to reference voltages before further signal processing in subsequent digital systems. Depending on how these functions are combined, different A/D converter architectures can be implemented with different requirements on each function. As discussed in Chapter 2, practical realizations show the trend that to a first order, converter power is directly proportional to sampling rate. However, power dissipation required becomes nonlinear as the speed capabilities of a process technology are pushed to the limit. Pipeline and two-step/multi-step converters tend to be the most efficient at achieving a given resolution and sampling rate specification.

This book is in a sense unique work as it covers the whole spectrum of design, test, debugging and calibration of multi-step A/D converters; it incorporates development of circuit techniques and algorithms to enhance the resolution and attainable sample rate of an A/D converter and to enhance testing and debugging

A. Zjajo and J. Pineda de Gyvez, *Low-Power High-Resolution Analog to Digital Converters*, Analog Circuits and Signal Processing, DOI 10.1007/978-90-481-9725-5_6, © Springer Science+Business Media B.V. 2011

potential to detect errors dynamically, to isolate and confine faults, and to recover and compensate for the errors continuously.

As presented in Chapter 3, the power proficiency for high resolution of multi-step converter by combining parallelism and calibration and exploiting low-voltage circuit techniques is demonstrated with a 1.8 V, 12-bit, 80 MS/s, 100 mW analog to-digital converter fabricated in five-metal layers 0.18-μm CMOS process. Lower power supply voltages significantly reduce noise margins and increase variations in process, device and design parameters. Consequently, it is steadily more difficult to control the fabrication process precisely enough to maintain uniformity. Micro-scopic particles present in the manufacturing environment and slight variations in the parameters of manufacturing steps can all lead to the geometrical and electrical properties of an IC to deviate from those generated at the end of the design process. Those defects can cause various types of malfunctioning, depending on the IC topology and the nature of the defect. To relief the burden placed on IC design and manufacturing originated with ever-increasing costs associated with testing and debugging of complex mixed-signal electronic systems, several circuit techniques and algorithms are developed and incorporated in the ATPG, DfT and BIST methodologies described in Chapter 4.

Process variation cannot be solved by improving manufacturing tolerances; variability must be reduced by new device technology or managed by design in order for scaling to continue. Similarly, within-die performance variation also imposes new challenges for test methods. This subject is treated in detail in Chapters 4 and 5. With the use of dedicated sensors, which exploit knowledge of the circuit structure and the specific defect mechanisms, the method described in this book facilitates early and fast identification of excessive process parameter variation effects. The expectation-maximization algorithm makes the estimation problem more tractable and also yields good estimates of the parameters for small sample sizes. To allow the test guidance with the information obtained through monitoring process variations implemented adjusted support vector machine clas-sifier simultaneously minimize the empirical classification error and maximize the geometric margin.

On a positive note, the use of digital enhancing calibration techniques reduces the need for expensive technologies with special fabrication steps. Indeed, the extra cost of digital processing is normally affordable as the use of submicron mixed signal technologies allows for efficient usage of silicon area even for relatively complex algorithms. Employed adaptive filtering algorithm for error estimation offers the small number of operations per iteration and does not require correlation function calculation nor matrix inversions. The presented foreground calibration algorithm does not need any dedicated test signal and does not require a part of the conversion time. It works continuously and with every signal applied to the A/D converter. The feasibility of the method for on-line and off-line debugging and calibration has been verified by experimental measurements from the silicon prototype fabricated in standard single poly, six metal 0.09-μm CMOS process.

6.2   Recommendations and Future Research

Even though the resolution of all today's A/D converter architectures is finally limited by jitter in the sampling clock, in low voltage multi-step A/D converters, the practical limitation for the conversion rate and resolution are imposed by full-resolution requirement on the time-interleaved front-end sample-and-hold stage, sub-D/A converter and residue signal. As a result, the design of fast and accurate multi-step A/D converters will become increasingly complex with the technology scaling. On the other hand, technology scaling allows an extensive use of digital signal processing to correct and compensate for the imperfections of the analog circuitry. Whether the evolution of the multi-stage architecture has reached saturation level, or will continue to be one of the dominant architectures of wide-band A/D converters, depends on the balance between these pros and contras of the development of the IC processes.

Appendix

A.1 Time Mismatch

Let the original sampled data sequence $S = [x(t_0), x(t_1), x(t_2), \ldots, x(t_m), \ldots, x(t_N), x(t_{N+1}), \ldots]$ be divided into N subsequences $S_0, S_1, S_2, \ldots, S_{N-1}$ as follows [133]:

$$S_0 = [x(t_0), x(t_N), x(t_{2N}), \ldots]$$

$$\vdots$$

$$S_n = [x(t_n), x(t_{N+n}), x(t_{2N+n}), \ldots] \tag{A.1}$$

$$\vdots$$

$$S_N = [x(t_{N-1}), x(t_{2N-1}), x(t_{3N-1}), \ldots$$

The S_n is obtained by uniformly sampling the signal $x(t + t_n)$ at the rate $1/NT$. Assume

$$\overline{S_n} = [x(t_n), 0, 0, \ldots (N - 1 \, zeros), x(t_{N+n}), 0, 0, \ldots] \tag{A.2}$$

the original sequence S, can be represented as

$$S = \sum_{n=0}^{N-1} \overline{S_n} z^{-n} \tag{A.3}$$

Then, the digital spectrum, $X(\omega)$, of S can be represented as

$$X(\omega) = \frac{1}{NT} \sum_{n=0}^{N-1} \left[\sum_{k=-\infty}^{\infty} X^a \left(\omega - \frac{2\pi k}{NT} \right) e^{j[\omega - (2\pi k/NT)] \times t_n} \right] \times e^{-jn\omega T} \tag{A.4}$$

Let $r_n T$, $n = 0, 1, N - 1$ be the sampling time offset encountered at the nth sample-hold unit (positive r_n means that the nth sample is delayed), and t_n real sampling time for nth sample-hold unit,

$$r_n T = nT - t_n \tag{A.5}$$

then (A.4) can be rewritten as

$$X(\omega) = \frac{1}{T} \sum_{k=-\infty}^{\infty} \left[\sum_{n=0}^{N-1} \frac{1}{N} e^{-j\left(\omega - k\frac{2\pi}{NT}\right) r_n T} e^{-jkn\frac{2\pi}{N}} \right] \times X^a \left(\omega - \frac{2\pi k}{NT} \right) \tag{A.6}$$

For a given sine wave $x(t) = sin(\omega_{in}t)$, the Fourier transform $X^a(\omega)$ is given by

$$X^a(\omega) = j\pi[\delta(\omega + \omega_{in}) - \delta(\omega - \omega_{in})] \qquad (A.7)$$

where δ is unit sample sequence $\delta[x] = 1$ at $x = 0$, and 0 elsewhere, (A.6) becomes [133]

$$X(\omega) = \frac{1}{T} \sum_{k=-\infty}^{\infty} \left[A(k)j\pi \left(\delta \left(\omega + \omega_{in} - k\frac{2\pi}{NT} \right) - \delta \left(\omega - \omega_{in} - k\frac{2\pi}{NT} \right) \right) \right] \qquad (A.8)$$

where

$$A(k) = \sum_{n=0}^{N-1} \left(\frac{1}{N} e^{-jr_n 2\pi f_{in}/f_s} \right) e^{-jkn(2\pi/N)} \qquad (A.9)$$

The digital spectrum given by (A.8) has N pairs of line spectra, each pairs centered at the fractional of the sampling frequency, such as $f_s/N, \ldots, (N-1)f_s/N$.

Fundamental corresponds to $k = 0$ while $k = 1, \ldots, N - 1$ corresponds to the distortion. The signal amplitude is determined by $A(0)$ while the distortion amplitudes are determined by $A(n)$, $n = 1, \ldots, N - 1$. $A(k)$ is a DFT of the sequence of $[(1/N)e^{-jr_n 2\pi f_0/f_s}, n = 0, 1, 2, \ldots, N - 1]$.

Assuming $r_n 2\pi f_{in}/f_s \ll 1$, the magnitude of the sidebands components can be expressed as:

$$\left| A(k) = \sum_{n=0}^{N-1} \left(e^{-jr_n 2\pi f_{in}/f_s} \right) e^{-jkn(2\pi/N)} \right| \approx \left| \frac{1}{N} \sum_{n=0}^{N-1} (1 - j2\pi r_n f_{in}/f_s) e^{-jkn(2\pi/N)} \right|$$

$$= \begin{cases} \frac{2\pi f_{in}}{Nf_s} \left| \sum_{n=0}^{N-1} r_n e^{-jkn(2\pi/N)} \right| & \text{for } k \neq 0, \pm N \\ \left| 1 - (j2\pi f_{in}/f_s) \left(\frac{1}{N} \sum_{n=0}^{N-1} r_n \right) \right| = 1 & \text{for } k = 0, \pm N \end{cases} \qquad (A.10)$$

The signal power is $A^2/2$ and power density of a spurious due to the time mismatch is expressed as

$$P_r^{spur}(k) = \frac{A^2}{2N^2} \left| \sum_{n=0}^{N-1} (1 - j2\pi r_n f_{in}/f_s) e^{-jkn(2\pi/N)} \right|^2 = \frac{A^2 4\pi^2 f_{in}^2}{2N^2 f_s^2} \sigma_r^2 \qquad (A.11)$$

A.2 Offset Mismatch

The offset mismatch can be modeled by adding a dc level with the input signal that is unique for each S/H unit. For a input signal $A\, sin(\omega_{in}t) + d_n$ with $n = 0, \ldots, N - 1$, the Fourier transform is given by

$$X^a(\omega) = j\pi A[\delta(\omega + \omega_{in}) - \delta(\omega - \omega_{in})] + 2\pi d_n \delta(\omega) \qquad (A.12)$$

If no timing error is assumed r_n is zero. Substituting previous equation in (A.8),

$$X(\omega) = \frac{1}{T} \sum_{k=-\infty}^{\infty} Aj\pi\delta\left[\left(\omega + \omega_{in} - \frac{2\pi k}{NT}\right) - \left(\omega - \omega_{in} - \frac{2\pi k}{NT}\right)\right] + \frac{1}{T}$$

$$\times \sum_{k=-\infty}^{\infty} A(k)2\pi\delta\left(\omega - \frac{2\pi k}{NT}\right) \qquad (A.13)$$

where

$$A(k) = \frac{1}{N} \sum_{n=0}^{N-1} d_n e^{-jkn(2\pi/N)} \qquad (A.14)$$

The first term of (A.12) corresponds to the input signal, while the second term corresponds to the distortion caused by channel offset. The distortion is not signal dependent and appears at nf_S/N, where $n = 0,1,\ldots,N - 1$. From previous equation can be seen that the distortion consists of a sum of impulses. Each impulse corresponds to a complex exponential signal in the time domain $e^{j\omega}$. The power of the exponential signal is 1. The factors $A(k)$ can be seen as the DFT of the sequence d_n/N, $n = 0, 1, \ldots, N - 1$. Power density of a spurious due to the offset mismatch is expressed as

$$P_r^{spur}(k) = \frac{1}{N^2} \left|\sum_{n=0}^{N-1} d_n e^{-jkn(2\pi/N)}\right|^2 = \frac{1}{N^2}\sigma_d^2 \qquad (A.15)$$

A.3 Gain Mismatch

To model gain mismatch, magnitude of one of the S/H unit input signals is different. The largest difference occurs at the peaks of the sine wave. The signal is $a_n\, sin(\omega_{in}t)$ with $n = 0, \ldots, N - 1$. The Fourier transform is given by

$$X^a(\omega) = j\pi a_n[\delta(\omega + \omega_{in}) - \delta(\omega - \omega_{in})] \qquad (A.16)$$

If no timing error is assumed r_n is zero. Substituting previous equation in (A.8),

$$X(\omega) = \frac{1}{T} \sum_{k=-\infty}^{\infty} \left[A(k)j\pi \left(\delta\left(\omega + \omega_{in} - k\frac{2\pi}{NT} \right) - \delta\left(\omega - \omega_{in} - k\frac{2\pi}{NT} \right) \right) \right] \quad \text{(A.17)}$$

where

$$A(k) = \frac{1}{N} \sum_{n=0}^{N-1} a_n e^{-jkn(2\pi/N)} \quad \text{(A.18)}$$

for the gain mismatches in N S/H units with spurious tones at $f_S/N \pm f_{in}$, $2f_S/N \pm f_{in}$, $\ldots ,(N-1)\, f_S/N \pm f_{in}$. Power density of a spurious due to the gain mismatch is expressed as

$$P_r^{\text{spur}}(k) = \frac{A^2}{2N^2} \left| \sum_{n=0}^{N-1} a_n e^{-jkn(2\pi/N)} \right|^2 = \frac{A^2}{2N^2} \sigma_a^2 \quad \text{(A.19)}$$

A.4 Bandwidth Mismatch

To model frequency dependent bandwidth mismatch, the S/H Amplifiers are approximated as the ideal one-pole amplifiers. For a one-pole system, $A(s) = A_0/(1 + s/\omega_0)$, the closed loop transfer function is

$$\frac{V_{out}}{V_{in}}(s) = \frac{A(s)}{1 + A(s)\beta} = \frac{A_0}{1 + A_0\beta + \frac{s}{\omega_0}} = \frac{\frac{A_0}{1 + A_0\beta}}{1 + \frac{s}{(1 + A_0\beta)\omega_0}} \quad \text{(A.20)}$$

Recognizing that $A_0\beta \gg 1$, the previous equation becomes

$$\frac{V_{out}}{V_{in}}(s) = \frac{1/\beta}{1 + \frac{s}{\beta A_0\omega_0}} = \frac{1/\beta}{1 + j\frac{f_{in}}{\beta A_0 f_0}} = \frac{1}{1 + j\frac{f_{in}}{f_1}} \approx 1 - j\frac{f_{in}}{f_1} = b_n \quad \text{(A.21)}$$

for $\beta = 1$. For a given sine wave $x(t) = sin(\omega_{in}t)$, the Fourier transform $X^a(\omega)$ is given by

$$X^a(\omega) = j\pi[\delta(\omega + \omega_{in}) - \delta(\omega - \omega_{in})] \quad \text{(A.22)}$$

and Eq. (A.8) with $r_n = 0$ becomes

$$X(\omega) = \frac{1}{T} \sum_{k=-\infty}^{\infty} \left[A(k)j\pi \left(\delta\left(\omega + \omega_{in} - k\frac{2\pi}{NT} \right) - \delta\left(\omega - \omega_{in} - k\frac{2\pi}{NT} \right) \right) \right] \quad \text{(A.23)}$$

where

$$A(k) = \frac{1}{N} \sum_{n=0}^{N-1} b_n e^{-jkn(2\pi/N)} \qquad (A.24)$$

with f_1 unity-gain frequency and b_n bandwidth offset experienced by the nth S/H amplifier with spurious tones at $f_S/N \pm f_{in}$, $2f_S/N \pm f_{in}$, ..., $(N-1)f_S/N \pm f_{in}$. The signal power is $A^2/2$ and power density of a spurious due to the gain-bandwidth mismatch is expressed as

$$P_r^{spur}(k) = \frac{A^2}{2N^2} \left| \sum_{n=0}^{N-1} b_n e^{-jkn(2\pi/N)} \right|^2 = \frac{A^2 f_{in}^2}{2N^2 f_1^2} \sigma_b^2 \qquad (A.25)$$

A.5 General Expression

The general expression for spurious-free dynamic range (SFDR) is given by

$$SFDR = 10\log_{10}\left(P_s/P_{spur}\right)$$

$$= 10\log_{10}\left(N^2 \lambda_x\right) \begin{array}{ll} \lambda_{off} = A^2/2\sigma_d^2 & \lambda_{time} = (f_s/(2f_{in}))^2/(\pi^2\sigma_r^2) \\ \lambda_{gain} = 1/\sigma_a^2 & \lambda_{bandwidth} = (f_1/f_{in})^2/(\sigma_b^2) \end{array} \qquad (A.26)$$

where the input signal power is defined as $P_s = A^2/2$.

B.1 Histogram Measurement of ADC Nonlinearities Using Sine Waves

The histogram or output code density is the number of times every individual code has occurred. For an ideal A/D converter with a full scale ramp input and random sampling, an equal number of codes is expected in each bin. The number of counts in the ith bin $H(i)$ divided by the total number of samples N_t, is the width of the bin as a fraction of full scale. By compiling a cumulative histogram, the cumulative bin widths are the transition levels.

The use of sine wave histogram tests for the determination of the nonlinearities of analog-to-digital converters (ADCs) has become quite common and is described in [336, 414]. When a ramp or triangle wave is used for histogram tests (as in [433]), additive noise has no effect on the results; however, due to the distortion or nonlinearity in the ramp, it is difficult to guarantee the accuracy.

For a differential nonlinearity test, a one percent change in the slope of the ramp would change the expected number of, codes by one percent. Since these errors would quickly accumulate, the integral nonlinearity test would become unfeasible. From brief consideration it is clear that the input source should have better precision than the converter being tested. When a sine wave is used, an error is produced, which becomes larger near the peaks. However, this error can be made as small and desired by sufficiently overdriving the A/D converter.

The probability density $p(V)$ for a function of the form $A \sin \omega t$ is

$$p(V) = \frac{1}{\pi\sqrt{A^2 - V^2}} \tag{B.27}$$

Integrating this density with respect to voltage gives the distribution function P (V_a, V_b)

$$P(V_a, V_b) = \frac{1}{\pi}\left\{\sin^{-1}\left[\frac{V_b}{A}\right] - \sin^{-1}\left[\frac{V_a}{A}\right]\right\} \tag{B.28}$$

which is in essence, the probability of a sample being in the range V_a to V_b. If the input has a *dc* offset, it has the form $V_o + A \sin \omega t$ with density

$$p(V) = \frac{1}{\pi\sqrt{A^2 - (V - V_o)^2}} \tag{B.29}$$

The new distribution is shifted by V_o as expected

$$P(V_a, V_b) = \frac{1}{\pi}\left\{\sin^{-1}\left[\frac{V_b - V_o}{A}\right] - \sin^{-1}\left[\frac{V_a - V_o}{A}\right]\right\} \tag{B.30}$$

The statistically correct method to measure the nonlinearities is to estimate the transitions from the data. The ratio of bin width to the ideal bin width $P(i)$ is the differential linearity and should be unity.

Subtracting on LSB gives the differential nonlinearity in LSBs

$$DNL(i) = \frac{H(i)/N_t}{P(i)} - 1 \tag{B.31}$$

Replacing the function $P(V_a, V_b)$ by the measured frequency of occurrence H/N_t, taking the cosine of both sides of (A.67) and solving for $\hat{V}_b$, which is an estimate of V_b, and using the following identities

$$\cos(\alpha - \beta) = \cos(\alpha)\cos(\beta) + \sin(\alpha)\sin(\beta) \tag{B.32}$$

$$\cos\left(\sin^{-1}\frac{V}{A}\right) = \frac{\sqrt{A^2 - V^2}}{A} \tag{B.33}$$

yields to

$$\hat{V}_b^2 - \left(2V_a\cos\left(\frac{\pi H}{N_t}\right)\right)\hat{V}_b - A^2\left(1 - \cos^2\left(\frac{\pi H}{N_t}\right)\right) + V_a^2 = 0 \tag{B.34}$$

In this consideration, the offset V_o is eliminated, since it does not effect the integral or differential nonlinearity. Solving for $\hat{V}_b$ and using the positive square root term as a solution so that $\hat{V}_b$ is greater than V_a

$$\hat{V}_b = V_a \cos\left(\frac{\pi H}{N_t}\right) + \sin\left(\frac{\pi H}{N_t}\right)\sqrt{A^2 - V_a^2} \tag{B.35}$$

This gives $\hat{V}_b$ in terms of V_a. $\hat{V}_k$ can be computed directly by using the boundary condition $V_o = -A$ and using

$$CH(k) = \sum_{i=0}^{k} H(i) \tag{B.36}$$

the estimate of the transition level $\hat{V}_b$ denoted as a T_k can be expressed as

$$T_k = -A\cos\left(\pi \frac{CH_{k-1}}{N_t}\right), \quad k = 1,\ldots,N-1 \tag{B.37}$$

A is not known, but being a linear factor, all transitions can be normalized to A so that the full range of transitions is ± 1.

B.2 Mean Square Error

As the probability density function associated with the input stimulus is known, the estimators of the actual transition level T_k and of the corresponding INL_k value expressed in least significant bits (LSBs) are represented as random variables defined, respectively, for a coherently sampled sinewave

$$s[m] = d + A\sin\left(2\pi\frac{D}{M}m + \theta_0\right) \quad m = 0, 1, \ldots, M-1 \tag{B.38}$$

$$T_k = d - A\cos\left(\pi\frac{CH_k}{M}\right), k = 1,\ldots,N-1 \, INL_k$$

$$= (T_k - T_k^i)/\Delta k = 1,\ldots,N-1 \tag{B.39}$$

where A, d, θ_0 are the signal amplitude, offset and initial phase, respectively, M is the number of collected data, D/M represents the ratio of the sinewave over the sampling frequencies. T_k^i is the ideal kth transition voltage, and $\Delta = FSR/2^B$ is the ideal code-bin width of the ADC under test, which has a full-scale range equal to FSR. A common model employed for the analysis of an analog-to digital converter affected by integral nonlinearities describes the quantization error ε as the sum of the quantization error of a uniform quantizer ε_q and the nonlinear behavior of the considered converter ε_n. For simplicity assuming that $/INL_k/ < \Delta/2$, we have:

$$\varepsilon_n = \sum_{k=1}^{N-1} \Delta \mathrm{sgn}(INL_k) i(s \in I_k) \tag{B.40}$$

where $sgn(.)$ and $i(.)$ represent the sign and the indicator functions, respectively, s denotes converter stimulus signal and the non-overlapping intervals I_k are defined as

$$I_k \hat{=} \begin{cases} (T_k^i - INL_k, T_k^i), & INL_k > 0 \\ (T_k^i, T_k^i - INL_k), & INL_k < 0 \end{cases} \tag{B.41}$$

The nonlinear quantizer mean-square-error, evaluated under the assumption of uniform stimulation of all converter output codes, is given by

$$mse = \int_{-\infty}^{\infty} [\varepsilon_q(s) + \varepsilon_n(s)]^2 f_s(s) ds \tag{B.42}$$

where f_s represent PDF of converter stimulus. Stimulating all device output codes with equal probability requires that

$$f_s(s) = \frac{1}{V_M - V_m} \cdot i(V_m \leqslant s < V_M) \tag{B.43}$$

Thus, mse becomes

$$mse = \frac{1}{V_M - V_m} \int_{V_m}^{V_M} [\varepsilon_q^2(s) + 2\varepsilon_q(s)\varepsilon_n(s) + \varepsilon_n^2(s)] ds \tag{B.44}$$

Assuming $\Delta = (V_M - V_m)/N$, and exploiting the fact the mse associated with the uniform quantization error sequence is $\Delta^2/12$

$$mse = \frac{\Delta^2}{12} + \frac{1}{N\Delta} \sum_{k=1}^{N-1} \int_{I_k} [2\Delta \mathrm{sgn}(INL_k)\varepsilon_q(s) + \Delta^2] ds \tag{B.45}$$

Since, for a rounding quantizer, $\varepsilon_q(s) = \Delta/2 - \Delta(s/\Delta - 1/2)$, it can be verified that $sgn(INL_k) \cdot \varepsilon_q(s) < 0$, so that

$$mse = \frac{\Delta^2}{12} + \frac{1}{N} \sum_{k=1}^{N-1} INL_k^2 \tag{B.46}$$

When characterizing A/D converters the SINAD is more frequently used than the *mse*. The SINAD is defined as

$$SINAD = 20\log_{10}\frac{rms_{(signal)}}{rms_{(noise)}}\,[dB] \tag{B.47}$$

Let the amplitude of the input signal be A_{dBFS}, expressed in *dB* relative full scale. Hence, the *rms* value is then

$$rms_{(signal)} = \frac{\Delta 10^{\frac{A_{dBFS}}{20}}2^{b-1}}{\sqrt{2}} \tag{B.48}$$

The $rms_{(noise)}$ amplitude is obtained from the *mse* expression above so that

$$rms_{(noise)} = \sqrt{mse}\,SINAD_{INL}$$

$$= 20b\log_{10}2 + 10\log_{10}\frac{3}{2} + A_{dBFS} - 10\log_{10}\left(\frac{mse}{\Delta^2/12}\right)[dB] \tag{B.49}$$

To calculate the effective number of bits ENOB, firstly express the SINAD for an ideal uniform ADC and than solve for *b*

$$SINAD_{(ideal)} = 20\log_{10}\left(\frac{\sqrt{6}A2^b}{FSR}\right)$$

$$ENOB = \frac{\log_2 10}{20}SINAD + \log_2\frac{FSR}{\sqrt{6}A} \tag{B.50}$$

Letting the amplitude $A = 10^{A(dBFS)/20}\,FSR/2$, and incorporating above equation, the ENOB can be expressed as

$$ENOB_{INL} = b - \frac{1}{2}\log_2\left(\frac{mse}{\Delta^2/12}\right)[dB] \tag{B.51}$$

B.3 Measurement Uncertainty

To estimate the uncertainty on the DNL and INL it is necessary to know the probability distribution of the cumulative probability Q_i to realize a measurement $V < UB_i$, with UB_i the uperbound of the *i*th level

$$Q_i = P(V<UB_i) = \int_{V_o-V}^{UB_i} p(V)dV \tag{B.52}$$

and using linear transformation

$$UB_i = -\cos \pi Q_i \tag{B.53}$$

The variance and cross-correlation of UB_i is derived using linear approximations. To realize the value Q_i, it is necessary to have N_i measurements with a value $<UB_i$, and $(N - N_i)$ measurements with a value $>UB_i$. The distribution of Q_i is a binomial distribution, which can be very well approximated by a normal distribution [414]

$$\begin{aligned} P(Q_i') &= C_N^{N_i} P(V<UB_i)^{N_i}(1 - P(V>UB_i))^{N-N_i} \\ &= C_N^{N_i} Q_i^{N_i}(1 - Q_i)^{N-N_i} \end{aligned} \tag{B.54}$$

with Q_i' the estimated value of Q_i. The mean and the standard deviation is given by

$$\mu_{Q_i'} = Q_i \quad \sigma_{Q_i'} = \sqrt{Q_i(1 - Q_i)/N} \tag{B.55}$$

which states that Q_i' is an unbiased estimate of Q_i. To calculate the covariance between Q_i and Q_j, firstly, let's define

$$\begin{aligned} Q_0 &= P(V>UB_j) \\ Q_{ij} &= P(UB_i<V<UB_j) = 1 - Q_i - Q_j \end{aligned} \tag{B.56}$$

and the relation

$$\begin{aligned} N_j &= N_i + N_{ij} & \sigma_{N_iN_j}^2 &= \sigma_{N_i}^2 + \sigma_{N_iN_{ij}}^2 \\ N_i + N_{ij} + N_0 &= N & \sigma_{N_0}^2 &= \sigma_{N_i}^2 + \sigma_{N_{ij}}^2 + 2\sigma_{N_iN_{ij}}^2 \end{aligned} \tag{B.57}$$

which leads to

$$\sigma_{N_iN_j}^2 = [\sigma_{N_i}^2 + \sigma_{N_0}^2 - \sigma_{N_{ij}}^2]/2 \tag{B.58}$$

with

$$\begin{aligned} \sigma_{N_i}^2 &= NQ_i(1 - Q_i) \\ \sigma_{N_0}^2 &= NQ_0(1 - Q_0) \\ \sigma_{N_{ij}}^2 &= NQ_{ij}(1 - Q_{ij}) \end{aligned} \tag{B.59}$$

or

$$\begin{aligned} \sigma_{N_iN_j}^2 &= NQ_iQ_0 = NQ_i(1 - Q_j) \\ \sigma_{Q_iQ_j}^2 &= Q_i(1 - Q_j)/N \end{aligned} \tag{B.60}$$

To calculate the variance $\sigma_{UB}{}^2$

$$\sigma_{UB_i}^2 = E[dUB_i dUB_i] = \pi^2 \sin^2 \pi Q_i \sigma_{Q_i}^2 = \pi^2 \sin^2 \pi Q_i Q_i (1 - Q_i)/N \qquad \text{(B.61)}$$

Similarly,

$$\sigma_{UB_i UB_j}^2 = E[dUB_i dUB_j] = \pi^2 \sin \pi Q_i \sin \pi Q_j Q_i (1 - Q_j)/N \qquad \text{(B.62)}$$

Since the differential nonlinearity of the ith level is defined as the ratio

$$DNL_i = \frac{UB_i - UB_{i-1}}{L_R} - 1 \qquad \text{(B.63)}$$

where L_R is the length of the record, the uncertainty in DNL_i and INL_i measurements can be expressed as

$$\sigma_{DNL_i}^2 = \sqrt{[\sigma_{UB_i}^2 + \sigma_{UB_{i-1}}^2 - 2\sigma_{UB_i UB_j}^2]}/L_R$$
$$\sigma_{INL_i}^2 = \sigma_{UB_i}/L_R \qquad \text{(B.64)}$$

The maximal uncertainty occurs for $Q_i = 0.5$, thus the previous equation can be approximated with

$$\sigma_{DNL_i}^2 \approx \sqrt{\pi/L_R} \times 1/\sqrt{N}$$
$$\sigma_{INL_i}^2 = \pi/2L_R \times 1/\sqrt{N} \qquad \text{(B.65)}$$

References

1. A. Yukawa, An 8-bit high-speed CMOS A/D converter. IEEE J. Solid-State Circuits **20**(3), 775–779 (1985)
2. T. Kumamoto, M. Nakaya, H. Honda, S. Asai, Y. Akasaka, Y. Horiba, An 8-bit high-speed CMOS A/D converter. IEEE J. Solid-State Circuits **21**(6), 976–982 (1986)
3. B. Peetz, B.D. Hamilton, J. Kang, An 8-bit 250 megasample per second analog-to-digital converter: operation without a sample and hold. IEEE J. Solid-State Circuits **21**(6), 997–1002 (1986)
4. Y. Akazawa, A. Iwata, T. Wakimoto, T. Kamato, H. Nakamura, H. Ikawa, A 400 MSPS 8 b flash AD conversion LSI. *IEEE International Solid-State Circuits Conference Digest of Technical Papers*, pp. 27–28, 1987
5. Y. Gendai, Y. Komatsu, S. Hirase, M. Kawata, An 8b 500MHz ADC. *IEEE International Solid-State Circuits Conference Digest of Technical Papers*, pp. 172–173, 1991
6. N. Shiwaku, Y. Tung, T. Hiroshima, K.-S. Tan, T. Kurosawa, K. McDonald, M. Chiang, A rail-to-rail video-band full Nyquist 8-bit A/D converter. *Proceedings of IEEE Custom Integrated Circuits Conference*, pp. 26.2.1–26.2.4, 1991
7. K.J. McCall, M.J. Demler, M.W.A. Plante, A 6-bit 125 MHz CMOS A/D converter. *Proceedings of IEEE Custom Integrated Circuits Conference*, pp. 16.8.1–16.8.4, 1992
8. I. Mehr, D. Dalton, A 500 Msample/s 6–bit Nyquist rate ADC for disk drive read channel applications. *Proceedings of IEEE European Solid-State Circuits Conference*, pp. 236–239, 1998
9. K. Yoon, S. Park, W. Kim, A 6 b 500 MSample/s CMOS flash ADC with a background interpolated auto-zeroing technique. *IEEE International Solid-State Circuits Conference Digest of Technical Papers*, pp. 326–327, 1999
10. B. Yu, W.C. Black Jr., A 900 MS/s 6b interleaved CMOS flash ADC. *Proceedings of IEEE Conference on Custom Integrated Circuits*, pp. 149–152, 2001
11. M. Choi, A.A. Abidi, A 6-b 1.3-Gsample/s A/D converter in 0.35-μm CMOS. IEEE J. Solid-State Circuits **36**(12), 1847–1858 (2001)
12. G. Geelen, A 6 b 1.1 GSample/s CMOS A/D converter. *IEEE International Solid-State Circuits Conference Digest of Technical Papers*, pp. 128–129, 2001
13. J. Lin, B. Haroun, An embedded 0.8 V/480 μW 6b/22 MHz flash ADC in 0.13 μm digital CMOS process using nonlinear double-interpolation technique. *IEEE International Solid-State Circuits Conference Digest of Technical Papers*, pp. 244–246, 2002
14. K. Uyttenhove, M. Steyaert, A 1.8–V, 6–bit, 1.3–GHz CMOS flash ADC in 0.25 μm CMOS. *Proceedings of IEEE European Solid-State Circuits Conference*, pp. 455–458, 2002
15. X. Jiang, Z. Wang, M.F. Chang, A 2 GS/s 6 b ADC in 0.18-μm CMOS. *IEEE International Solid-State Circuits Conference Digest of Technical Papers*, pp. 322–323, 2003

16. C. Sandner, M. Clara, A. Santner, T. Hartig, F. Kuttner, A 6bit, 1.2GSps low-power flash-ADC in 0.13 μm digital CMOS. *Proceedings of IEEE European Solid-State Circuits Conference*, pp. 339–342, 2004

17. C. Paulus, H.-M. Bluthgen, M. Low, E. Sicheneder, N. Bruls, A. Courtois, M. Tiebout, R. Thewes, A 4GS/s 6b flash ADC in 0.13-μm CMOS. *IEEE Symposium on VLSI Circuits Digest of Technical Papers*, pp. 420–423, 2004

18. C.-C. Huang, J.-T. Wu, A Background Comparator Calibration Technique for Flash Analog-to-Digital Converters. IEEE Trans. Circuits Syst. **52**(9), 1732–1740 (2005)

19. O. Viitala, S. Lindfors, K. Halonen, A 5-bit 1-GS/s Flash-ADC in 0.13-μm CMOS Using Active Interpolation. *Proceedings of IEEE European Solid-State Circuits Conference*, pp. 412–415, 2006

20. S. Park, Y. Palaskas, M.P. Flynn, A 4-GS/s 4-bit flash ADC in 0.18-μm CMOS. IEEE J. Solid-State Circuits **42**(9), 1865–1872 (2007)

21. I.-H. Wang, S.-I. Liu, A 1V 5-Bit 5GSample/sec CMOS ADC for UWB receivers. *Proceedings of IEEE International Symposium on VLSI Design, Automation and Test*, pp. 1–4, 2007

22. M. Ishikawa, T. Tsukahara, An 8-bit 50-MHz CMOS subranging A/D converter with pipelined wide-band S/H. IEEE J. Solid-State Circuits **24**(6), 1485–1491 (1989)

23. S. Hosotani, T. Miki, A. Maeda, N. Yazawa, An 8bit 20MS/s CMOS A/D converter with 50-mW power consumption. IEEE J. Solid-State Circuits **25**(1), 167–172 (1990)

24. T. Matsuura, H. Kojima, E. Imaizumi, K. Usui, S. Ueda, An 8b 50 MHz 225 mW submicron CMOS ADC using saturation eliminated comparators. *Proceedings of IEEE Custom Integrated Circuits Conference*, pp 6.4.1–6.4.4, 1990

25. B. Razavi, B.A. Wooley, A 12b 5MSample/s two-step CMOS A/D converter. IEEE J. Solid-State Circuits **27**(12), 1667–1678 (1992)

26. S.-H. Lee, B.-S. Song, Digital-domain calibration of multistep analog-to-digital converters. IEEE J. Solid-State Circuits **27**(12), 1679–1688 (1992)

27. T. Miki, H. Kouno, T. Kumamoto, Y. Kinoshita, T. Igarashi, K. Okeda, A 10-b 50 MS/s 500-mW A/D converter using a differential-voltage subconverter. IEEE J. Solid-State Circuits **29**(4), 516–522 (1994)

28. M. Ito, T. Miki, S. Hosotani, T. Kumamoto, Y. Yamashita, M. Kijima, T. Okuda, K. Okada, A 10 bit 20MS/s 3V Supply CMOS A/D Converter. IEEE J. Solid-State Circuits **29**(12), 1531–1531 (1994)

29. S.-U. Kwak, B.-S. Song, K. Bacrania, A 15 b 5 MSample/s low-spurious CMOS ADC. *IEEE International Solid-State Circuits Conference Digest of Technical Papers*, pp. 146–147, 1997

30. H. van der Ploeg, R. Remmers, A 3.3-V, 10-b, 25-MSample/s two-step ADC in 0.35-μm CMOS. IEEE J. Solid-State Circuits **34**(12), 1803–1811 (1999)

31. C. Moreland, F. Murden, M. Elliott, J. Young, M. Hensley, R. Stop, A 14-bit 100-MSample/s subranging ADC. IEEE J. Solid-State Circuits **35**(7), 1791–1798 (2000)

32. P. Hui, M. Segami, M. Choi, C. Ling, A.A. Abidi, A 3.3-V 12-b 50-MS/s A/D converter in 0.6-μm CMOS with over 80-dB SFDR. IEEE J. Solid-State Circuits **35**(12), 1769–1780 (2000)

33. M.-J. Choe, B.-S. Song, K. Bacrania, A 13-b 40-MSamples/s CMOS pipelined folding ADC with background offset trimming. IEEE J. Solid-State Circuits **35**(6), 1781–1790 (2000)

34. H. van der Ploeg, G. Hoogzaad, H.A.H. Termeer, M. Vertregt, R.L.J. Roovers, A 2.5-V 12-b 54-Msample/s 0.25-μm CMOS ADC in 1-mm² with mixed-signal chopping and calibration. IEEE J. Solid-State Circuits **36**(12), 1859–1867 (2001)

35. M. Clara, A. Wiesbauer, F. Kuttner, A 1.8 V Fully Embedded 10 b 160 MS/s Two-Step ADC in 0.18 μm CMOS. *Proceedings of IEEE Custom Integrated Circuit Conference*, pp. 437–440, 2002

36. T.-C. Lin, J.-C. Wu, A two-step A/D converter in digital CMOS processes. *Proceedings of IEEE Asia-Pacific Conference on ASIC*, pp. 177–180, 2002

37. A. Zjajo, H. van der Ploeg, M. Vertregt, A 1.8V 100mW 12-bits 80Msample/s two-step ADC in 0.18-μm CMOS. *Proceedings of IEEE European Solid-State Circuits Conference*, pp. 241–244, 2003

38. H. van der Ploeg, M. Vertregt, M. Lammers, A 15-bit 30 MS/s 145 mW three-step ADC for imaging applications. *Proceedings of IEEE European Solid-State Circuits Conference*, pp. 161–164, 2005

39. N. Ning, F. Long, S.-Y. Wu, Y. Liu, G.-Q. Liu, Q. Yu, M.-H. Yang, An 8-Bit 250MSPS modified two-step ADC. *Proceedings of IEEE International Conference on Communications, Circuits and Systems*, pp. 2197–2200, 2006

40. T. Sekino, M. Takeda, K. Koma, A Monolithic 8b Two-Step Parallel ADC without DAC and Subtractor Circuits. *IEEE International Solid-State Circuits Conference Digest of Technical Papers*, pp. 46–47, 1982

41. A.G.F. Dingwall, V. Zazzu, An 8-MHz CMOS subranging 8-bit A/D converter. IEEE J. Solid-State Circuits **20**(6), 1138–1143 (1985)

42. J. Fernandes, S.R. Lewis, A.M. Mallinson, G.A. Miller, A 14-bit 10-μs subranging A/D converter with S/H. IEEE J. Solid-State Circuits **23**(6), 1309–1315 (1988)

43. M. Ishikawa, T. Tsukahara, An 8-bit 50-MHz CMOS subranging A/D converter with pipelined wide-band S/H. IEEE J. Solid-State Circuits **24**(6), 1485–1491 (1989)

44. R. Petschacher, B. Zojer, B. Astegher, H. Jessner, A. Lechner, A 10-b 75-MSPS subranging A/D converter with integrated sample and hold. IEEE J. Solid-State Circuits **25**(6), 1339–1346 (1990)

45. G. Sou, G.-N. Lu, G. Klisnick, M. Redon, A 10-bit 20-Msample/s Sub-Ranging ADC for Low-Power and Low-Voltage Applications. *Proceedings of IEEE European Solid-State Circuits Conference*, pp. 424–427, 1998

46. A. Wiesbauer, M. Clara, M. Harteneck, T. Potscher, C. Fleischhacker, G. Koder, C. Sandner, a fully integrated analog front-end macro for cable modem applications in 0.18-μm CMOS. *Proceedings of IEEE European Solid-State Circuits Conference*, pp. 245–248, 2001

47. R.C. Taft, M.R. Tursi, A 100-MS/s 8-b CMOS subranging ADC with sustained parametric performance from 3.8 V down to 2.2 V. IEEE J. Solid-State Circuits **36**(3), 331–338 (2001)

48. J. Mulder, C.M. Ward, C.-H. Lin, D. Kruse, J.R. Westra, M. Lughtart, E. Arslan, R.J. van de Plassche, K. Bult, F.M.L. van der Goes, A 21-mW 8-b 125-MSample/s ADC in 0.09-mm2 0.13-μm CMOS. IEEE J. Solid-State Circuits **39**(5), 2116–2125 (2004)

49. P.M. Figueiredo, P. Cardoso, A. Lopes, C. Fachada, N. Hamanishi, K. Tanabe, J. Vital, A 90nm CMOS 1.2V 6b 1GS/s Two-Step Subranging ADC. *IEEE International Solid-State Circuits Conference Digest of Technical Papers*, pp. 568–569, 2006

50. Y. Shimizu, S. Murayama, K. Kudoh, H. Yatsuda, A 30mW 12b 40MS/s subranging ADC with a high-gain offset-canceling positive-feedback amplifier in 90 nm digital CMOS. *IEEE International Solid-State Circuits Conference Digest of Technical Papers*, pp. 216–217, 2006

51. J. Huber, R.J. Chandler, A.A. Abidi, A 10b 160MS/s 84mW 1V subranging ADC in 90nm CMOS. *IEEE International Solid-State Circuits Conference Digest of Technical Papers*, pp. 454–455, 2007

52. C. Cheng, Y. Jiren, A 10-bit 500-MS/s 124-mW subranging folding ADC in 0.13 μm CMOS. *Proceedings of IEEE International Symposium on Circuits and Systems*, pp. 1709–1712, 2007

53. Y. Shimizu, S. Murayama, K. Kudoh, H. Yatsuda, A Split-Load Interpolation-amplifier-array 300MS/s 8b subranging ADC in 90 nm CMOS. *IEEE International Solid-State Circuits Conference Digest of Technical Papers*, pp. 552–553, 2008

54. S.H. Lewis, P.R. Gray, A pipelined 5-Msample/s 9-bit analog-to-digital converter. IEEE J. Solid-State Circuits **22**(4), 954–961 (1987)

55. S. Sutarja, P. Gray, A pipelined 13-bit, 250-ks/s, 5-V analog-to-digital converter. IEEE J. Solid-State Circuits **23**(6), 1316–1323 (1988)

56. B.-S. Song, M.F. Tompsett, K.R. Lakshmikumar, A 12 bit 1 MHz capacitor error averaging pipelined A/D. IEEE J. Solid-State Circuits **23**(10), 1324–1333 (1988)

57. Y.-M. Lin, B. Kim, P.R. Gray, A 13-b 2.5-MHz self-calibrated pipelined A/D converter in 3-μm CMOS. IEEE J. Solid-State Circuits **26**(5), 628–635 (1991)

58. T. Matsuura, M. Hotta, K. Usui, E. Imaizumi, S. Ueda, A 95 mW, 10b 15 MHz low-power CMOS ADC using analog double-sampled pipelining scheme. *Proceedings of IEEE Symposium on VLSI Circuits*, pp. 98–99, 1992

59. S.H. Lewis, H.S. Fetterman, G.F. Gross, R. Ramachandran, T.R. Viswanathan, A 10-b 20-Msample/s analog-to-digital converter. IEEE J. Solid-State Circuits 27(3), 351–358 (1992)

60. C. Mangelsdorf, S.-H. Lee, M. Martin, H. Malik, T. Fukuda, H. Matsumoto, Design for testability in digitally-corrected ADCs. *IEEE International Solid-State Circuits Conference Digest of Technical Papers*, pp. 70–71, 1993

61. C. Mangelsdorf, H. Malik, S.-H. Lee, S Hisano, M. Martin, A two residue architecture for mulitstage ADCs. *IEEE International Solid-State Circuits Conference Digest of Technical Papers*, pp. 64–65, 1993

62. K. Kusumoto, A. Matsuzawa, K. Murata, A 10 b 20MHz 30 mW pipelined interpolating CMOS ADC. IEEE J. Solid-State Circuits 28(12), 1200–1206 (1993)

63. A.N. Karanicolas, H.-S. Lee, K.L. Bacrania, A 15-b 1-Msample/s digitally self-calibrated pipeline ADC. IEEE J. Solid-State Circuits 28(12), 1207–1215 (1993)

64. W.T. Colleran, A.A. Abidi, A 10-b, 75 Ms/s two stage pipelined bipolar A/D converter. IEEE J. Solid-State Circuits 28(12), 1187–1199 (1994)

65. D.A. Mercer, A 14-b, 2.5 MSPS pipelined ADC with on-chip EPROM. IEEE J. Solid-State Circuits 31(1), 70–76 (1996)

66. I. Opris, L. Lewicki, B. Wong, A single-ended 12-bit 20 MSample/s self-calibrating pipeline A/D converter. IEEE J. Solid-State Circuits 33(11), 1898–1903 (1998)

67. A.M. Abo, P.R. Gray, A 1.5-V, 10-bit, 14.3-MS/s CMOS pipeline analog-to-digital converter. IEEE J. Solid-State Circuits 34(5), 599–606 (1999)

68. H.-S. Chen, K. Bacrania, B.-S. Song, A 14b 20MSample/s CMOS pipelined ADC. *IEEE International Solid-State Circuits Conference Digest of Technical Papers*, pp. 46–47, 2000

69. I. Mehr, L. Singer, A 55-mW, 10-bit, 40-Msample/s Nyquist-rate CMOS ADC. IEEE J. Solid-State Circuits 35(3), 70–76 (2000)

70. Y. Chiu, Inherently linear capacitor error-avaraging techniques for pipelined A/D conversion. IEEE Trans. Circuits Syst.–II 47, 229–232 (2000)

71. B.W. Lee, G.H. Cho, A CMOS 10 bit 37MS/s pipelined A/D converter with code regeneration and averaging. *Proceedings of IEEE European Solid-State Circuits Conference*, pp. 314–317, 2000

72. J. Goes, J.C. Vital, L. Alves, N. Ferreira, P. Ventura, E. Bach, J.E. Franca, R. Koch, A low-power 14-b 5 MS/s CMOS pipeline ADC with background analog self-calibration. *Proceedings of IEEE European Solid-State Circuits Conference*, pp. 172–175, 2000

73. A. Loloee, A. Zanchi, J. Huawen, S. Shehata, E. Bartolome, A 12b 80MSps pipelined ADC core with 190 mW consumption from 3 V in 0.18 µm digital CMOS. *Proceedings of IEEE European Solid-State Circuits Conference*, pp. 467–470, 2002

74. B.-M. Min, P. Kim, F.W. Bowman, D.M. Boisvert, A.J. Aude, A 69-mW 10-bit 80-MSample/s pipelined CMOS ADC. IEEE J. Solid-State Circuits 38(12), 1187–1199 (2003)

75. T.N. Andersen, A. Briskemyr, F. Telstø, J. Bjørnsen, T. E. Bonnerud, B. Hernes, Ø. Moldsvor, A 97 mW 100 MS/s 12b pipeline ADC implemented in 0.18 µm digital CMOS. *Proceedings of IEEE European Solid-State Circuits Conference*, pp. 247–250, 2004

76. J. Li, U.-K. Moon, A 1.8-V 67-mW 10-bit 100-MS/s pipelined ADC using time-shifted CDS technique. IEEE J. Solid-State Circuits 39(9), 1468–1476 (2004)

77. X. Wang, P.J. Hurst, S.H. Lewis, A 12-bit 20-Msample/s pipelined analog-to-digital converter with nested digital background calibration. IEEE J. Solid-State Circuits 39(11), 1799–1808 (2004)

78. D. Kurose, T. Ito, T. Ueno, T. Yamaji, T. Itakura, 55-mW 200-MSPS 10-bit pipeline ADCs for wireless receivers. *Proceedings of IEEE European Solid-State Circuits Conference*, pp. 527–530, 2005

79. C.T. Peach, A. Ravi, R. Bishop, K. Soumyanath, D.J. Allstot, A 9-b 400 Msample/s pipelined analog-to-digital converter in 90 nm CMOS. *Proceedings of IEEE European Solid-State Circuits Conference*, pp. 535–538, 2005

80. A.M.A. Ali, C. Dillon, R. Sneed, A.S. Morgan, S. Bardsley, J. Kornblum, L. Wu, A 14-bit 125 MS/s IF/RF sampling pipelined ADC with 100 dB SFDR and 50 fs jitter. IEEE J. Solid-State Circuits **41**(8), 1846–1855 (2006)

81. M. Daito, H. Matsui, M. Ueda, K. Iizuka, A 14-bit 20-MS/s pipelined ADC with digital distortion calibration. IEEE J. Solid-State Circuits **41**(11), 2417–2423 (2006)

82. T. Ito, D. Kurose, T. Ueno, T. Yamaji, T. Itakura, 55-mW 1.2-V 12-bit 100-MSPS pipeline ADCs for wireless receivers. *Proceedings of IEEE European Solid-State Circuits Conference*, pp. 540–543, 2006

83. J. Treichler, Q. Huang, T. Burger, A 10-bit ENOB 50-MS/s pipeline ADC in 130-nm CMOS at 1.2 V supply. *Proceedings of IEEE European Solid-State Circuits Conference*, pp. 552–555, 2006

84. I. Ahmed, D.A. Johns, An 11-bit 45MS/s pipelined ADC with rapid calibration of DAC errors in a multi-bit pipeline stage. *Proceedings of IEEE European Solid-State Circuits Conference*, pp. 147–150, 2007

85. S.-C. Lee, Y.-D. Jeon, J.-K. Kwon, J. Kim, A 10-bit 205-MS/s 1.0- mm^2 90-nm CMOS pipeline ADC for flat panel display applications. IEEE J. Solid-State Circuits **42**(12), 2688–2695 (2007)

86. J. Li, R. Leboeuf, M. Courcy, G. Manganaro, A 1.8V 10b 210MS/s CMOS pipelined ADC featuring 86 dB SFDR without calibration. *Proceedings of IEEE Custom Integrated Circuits Conference*, pp. 317–320, 2007

87. M. Boulemnakher, E. Andre, J. Roux, F. Paillardet, A 1.2V 4.5 mW 10 b 100 MS/s pipeline ADC in a 65nm CMOS. *IEEE International Solid-State Circuits Conference Digest of Technical Papers*, pp. 250–251, 2008

88. Y.-S. Shu, B.-S. Song, A 15-bit linear 20-MS/s pipelined ADC digitally calibrated with signal-dependent dithering. IEEE J. Solid-State Circuits **43**(2), 342–350 (2008)

89. J. Shen, P.R. Kinget, A 0.5-V 8-bit 10-Ms/s pipelined ADC in 90-nm CMOS. IEEE J. Solid-State Circuits **43**(4), 1799–1808 (2008)

90. W.C. Black Jr., D.A. Hodges, Time interleaved converter arrays. IEEE J. Solid-State Circuits **15**(6), 1022–1029 (1980)

91. C.S.G. Conroy, D.W. Cline, P.R. Gray, A high-speed parallel pipelined ADC technique in CMOS. *Proceedings of IEEE Symposium on VLSI Circuits*, pp. 96–97, 1992

92. M. Yotsuyanagi, T. Etoh, K. Hirata, A 10 b 50 MHz pipelined CMOS A/D converter with S/H. IEEE J. Solid-State Circuits **28**, 292–300 (1993)

93. K. Nagaraj, H. Fetterman, J. Anidjar, S. Lewis, R. Renninger, An 8-b 50-Msamples/s pipelined A/D converter with an area and power efficient architecture. *Proceedings of IEEE Custom Integrated Circuits Conference,* pp. 423–426, 1996

94. W. Bright, 8 b 75 M Sample/s 70 mW parallel pipelined ADC incorporating double sampling. *IEEE International Solid-State Circuits Conference Digest of Technical Papers*, pp. 146–147, 1998

95. K. Dyer, D. Fu, S. Lewis, P. Hurst, Analog background calibration technique for time-interleaved analog-to-digital converters. IEEE J. Solid-State Circuits **33**(12), 1912–1919 (1998)

96. D. Fu, K.C. Dyer, S.H. Lewis, P.J. Hurst, A digital background calibration technique for time-interleaved analog-to-digital converters. IEEE J. Solid-State Circuits **33**(12), 1904–1911 (1998)

97. L. Sumanen, M. Waltari, K.A.I. Halonen, A 10-bit 200-MS/s CMOS parallel pipeline A/D converter. *Proceedings of IEEE European Solid-State Circuits Conference*, pp. 439–442, 2000

98. S.M. Jamal, D. Fu, N.C.-J. Chang, P.J. Hurst, S.H. Lewis, A 10-b 120-MSample/s time-interleaved analog-to-digital converter with digital background calibration. IEEE J. Solid-State Circuits **37**(12), 1618–1627 (2002)

99. J. Talebzadeh, M.R. Hasanzadeh, M. Yavari, O. Shoaei, A 10-bit 150-MS/s parallel pipeline A/D converter in 0.6-μm CMOS. *Proceedings of IEEE International Symposium on Circuits and Systems*, pp. 133–136, 2002

100. D. Subiela, S. Engels, L. Dugoujon, R. Esteve-Bosch, B. Mota, L. Musa, A. Jimenez-de-Parga, A low-power 16-channel AD converter and digital processor ASIC. *Proceedings of IEEE European Solid-State Circuits Conference*, pp. 259–262, 2002

101. D. Miyazaki, M. Furuta, S. Kawahito, A 75mW 10bit 120MSample/s parallel pipeline ADC. *Proceedings of IEEE European Solid-State Circuits Conference*, pp. 719–722, 2003

102. B. Xia, A. Valdes-Garcia, E. Sanchez-Sinencio, A configurable time-interleaved pipeline ADC for multi-standard wireless receivers. *Proceedings of IEEE European Solid-State Circuits Conference*, pp. 259–262, 2004

103. S.-C. Lee, G.-H. Kim, J.-K. Kwon, J. Kim, S.-H. Lee, Offset and dynamic gain-mismatch reduction techniques for 10 b 200 Ms/s parallel pipeline ADCs. *Proceedings of IEEE European Solid-State Circuits Conference*, pp. 531–534, 2005

104. S. Limotyrakis, S.D. Kulchycki, D.K. Su, B.A. Wooley, A 150-MS/s 8-b 71-mW CMOS time-interleaved ADC. IEEE J. Solid-State Circuits **40**(5), 1057–1067 (2005)

105. C.-C. Hsu, F.-C. Huang, C.-Y. Shih, C.-C. Huang, Y.-H. Lin, C.-C. Lee, B. Razavi, An 11b 800 MS/s time-interleaved ADC with digital background calibration. *IEEE International Solid-State Circuits Conference Digest of Technical Papers*, pp. 464–465, 2007

106. Z.-M. Lee, C.-Y. Wang, J.-T. Wu, A CMOS 15-bit 125-MS/s time-interleaved ADC with digital background calibration. IEEE J. Solid-State Circuits **42**(10), 2149–2160 (2007)

107. R.H. Walden, Analog-to-digital converter survey and analysis. IEEE J. Sel. Areas Commun. **17**(4), 539–550 (1999)

108. K. Kattmann, J. Barrow, A technique for reducing differential nonlinearity errors in flash A/D converters. *IEEE International Solid-State Circuits Conference Digest of Technical Papers*, pp. 170–171, 1991

109. K. Bult, A. Buchwald, An embedded 240-mW 10-b 50-MS/s CMOS ADC in 1 mm^2. IEEE J. Solid-State Circuits **32**(4), 1887–1895 (1997)

110. H. Kimura, A. Matsuzawa, T. Nakamura, S. Sewada, A 10-b 300-MHz interpolated-parallel A/D converter. IEEE J. Solid-State Circuits **28**(5), 438–446 (1993)

111. B. Ginetti, P. Jespers, A 1.5 MS/s 8-bit pipelined RSD A/D converter. *Proceedings of IEEE European Solid-State Circuits Conference*, pp. 137–140, 1990

112. B. Ginetti, P.G.A. Jespers, A. Vandemeulebroecks, A CMOS 13-b cyclic RSD A/D converter. IEEE J. Solid-State Circuits **27**(8), 957–965 (1992)

113. E.G. Soenen, R.L. Geiger, An architecture and an algorithm for fully digital correction of monolithic pipelined ADC's. IEEE Trans. Circuits Syst.–II **42**, 143–153 (1995)

114. S.-H. Lee, B.-S. Song, Digital-domain calibration for multistep analog-to-digital converters. IEEE J. Solid-State Circuits **27**(5), 1679–1688 (1992)

115. P.W. Li, M.J. Chin, P.R. Gray, R. Castello, A ratio-independent algorithmic analog-to-digital conversion technique. IEEE J. Solid-State Circuits **19**(8), 828–836 (1984)

116. T. Matsuura, T. Nara, T. Komatsu, E. Imaizumi, T. Matsutsuru, R. Horita, H. Katsu, S. Suzumura, K. Sato, A 240-Mbps, 1-W CMOS EPRML read-channel LSI chip using an interleaved subranging pipeline A/D converter. IEEE J. Solid-State Circuits **33**(4), 1840–1850 (1998)

117. K.Y. Kim, N. Kusayanagi, A.A. Abidi, A 10-b, 100-MS/s CMOS A/D converter. IEEE J. Solid-State Circuits **32**(8), 302–311 (1997)

118. S.H. Lewis, R. Ramachandran, W.M. Snelgrove, Indirect testing of digital-correction circuits in analog-to-digital converters with redundancy. IEEE Trans. Circuits Syst.–II: Analog Digital Signal Process **42**(7), 437–445 (1995)

119. H.P. Tuinhout, G. Hoogzaad, M. Vertregt, R.L.J. Roovers, C. Erdmann, Design and characterization of a high-precision resistor ladder test structure. IEEE Trans. Semicond. Manuf. **16**, 187–193 (2003)

120. J.M. Rabaey, A. Chandrakasan, B. Nikolic, *Digital Integrated Circuits: A Design Perspective*, 2nd edn. (Prentice Hall, New Jersey, 2003)

121. A.A. Abidi, High-frequency noise measurements on FETs with small dimensions. IEEE Trans. Electron Devices **33**(11), 1801–1805 (1986)

122. C. Enz, Y. Cheng, MOS transistor modeling for RF IC design. IEEE J. Solid-State Circuits **35**(2), 186–201 (2000)

123. G. Wegmann, E.A. Vittoz, F. Rahali, Charge injection in analog MOS switches. IEEE J. Solid-State Circuits **22**(5), 1091–1097 (1987)

124. J. Sheu, C. Hu, Switch-induced error voltage on a switched capacitor. IEEE J. Solid-State Circuits **19**(4), 519–525 (1984)

125. D.G. Haigh, B. Singh, A switching scheme for switched capacitor filters which reduces the effect of parasitic capacitances associated with switch control terminals. Proc. IEEE Int. Symp. Circuits Syst. **2**(7), 586–589 (1983)

126. K. Nagaraj, D.A. Martin, M. Wolfe, R. Chattopadhyay, S. Pavan, J. Cancio, T.R. Viswanathan, A dual-mode 700-Msamples/s 6-bit 200-Msamples/s 7-bit A/D converter in a 0.25-μm digital CMOS process. IEEE J. Solid-State Circuits **35**(12), 1760–1768 (2000)

127. P.E. Allen, D.R. Holberg, *CMOS Analog Circuit Design* (Oxford University Press, New York, 2002)

128. J. Shieh, M. Patil, B. Scheu, Measurement and analysis of charge injection in MOS analog switches. IEEE J. Solid-State Circuits **22**, 277–281 (1987)

129. J. Kuo, R. Dutton, B. Wooley, MOS pass transistor turn-off transient analysis. IEEE Trans Electron Device **33**(6), 1545–1555 (1986)

130. M. Pelgrom, A. Duinmaijer, A. Welbers, Matching properties of MOS transistors. IEEE J. Solid-State Circuits **24**(5), 1433–1439 (1989)

131. K. Lakshmikumar, R. Hadaway, M. Copeland, Characterization and modeling of mismatch in MOS transistors for precision analog design. IEEE J. Solid-State Circuits **21**(8), 1057–1066 (1986)

132. J. McCreary, Matching properties, and voltage and temperature dependence of MOS capacitors. IEEE J. Solid-State Circuits **6**(6), 608–616 (1981)

133. Y.C. Jenq, Digital spectra of nonuniformly sampled signals: fundamentals and high-speed waveform digitizers. IEEE Trans. Instrum. Meas. **37**(2), 245–251 (1988)

134. A. Petraglia, S.K. Mitra, Analysis of mismatch effects among A/D converters in a time-interleaved waveform digitizer. IEEE Trans. Instrum. Meas. **40**(5), 831–835 (1991)

135. Y.C. Jenq, Perfect reconstruction of digital spectrum from nonuniformly sampled signals. IEEE Trans. Instrum. Meas. **46**(3), 649–652 (1997)

136. N. Kurosawa, H. Kobayashi, K. Maruyama, H. Sugawara, K. Kobayashi, Explicit analysis of channel mismatch effects in time-interleaved ADC systems. IEEE Trans. Circuits Syst. I: Fund. Theor. Appl. **48**(3), 261–271 (2001)

137. H. Ohara, H.X. Ngo, M.J. Armstrong, C.F. Rahim, P.R. Gray, A CMOS programmable self-calibrating 13-bit eight-channel data acquisition peripheral. IEEE J. Solid-State Circuits **22**(6), 930–938 (1987)

138. B.J. Hosticka, Improvement of the gain of MOS amplifiers. IEEE J. Solid-State Circuits **14**(6), 1111–1114 (1979)

139. E. Sackinger, W. Guggenbuhl, A high-swing high-impedance MOS cascode circuit. IEEE J. Solid-State Circuits **25**(1), 289–298 (1990)

140. K. Bult, G. Geelen, A fast-settling CMOS Op Amp for SC circuits with 90-dB DC gain. IEEE J. Solid-State Circuits **25**(6), 1379–1384 (1990)

141. J. Lloyd, H.-S. Lee, A CMOS Op Amp with fully-differential gain-enhancement. IEEE Trans. Circuits Syst. II: Analog Digital Signal Processing **41**(3), 241–243 (1994)

142. C.A. Laber, P.R. Gray, A positive-feedback transconductance amplifier with applications to high frequency high Q CMOS switched capacitor filters. IEEE J. Solid-State Circuits **13**(6), 1370–1378 (1988)

143. A.A. Abidi, An analysis of bootstrapped gain enhancement techniques. IEEE J. Solid-State Circuits **22**(6), 1200–1204 (1987)
144. A. de la Plaza, High frequency switched capacitor filter using unity-gain buffers. IEEE J. Solid-State Circuits **21**(3), 470–477 (1986)
145. P.C. Yu, H.-S. Lee, A high swing 2-V CMOS operational amplifier with replica-Amp gain enhancement. IEEE J. Solid-State Circuits **28**(12), 1265–1272 (1993)
146. D. Wan, M. Franklin, Asynchronous and clocked control structures for VLSI based interconnection networks. IEEE Trans. Comput. **32**(3), 284–293 (1983)
147. K. Poulton, R. Neff, A. Muto, W. Liu, A. Burnstein, M. Heshami, A 4G sample/s 8b ADC in 0.35 μm CMOS. *IEEE International Solid-State Circuits Conference Digest of Technical Papers*, pp. 166–167, 2002
148. S.M. Louwsma, E.J.M. van Tuijl, M. Vertregt, B. Nauta, A 1.35 GS/s, 10b, 175 mW time-interleaved AD converter in 0.13 μm CMOS. *Proceedings of IEEE Symposium on VLSI Circuits*, pp. 62–63, 2007
149. C.T. Chuang, Analysis of the settling behavior of an operational amplifier. IEEE J. Solid-State Circuits **17**(1), 74–80 (1982)
150. G. Nicollini, P. Confalonieri, D. Senderowicz, A fully differential sample-and-hold circuit for high-speed applications. IEEE J. Solid-State Circuits **24**(5), 1461–1465 (1989)
151. K. Gulati, H.-S. Lee, A high-swing CMOS telescopic operational amplifier. IEEE J. Solid-State Circuits **33**(12), 2010–2019 (1998)
152. T.C. Choi, R.T. Kaneshiro, W. Brodersen, P.R. Gray, W.B. Jett, M. Wilcox, High-frequency CMOS switched-capacitor filters for communications application. IEEE J. Solid-State Circuits **18**, 652–664 (1983)
153. R. Harjani, R. Heineke, F. Wang, An integrated low-voltage class AB CMOS OTA. IEEE J. Solid-State Circuits **34**(2), 134–142 (1999)
154. R. Hogervorst, J.H. Huijsing, *Design of Low-Voltage Low-Power Operational Amplifier Cells* (Kluwer, Dordrecht, 1999)
155. B.K. Ahuja, An improved frequency compensation technique for CMOS operational amplifiers. IEEE J. Solid-State Circuits **18**(6), 629–633 (1983)
156. J.H. Huijsing, F. Tol, Monolithic operational amplifier design with improved HF behavior. IEEE J. Solid-State Circuits **11**(2), 323–327 (1976)
157. W. Sansen, Z.Y. Chang, Feedforward compensation techniques for high-frequency CMOS amplifiers. IEEE J. Solid-State Circuits **25**(6), 1590–1595 (1990)
158. F. Op't Eynde, W. Sansen, A CMOS wideband amplifier with 800 MHz gain-bandwidth. *Proceedings of IEEE Custom Integrated Circuits Conference*, pp. 9.1.1–9.1.4, 1991
159. S. Setty, C. Toumazou, Feedforward Compensation techniques in the design of low voltage Opamps and OTAs. Proc. IEEE Int. Symp. Circuits Syst. **1**, 464–467 (1998)
160. B.Y. Kamath, R.G. Meyer, P.R. Gray, Relationship between frequency response and settling time of operational amplifiers. IEEE J. Solid-State Circuits **9**(6), 347–352 (1974)
161. R.E. Vallee, E.I. El-Masry, A very high-frequency CMOS complementary folded cascode amplifier. IEEE J. Solid-State Circuits **29**(2), 130–133 (1994)
162. D.A. Johns, K. Martin, *Analog Integrated Circuit Design* (Wiley, New York, 1997)
163. A.A. Abidi, On the operation of cascode gain stages. IEEE J. Solid-State Circuits **23**(6), 1434–1437 (1988)
164. B.J. Hosticka, Improvement of the gain of MOS amplifiers. IEEE J. Solid-State Circuits **14**(6), 1111–1114 (1979)
165. E. Säckinger, W. Guggenbühl, A high-swing, high-impedance MOS cascode circuit. IEEE J. Solid-State Circuits **25**(1), 289–297 (1990)
166. U. Gatti, F. Maloberti, G. Torelli, A novel CMOS linear transconductance cell for continuous-time filters. *Proceedings of IEEE International Symposium on Circuits and Systems*, pp. 1173–1176, 1990.
167. B.J. Hosticka, Dynamic CMOS amplifiers. IEEE J. Solid-State Circuits **15**(5), 881–886 (1980)

168. M. Steyaert, R. Roovers, J. Craninckx, A 100 MHz 8 bit CMOS interpolating A/D converter. *Proceedings of IEEE Custom Integrated Circuit Conference*, pp. 28.1.1–28.1.4, 1993

169. R. van de Plassche, P. Baltus, An 8 b 100 MHz folding ADC. *IEEE International Solid-State Circuits Conference Digest of Technical Papers*, pp. 222–223, 1988

170. P. Scholtens, M. Vertregt, A 6-b 1.6-Gsample/s flash ADC in 0.18-µm CMOS using averaging termination. IEEE J. Solid-State Circuits 37(12), 1599–1609 (2002)

171. R. Ockey, M. Syrzycki, Optimization of a latched comparator for high-speed analog-to-digital converters. Proc. IEEE Canadian Conf Elect Comput Eng 1, 403–408 (1999)

172. P.M. Figueiredo, J.C. Vital, Low kickback noise techniques for CMOS latched comparators. Proc. IEEE Int. Symp. Circuits Syst. 1, 537–540 (2004)

173. F. Murden, R. Gosser, 12b 50 M Sample/s two-stage A/D converter. *IEEE International Solid-State Circuits Conference Digest of Technical Papers*, pp. 278–279, 1995

174. J. Robert, G.C. Temes, V. Valencic, R. Dessoulavy, D. Philippe, A 16-bit low-voltage CMOS A/D converter. IEEE J. Solid-State Circuits 22(2), 157–263 (1987)

175. T.B. Cho, P.R. Gray, A 10 b, 20 Msample/s, 35 mW pipeline A/D converter. IEEE J. Solid-State Circuits 30(3), 166–172 (1995)

176. L. Sumanen, M. Waltari, K. Halonen, A mismatch insensitive CMOS dynamic comparator for pipeline A/D converters. *Proceedings of the IEEE International Conference on Circuits and Systems*, pp. 32–35, 2000

177. T. Kobayashi, K. Nogami, T. Shirotori, Y. Fujimoto, A current-controlled latch sense amplifier and a static power-saving input buffer for low-power architecture. IEEE J. Solid-State Circuits 28(4), 523–527 (1993)

178. B. Nauta, A.G.W. Venes, A 70-MS/s 110-mW 8-b CMOS folding and interpolating A/D converter. IEEE J. Solid-State Circuits 30(12), 1302–1308 (1995)

179. G.M. Yin, F. op't Eynde, W. Sansen, A high-speed CMOS comparator with 8-b resolution. IEEE J. Solid-State Circuits 27(2), 208–211 (1992)

180. J. van Valburg, R.J. van de Plassche, An 8-bit 650-MHz folding ADC. IEEE J. Solid-State Circuits 27(12), 1662–1666 (1992)

181. R.J. van de Plassche, R.J. Grift, A high-speed 7-b A/D converter. IEEE J. Solid-State Circuits 14(6), 938–943 (1979)

182. M. Flynn, D. Allstot, CMOS folding ADC's with current-mode interpolation. IEEE J. Solid-State Circuits 31(9), 1248–1257 (1996)

183. A.G.W. Venes, R.J. van de Plassche, A 80-MHz, 8-b CMOS folding A/D converter. IEEE J. Solid-State Circuits 31(12), 1846–1853 (1996)

184. P. Vorenkamp, R. Roovers, A 12-bit, 60-MSample/s cascaded folding and interpolating ADC. IEEE J. Solid-State Circuits 32(12), 1876–1886 (1997)

185. B.S. Song, P. Rakers, S. Gillig, A 1-V 6-b 50-MS/s current-interpolating CMOS ADC. IEEE J. Solid-State Circuits 35(4), 647–651 (2000)

186. W. An, C.A.T. Salama, An 8-bit, 1-Gsample/s folding-interpolating analog-to-digital converter. *Proceedings of IEEE European Solid-State Circuits Conference*, pp. 228–231, 2000

187. R. Taft, C. Menkus, M.R. Tursi, O. Hidri, V.Pons, A 1.8V 1.6GS/s 8b Self-calibrating folding ADC with 7.26 ENOB at Nyquist frequency. *IEEE International Solid-State Circuits Conference Digest of Technical Papers*, pp. 252–256, 2004

188. S. Hwang, J. Moon, S. Jung, M. Song, Design of a 1.8V 6-bit 100MSPS 5mW CMOS A/D converter with low power folding-interpolation techniques. *Proceedings of IEEE European Solid-State Circuits Conference*, pp. 548–551, 2006

189. C. Yihui, H. Qiuting, T. Burger, A 1.2V 200-MS/s 10-bit folding and interpolating ADC in 0.13-µm CMOS. *Proceedings of IEEE European Solid-State Circuits Conference*, pp. 155–158, 2007

190. C.-C. Hsu, C.-C. Huang, Y.-H. Lin, C.-C. Lee, A 10b 200MS/s pipelined folding ADC with offset calibration. *Proceedings of IEEE European Solid-State Circuits Conference*, pp. 151–154, 2007

191. S. Limotyrakis, K. Nam, B. Wooley, Analysis and simulation of distortion in folding and interpolating A/D converters. IEEE Transac Circuits Syst-II: Analog Digital Signal Process **49**(3), 161–169 (2002)

192. R. Roovers, M.S.J. Steyaert, A 175 Ms/s, 6 b, 160 mW, 3.3 V CMOS A/D converter. IEEE J. Solid-State Circuits **31**(7), 938–944 (1996)

193. X. Jiang, Y. Wang, A.N. Willson Jr., A 200 MHz 6-bit folding and interpolating ADC in 0.5-μm CMOS. Proc. IEEE Int. Symp. Circuits Syst. **1**, 5–8 (1998)

194. M.P. Flynn, B. Sheahan, A 400-Msample/s, 6-b CMOS folding and interpolating ADC. IEEE J. Solid-State Circuits **33**(12), 1932–1938 (1998)

195. C. Lane, A 10-bit 60-MSPS flash ADC. *IEEE Bipolar Circuits and Technology Meeting Digest of Technical Papers,* pp. 44–47, 1989

196. J. Guilherme, P. Figueiredo, P. Azevedo, G. Minderico, A. Leal, J. Vital, J. Franca, A pipeline 15-b 10-Msample/s analog-to-digital converter for ADSL applications. *Proceedings of the IEEE International Symposium on Circuits and Systems,* pp. 396–399, 2001

197. C.M. Hammerschmied, Q. Huang, Design and implementation of an untrimmed MOSFET-Only 10-bit A/D converter with -79-dB THD. IEEE J. Solid-State Circuits **33**(8), 1148–1057 (1998)

198. A.G.F. Dingwall, Monolithic expandable 6 bit 20 MHz CMOS/SOS A/D converter. IEEE J. Solid-State Circuits **14**(6), 926–932 (1979)

199. O. Moldsvor, G.S. Ostrem, An 8-bit, 200 MSPS folding and interpolating ADC. Analog Integr. Circuits Signal Process. **15**(1), 37–47 (1998)

200. B. Razavi, *Principles of Data Conversion System Design* (IEEE Press, Piscataway, 1995)

201. T. Shimizu, M. Hotta, K. Maio, S. Ueda, A 10-bit 20-MHz two-step parallel A/D converter with internal S/H. IEEE J. Solid-State Circuits **24**(1), 13–20 (1989)

202. P. Vorenkamp, J. Verdaasdonk, A 10b 50MS/s pipelined ADC. *IEEE International Solid-State Circuit Conference Digest of Technical Papers*, pp. 32–33, 1992

203. G. Erdi, A precision trim technique for monolithic analog circuits. IEEE J. Solid-State Circuits **10**(6), 412–416 (1975)

204. M. Mayes, S. Chin, L. Stoian, A low-power 1 MHz 25 mW 12-bit time-interleaved analog-to-digital converter. IEEE J. Solid-State Circuits **31**(2), 169–178 (1996)

205. H.-S. Lee, D. Hodges, P. Gray, A self-calibrating 15 bit CMOS A/D converter. IEEE J. Solid-State Circuits **19**(6), 813–819 (1984)

206. P. Yu, S. Shehata, A. Joharapurkar, P. Chugh, A. Bugeja, X. Du, S.-U. Kwak, Y. Panantono-poulous, T. Kuyel, A 14b 40MSample/s pipelined ADC with DFCA. *IEEE International Solid-State Circuit Conference Digest of Technical Papers*, pp. 136–137, 2001

207. I. Galton, Digital cancellation of D/A converter noise in pipelined A/D converters. IEEE Trans. Circuits Syst. **47**(3), 185–196 (2000)

208. J.M. Ingino, B.A. Wooley, A continuously calibrated 12-b, 10-MS/s, 3.3-V A/D converter. IEEE J. Solid-State Circuits **33**(12), 1920–1931 (1998)

209. O.E. Erdogan, P.J. Hurst, S.H. Lewis, A 12-b digital-background-calibrated algorithmic ADC with -90-dB THD. IEEE J. Solid-State Circuits **34**(12), 1812–1820 (1999)

210. U.-K. Moon, B.-S. Song, Background digital calibration techniques for pipelined ADC's. IEEE Trans. Circuits Syst.–II **44**(2), 102–109 (1997)

211. T.-H. Shu, B.-S. Song, K. Bacrania, A 13-b 10-Msample/s ADC digitally calibrated with oversampling delta-sigma converter. IEEE J. Solid-State Circuits **30**(4), 443–452 (1994)

212. C.C. Enz, G.C. Temes, Circuit techniques for reducing the effects of Opamp imperfections: autozeroing, correlated double sampling, and chopper stabilization. Proc. IEEE **84**(11), 1584–1614 (1996)

213. G.C.M. Meijer, Concepts and focus point for intelligent sensor systems. Sens. Actuat. **41**, 183–191 (1994)

214. J.J.F. Rijns, CMOS low-distortion high-frequency variable-gain amplifier. IEEE J. Solid-State Circuits **31**(7), 1029–1034 (1996)

215. D.K. Su, M.J. Loinaz, S. Masui, B.A. Wooley, Experimental results and modeling techniques for substrate noise in mixed-signal integrated circuits. IEEE J. Solid-State Circuits **28**(4), 420–430 (1993)
216. M. Shinagawa, Y. Akazawa, T. Wakimoto, Jitter analysis of high-speed sampling systems. IEEE J. Solid-State Circuits **25**(5), 220–224 (1990)
217. J. Doernberg, P.R. Gray, D.A. Hodges, A 10-bit 5-Msample/s CMOS two-step flash ADC. IEEE J. Solid-State Circuits **24**(2), 241–249 (1989)
218. T. M. Sounder, G.N. Stenbakken, A comprehensive approach for modeling and testing analog and mixed-signal devices. *Proceedings of IEEE International Test Conference*, pp. 169–176, 1990
219. N. Nagi, A. Chatterjee, A Balivada, J.A. Abraham, Fault-based automatic test generator for linear analog circuits. *Proceedings of IEEE International Conference on Computer Aided Design*, pp. 88–91, 1993
220. T. Koskinen, P.Y.K. Cheung, Hierarchical tolerance analysis using statistical behavioral models. IEEE Trans. Comput. Aided Design **15**(5), 506–516 (1996)
221. S.J. Spinks, C.D. Chalk, I.M. Bell, M. Zwolinski, Generation and verification of tests for analog circuits subject to process parameter deviations. J. Electron. Test.: Theor. Appl. **20**, 11–23 (2004)
222. A. Zjajo, J. Pineda de Gyvez, Evaluation of signature-based testing of RF/analog circuits. *Proceedings of IEEE European Test Symposium*, pp. 62–67, 2005
223. R. Voorakaranam, S.S. Akbay, S. Bhattacharya, S. Cherubal, A. Chatterjee, Signature testing of analog and RF circuits: algorithms and methodology. IEEE Trans. Circuits Syst.-I: Fund. Theor. Appl. **54**, 1018–1031 (2007)
224. A. McKeon, A. Wakeling, Fault diagnosis in analog circuit using AI techniques. *Proceedings of IEEE International Test Conference*, pp. 118–123, 1989
225. L. Milor, V. Visvanathan, Detection of catastrophic faults in analog integrated circuits. IEEE Trans. Comput.-Aided Design **8**(2), 114–130 (1989)
226. G. Devarayanadurg, M. Soma, Analytical fault modeling and static test generation for analog ICs. *Proceedings of IEEE/ACM International Conference on Computer-Aided Design*, pp. 44–47, 1994
227. Z. Wang, G. Gielen, W. Sansen, A novel method for the fault detection of analog integrated circuits. Proc. IEEE Int. Symp. Circuits Syst. **1**, 347–350 (1994)
228. K. Saab, N. Ben-Hamida, B. Kaminska, Parametric fault simulation and test vector generation. *Proceedings of IEEE Design, Automation and Test in Europe Conference*, pp. 650–656, 2000
229. F. Liu, P.K. Nikolov, S. Ozev, Parametric fault diagnosis for analog circuits using a Bayesian framework. *Proceedings of IEEE VLSI Test Symposium*, pp. 272–277, 2006
230. J. Neyman, E. Pearson, On the problem of the most efficient tests of statistical hypotheses. Philos. Trans. R. Soc. Lond. **A 231**, 289–337 (1933)
231. A. Zjajo, J. Pineda de Gyvez, Analog automatic test pattern generation for quasi-static structural test. *IEEE Transaction on Very Large Scale Integration (VLSI) Systems*, **17**, 1383–1391 (2009)
232. E. Silva, J. Pineda de Gyvez, G. Gronthoud, Functional vs. multi-VDD testing of RF circuits. *Proceedings of IEEE International Test Conference*, 2005
233. M. Loève, *Probability Theory* (D. Van Nostrand, Princeton, NJ, 1960)
234. J. Vlach, K. Singhal, *Computer Methods for Circuit Analysis and Design* (Van Nostrand Reinhold, New York, 1983)
235. K.E. Brenan, S.L. Campbell, L.R. Petzold, *The Numerical Solution of Initial Value Problems in Ordinary Differential-Algebraic Equations* (North Holland, New York, 1989)
236. J. Butcher, P. Chartier, Parallel general linear methods for stiff ordinary differential and differential algebraic equations. Appl. Numer. Math. **17**, 213–222 (1995)
237. P.J. Rrabier, W.C. Rheinboldt, Techniques of scientific computing (part 4), in *Handbook of Numerical Analysis*, ed. by P.G. Ciarlet, vol. 8 (North Holland/Elsevier, Amsterdam, 2002), pp. 183–540

238. M. Gunther, U. Feldmann, CAD based electric modeling in Industry. Math. Comput. Simul. **39**, 573–582 (1995)
239. M. Gunther, U. Feldmann, CAD based electric modeling in industry, Part I: mathematical structure and index of network equations. Surv. Math. Indus. **8**, 97–129 (1999)
240. C. Tischendorf, Topological index calculation of DAEs in circuit simulation. Surv. Math. Indus. **8**, 187–199 (1999)
241. G. Ali, A. Bartel, M. Günther, Parabolic differential-algebraic models in electrical network design. Soc. Indus. Appl. Math. – J. Multiscale Model. Simul. **4**, 813–838 (2005)
242. H.P. Tuinhout, S. Swaving, J. Joosten, A fully analytical MOSFET model parameter extraction approach. IEEE Proc. Microelectron. Test Struct. **1**(1), 79–84 (1988)
243. T.L. Chen, G. Gildenblat, Symmetric bulk charge linearization in the charge-sheet model. IEEE Electron. Lett. **37**, 791–793 (2001)
244. R. van Langevelde, A.J. Scholten, D.B.M. Klassen, *MOS Model 11: Level 1102*. Philips Research Technical Report 2004/85
245. M. Grigoriu, On the spectral representation method in simulation. Probab. Eng. Mech. **8**, 75–90 (1993)
246. H. Stark, W.J. Woods, *Probability, Random Process and Estimation Theory for Engineers* (Prentice-Hall, Englewood Cliffs, NJ, 1994)
247. R. Ghanem, P.D. Spanos, *Stochastic Finite Element: A Spectral Approach* (Springer, New York, 1991)
248. P. Friedberg, Y. Cao, J. Cain, R. Wang, J. Rabaey, C. Spanos, Modeling within-die spatial correlation effects for process-design co-optimization. *Proceedings of IEEE International Symposium on Quality of Electronic Design*, pp. 516–521, 2005
249. J. Xiong, V. Zolotov, L. He, Robust extraction of spatial correlation. *Proceedings of IEEE International Symposium on Physical Design*, pp. 2–9, 2006
250. B.E. Stine, D.S. Boning, J.E. Chung, Analysis and decomposition of spatial variation in integrated circuit process and devices. *IEEE Transaction on Semiconductor Manufacturing*, pp. 24–41, 1997
251. A. Zjajo, J. Pineda de Gyvez, G. Gronthoud, Structural fault modeling and fault detection through Neyman-Pearson decision criteria for analog integrated circuits. J. Electron. Test.: Theor. Appl. **22**, 399–409 (2006)
252. S.D. Huss, R.S. Gyurcsik, Optimal ordering of analog integrated circuit tests to minimize test time. *Proceedings of Design Automation Conference*, pp. 494–499, 1991
253. IEEE Std. 1149.4-1999, Test Technology Technical Committee of the IEEE Computer Society, *IEEE Standard for a Mixed-Signal Test Bus*. Institute of Electrical and Electronic Engineers Inc.
254. G. Schafer, H. Sapotta, W. Dennerm, Block-oriented test strategy for analog circuits. *Proceedings of IEEE European Solid-State Circuit Conference*, pp. 217–220, 1991
255. M. Soma, A design for test methodology for active analog filters. *Proceedings of IEEE International Test Conference*, pp. 183–192, 1990
256. A.H. Bratt, R.J. Harvey, A.P. Dorey, A.M.D. Richardson, Design for test structure to facilitate test vector application with low performance loss in non-test mode. Electron. Lett. **4**(4), 299–313 (1993)
257. D. Vazquez, J.L. Huertas, A. Rueda, A new strategy for testing analog filters. *Proceedings of IEEE VLSI Test Symposium*, pp. 36–41, 1994
258. D. Vazquez, J.L. Huertas, A. Rueda, Reducing the impact of DfT on the performance of analog integrated circuits: improved SW-OPAMP design. *Proceedings of IEEE VLSI Test Symposium*, pp. 42–48, 1996
259. B. Vinnakota, R. Harjani, DFT for digital detection of analog parametric faults in SC filters. IEEE Trans. Comput.-Aided Design Integr. Circuits Syst. **19**(7), 789–798 (2000)
260. E. Peralias, A. Rueda, J.L. Huertas, A DFT technique for analog-to-digital converters with digital correction. *Proceedings of IEEE VLSI Test Symposium*, pp. 302–307, 1997

261. J. Pineda de Gyvez, G. Gronthoud, R. Amine, VDD Ramp Testing for RF Circuits. *Proceedings of IEEE International Test Conference*, pp. 651–658, 2003
262. A. Zjajo, J. Pineda de Gyvez, G. Gronthoud, A DC approach for detection and simulation of parametric faults in analog and mixed-signal circuits. *Proceedings of IEEE International Mixed-Signal Testing Workshop*, pp. 155–164, 2005
263. A. Zjajo, H.J. Bergveld, R. Schuttert, J. Pineda de Gyvez, Power-scan chain: design for analog testability. *Proceedings of International Test Conference*, 2005
264. S. Somayayula, E. Sanchez-Sinencio, J. Pineda de Gyvez, Analog fault diagnosis based on ramping power supply current signature. IEEE Trans. Circuits Syst.-II **43**(10), 703–712 (1996)
265. S. Tabatabaei, A. Ivanov, A built-in current monitor for testing analog circuit blocks. Proc. IEEE Int. Symp. Circuits Syst. **2**, 109–114 (1999)
266. J.R. Vazquez, J. Pineda de Gyvez, Built-in current sensor for ΔI_{DDQ} testing. IEEE J. Solid-State Circuits **39**(3), 511–518 (2004)
267. IEEE Std. 1149.1-2001, Test Technology Technical Committee of the IEEE Computer Society, *IEEE Standard Test Access Port and Boundary-Scan Architecture*. Institute of Electrical and Electronic Engineers Inc.
268. J. Galan, R.G. Carvajal, A. Torralba, F. Munoz, J. Ramirez-Angulo, A low-power low-voltage OTA-C sinusoidal oscillator with a large tuning range. IEEE Trans. Circuits Syst. **52** (2), 283–291 (2005)
269. G. Chang, A. Rofougaran, K. Mong-Kai, A.A. Abidi, H. Samueli, A low-power cmos digitally-synthesized 0–13 MHZ sinewave generator. *IEEE International Solid-State Circuits Conference Digest of Technical Papers*, pp. 32–33, 1994
270. B. Dufort, G.W. Roberts, On-chip analog signal generation for mixed-signal built-in-self-test. IEEE J. Solid-State Circuits **33**(3), 318–330 (1999)
271. J. Huang, K. Cheng, A sigma-delta modulation based BIST scheme for mixed-signal circuits. *Proceedings of IEEE Design Automation Conference*, pp. 605–610, 2000
272. A.K. Lu, G.W. Roberts, D. Johns, A high quality analog oscillator using oversampling D/A conversion techniques. IEEE Trans. Circuits Syst. II **41**(7), 437–444 (1994)
273. M.J. Barragan, D. Vazquez, A. Rueda, J.L. Huertas, On-chip analog sinewave generator with reduced circuitry resources. *Proceedings of IEEE Midwest Symposium on Circuits and Systems*, pp. 638–642, 2006
274. Y.P. Tsividis, Integrated continuous-time filter design-an overview. IEEE J. Solid-State Circuits **29**(3), 166–176 (1994)
275. A.M. Durham, J.B. Hughes, W. Redman-White, Circuit architectures for high linearity monolithic continuous-time filtering. IEEE Trans. Circuits Syst.–II **39**(9), 651–657 (1992)
276. R. Gharpurey, N. Yanduru, F. Dantoni, P. Litmanen, G. Sirna, T. Mayhugh, C. Lin, I. Deng, P. Fontaine, F. Lin, A direct conversion receiver for the 3G WCDMA standard. *Proceedings of the IEEE Custom Integrated Circuits Conference*, pp. 239–242, 2002
277. S. Lindfors, J. Jussila, K. Halonen, L. Siren, A 3-V continuous-time filter with on-chip tuning for IS-95. IEEE J. Solid-State Circuits **34**(8), 1150–1154 (1999)
278. J.K. Pyykönen, A low distortion wideband active-RC filter for a multicarrier base station transmitter. *Proceedings of the IEEE International Symposium on Circuits and Systems*, pp. 244–247, 2001
279. H. Khorramabadi, M.J. Tarsia, N.S. Woo, Baseband filters for IS-95 CDMA receiver applications featuring digital automatic frequency tuning. *IEEE International Solid-State Circuits Conference Digest of Technical Papers*, pp. 172–173, 1996
280. T. Salo, S. Lindfors, T. Hollman, K. Halonen, Programmable direct digital tuning circuit for a continuous-time filter. *Proceedings of the European Solid-State Circuits Conference*, pp. 168–171, 2000
281. S. Lindfors, T. Hollman, T. Salo, K. Halonen, A 2.7V CMOS GSM/WCDMA continuous-time filter with automatic tuning. *Proceedings of the IEEE Custom Integrated Circuits Conference*, pp. 9–12, 2001

282. Y. Tsividis, Continuous-time filters in telecommunications chips. *Proceedings of IEEE Communications Magazine*, pp. 132–137, 2001

283. Z. Czarnul, Modification of Banu-Tsividis continuous-time integrator structure. IEEE Trans. Circuits Syst. **33**(7), 714–716 (1986)

284. U.-K. Moon, B.-S. Song, Design of a low-distortion 22-kHz fifth-order Bessel filter. IEEE J. Solid-State Circuits **28**(12), 1254–1264 (1993)

285. A. Yoshizawa, Y.P. Tsividis, Anti-blocker design techniques for MOSFET-C filters for direct conversion receivers. IEEE J. Solid-State Circuits **37**(3), 357–364 (2002)

286. A. Yoshizawa, Design considerations for large dynamic range MOSFET-C filters for direct conversion receivers. *Proceedings of the European Solid-State Circuits Conference*, pp. 655–658, 2002

287. B. Nauta, A CMOS transconductance-C filter technique for very high frequencies. IEEE J. Solid-State Circuits **27**(2), 142–153 (1992)

288. S. Lindfors, K. Halonen, M. Ismail, A 2.7-V elliptical MOSFET-Only gmC-OTA filter. IEEE Trans. Circuits Syst.−II **47**(2), 89–95 (2000)

289. D. Python, A.-S. Porret, C. Enz, A 1V 5th-order bessel filter dedicated to digital standard processes. *Proceedings of the IEEE Custom Integrated Circuits Conference*, pp. 505–508, 1999

290. T. Itakura, T. Ueno, H. Tanimoto, A. Yasuda, R. Fujimoto, T. Arai, H. Kokatsu, A 2.7-V, 200-kHz, 49-dBm, stopband-IIP3, low-noise, fully balanced Gm-C filter IC. IEEE J. Solid-State Circuits **34**(8), 1155–1159 (1999)

291. T.C. Kuo, B.B. Lusignan, A very low power channel select filter for IS-95 CDMA receiver with on-chip tuning. *IEEE Symposium on VLSI Circuits Digest of Technical Papers*, pp. 244–247, 2000

292. K. Halonen, S. Lindfors, J. Jussila, L. Siren, A 3V GmC-filter filter with on-chip tuning for CDMA. *Proceedings of the IEEE Custom Integrated Circuits Conference*, pp. 83–86, 1997

293. C.A. Laber, P.R. Gray, A 20-MHz sixth-order BiCMOS Parasitic-insensitive continuous-time filter and second-order equalizer optimized for disk-drive read channels. IEEE J. Solid-State Circuits **27**(4), 462–470 (1993)

294. B. Razavi, *Design of Analog CMOS Integrated Circuits* (McGraw-Hill, New York, 2001)

295. D.G. Haigh, B. Singh, A switching scheme for switched capacitor filters which reduces the effect of parasitic capacitances associated with switch control terminals. *Proceedings of IEEE International Symposium on Circuits and Systems*, pp. 586–589, 1983

296. R.W. Brodersen, P.R. Gray, D.A. Hodges, MOS switched-capacitor filters. Proc. IEEE **67**, 61–75 (1979)

297. G.M. Jacobs, D.J. Allstot, R.W. Brodersen, P.R. Gray, Design techniques for MOS switched-capacitor ladder filters. IEEE Trans. Circuits Syst. **25**(12), 1014–1021 (1978)

298. A. Fettweis, D. Herbst, B. Hoefflinger, J. Pandel, R. Schweer, MOS switched capacitor filters using voltage inverter switches. IEEE Trans. Circuits Syst. **27**(6), 527–538 (1980)

299. F. Montecchi, On design of switched-capacitor filters with the voltage-inverter switch approach. Proc. IEEE Int. Symp. Circuits Syst. **2**, 1479–1482 (1988)

300. R. Gregorian, G.C. Temes, *Analog MOS Integrated Circuits for Signal Processing* (Wiley, New York, 1986)

301. K. Martin, A.S. Sedra, Exact design of switched-capacitor bandpass filters using coupled-biquad structures. IEEE Trans. Circuits Syst. **27**(6), 469–478 (1980)

302. K. Martin, A.S. Sedra, Effects of the Op-Amp finite gain and bandwidth on the performance of switched-capacitor filters. IEEE Trans. Circuits Syst. **28**(8), 822–829 (1981)

303. A. Baschirotto, F. Severi, R. Castello, A 200-Ms/s 10-mW switched-capacitor filter in 0.5-μm CMOS technology. IEEE J. Solid-State Circuits **35**(8), 1215–1219 (2000)

304. A.D. Plaza, High-frequency switched-capacitor filter using unity-gain buffers. IEEE J. Solid-State Circuits **21**(8), 470–477 (1986)

305. C.Y. Wu, P.H. Lu, M.K. Tsai, Design techniques for high-frequency CMOS switched-capacitor filters using non-Op-Amp-based unity-gain amplifiers. IEEE J. Solid-State Circuits **26**(4), 1460–1466 (1991)

306. B.K. Thandri, S.J. Silva-Martinez, F. Maloberti, A feedforward compensation scheme for high gain wideband amplifiers. *Proceedings of IEEE International Conference on Electronics, Circuits and Systems*, pp. 1115–1118, 2001

307. S. Pavan, Y. Tsividis, *High Frequency Continuous Time Filters in Digital CMOS Processes* (Kluwer, Boston, 2000)

308. Y. Tsividis, Y. Papananos, Continuous-time filters using buffer with gain lower than unity. IEEE Electron Lett. **30**(8), 629–630 (1994)

309. J.S. Silva-Martinez, M. Steyaert, W. Sansen, *High-Performance CMOS Continuous-Time Filters* (Kluwer, Boston, 1993)

310. B. Nauta, *Analog CMOS Filters for Very High Frequencies* (Kluwer, Boston, 1993)

311. F. Krummenacher, N. Joehl, A 4 MHz CMOS continuous-time filter with On-Chip automatic tuning. IEEE J. Solid-State Circuits **23**, 750–758 (1988)

312. R. Schaumann, M.E. Valkenburg, *Design of Analog Filters* (Oxford University Press, New York, 2001)

313. R. Schaumann, M.S. Ghausi, K.R. Laker, *Design of Analog Filters* (Prentice-Hall, Englewood Cliffs, NJ, 1990)

314. F. Maloberti, F. Montecchi, G. Torelli, E. Halasz, Bilinear design of fully differential switched-capacitor ladder filters. IEE Proc. Electro Circuits Syst. **132**, 266–272 (1985)

315. M. Maymandi-Nejad, M. Sachdev, Continuous-time common mode feedback technique for sub 1V analogue circuits. IEEE Electron Lett. **38**, 1408–1409 (2002)

316. C. Yoo, S. Lee, W. Kim, A +/− 1.5-V, 4-MHz CMOS continuous-time filter with a single-integrator based tuning. IEEE J. Solid-State Circuits **33**(4), 18–27 (1998)

317. K.-H. Loh, D.L. Hiser, W.J. Adams, R.L. Geiger, A versatile digitally controlled continuous-time filter structure with wide-range and fine resolution capability. IEEE Trans Circuit Syst II **39**(7), 265–276 (1992)

318. A. Brandolini, A. Gandelli, Testing methodologies for analog-to-digital converters. *IEEE Transactions on Instrumentation and Measurement*, pp. 595–603, 1993

319. T. Yamaguchi, Static testing of ADCs using wavelet transforms. *Proceedings of IEEE Asian Test Symposium*, pp. 188–193, 1997

320. M.F. Toner, G.W. Roberts, A BIST scheme for an SNR test of a sigma-delta ADC. *Proceedings of IEEE International Test Conference*, pp. 805–14, 1993

321. A.C. Serra, P.S. Girao, Static and dynamic testing of A/D converter using a VXI based system. *Proceedings of IEEE Instrumentation and Measurement Technology Conference*, pp. 903–906, 1994

322. G. Chiorboli, G. Franco, C. Morandi, Analysis of distortion in A/D converters by time-domain and code-density techniques. *Proceedings of IEEE Transactions on Instrumentation and Measurement*, pp. 45–49, 1996

323. K. Arabi, B. Kaminska, J. Rzeszut, A new built-in self-test approach for digital-to-analog and analog-to-digital converters. *Proceedings of IEEE International Conference on Computer Aided Design*, pp. 491–494, 1994

324. K. Arabi, B. Kaminska, Efficient and accurate testing of analog-to-digital converters using oscillation-test method. *Proceedings of IEEE European Design and Test Conference*, pp. 348–352, 1997

325. R. de Vries, T. Zwemstra, E. Bruls, P. Regtien, Built-in self-test methodology for A/D converters. *Proceedings of IEEE European Design and Test Conference*, pp. 353–358, 1997

326. S.K. Sunter, N. Nagi, A simplified polynomial-fitting algorithm for DAC and ADC BIST. *Proceedings of IEEE International Test Conference*, pp. 389–395, 1997

327. F. Azais, S. Bernard, Y. Bertrand, M. Renovell, Towards an ADC BIST scheme using the histogram test technique. *Proceedings of IEEE European Test Workshop*, pp. 53–58, 2000

328. E. Peralias, A. Rueda, J.A. Prieto, J.L. Huertas, DfT & On-line test of high-performance data converters: a practical case. *Proceedings of International Test Conference*, pp. 534–540, 1998

329. K. Arabi, B. Kaminska, J. Rzeszut, BIST for D/A and A/D converters. *Proceedings of IEEE Design and Test of Computers*, pp. 40–49, 1996

330. K. Arabi, B. Kaminska, J. Rzeszut, A new built-in-self-test approach for digital-to-analog and analog-to-digital converters. *Proceedings of IEEE International Conference on Computer Aided Design*, pp. 491–494, 1994

331. Y.-C. Wen, K.-J. Lee, BIST structure for DAC testing. IEE Electron Lett. **34**(12), 1173–1174 (1998)

332. S.J. Chang, C.L. Lee, J.E. Chen, BIST scheme for DAC testing. IEE Electron Lett. **38**(15), 776–777 (2002)

333. *IEEE 1057 Standard for Digitizing Waveform Recorders*, 1994

334. *IEEE 1241 Standard for Analog-to-Digital Converters*, 2000

335. M. Vanden Bossche, J. Schoukens, J. Eenneboog, Dynamic testing and diagnostics of A/D converters. IEEE Trans. Circuits Syst. **33**(8), 775–785 (1986)

336. J. Doernberg, H.-S. Lee, D.A. Hodges, Full-speed testing of A/D converters. IEEE J. Solid-State Circuits **19**(6), 820–827 (1984)

337. N. Giaquinto, A. Trotta, Fast and accurate ADC testing via an enhanced sine wave fitting algorithm. IEEE Trans. Instrum. Meas. **46**(2), 1020–1024 (1997)

338. F. Alegria, P. Arpaia, A.M. da Cruz Serra, P. Daponte, ADC histogram test by triangular small-waves. Proc IEEE Instrum Meas Technol Conf **3**, 1690–1695 (2001)

339. L. Jin, K. Parthasarathy, T. Kuyel, D. Chen, R.L. Geiger, Linearity testing of precision analog-to-digital converters using stationary nonlinear inputs. *Proceedings of IEEE International Test Conference*, pp. 218–227, 2003

340. L. Jin, D. Chen, R. Geiger, SEIR linearity testing of precision A/D converters in nonstationary environments with center-symmetric interleaving. IEEE Trans. Instrum. Meas. **56**(5), 1776–1785 (2007)

341. E. Korhonen, J. Häkkinen, J. Kostamovaara, A robust algorithm to identify the test stimulus in histogram-based A/D converter testing. IEEE Trans. Instrum. Meas. **56**(6), 2369–2374 (2007)

342. H. Shin, J. Park, J.A. Abraham, A statistical digital equalizer for loopback-based linearity test of data converters. *Proceedings of IEEE Asian Test Conference*, pp. 245–250, 2006

343. H. Shin, B. Kim, J.A. Abraham, Spectral prediction for specification-based loopback test of embedded mixed-signal circuits. *Proceedings of IEEE VLSI Test Symposium*, pp. 412–417, 2006

344. X. Sheng, H. Kerkhoff, A. Zjajo, G. Gronthoud, Exploring dynamics of embedded ADC through adapted digital input stimuli. *Proceedings of IEEE International Workshop on Mixed-Signals, Sensors, and Systems Test*, pp. 1–7, 2008

345. X. Sheng, H. Kerkhoff, A. Zjajo, G. Gronthoud, Time-modulo reconstruction algorithms for cost-efficient A/D converter multi-site test. *IEEE European Test Symposium*, 2009, accepted for publication

346. F.H. Irons, D.M. Hummels, The modulo time plot-a useful data acquisition diagnostic tool. IEEE Trans. Instrum. Meas. **45**(3), 734–738 (1996)

347. K. Bowman, J. Meindl, Impact of Within-die parameter fluctuations on the future maximum clock frequency distribution. *Proceedings of IEEE Custom Integrated Circuits Conference*, pp. 229–232, 2001

348. C. Michael, M. Ismail, *Statistical Modeling for Computer-Aided Design of MOS VLSI Circuits* (Kluwer, Boston, 1993)

349. P.R. Gray, R.G. Meyer, *Analysis and Design of Analog Integrated Circuits* (Wiley, New York, 1984)

350. A. Demir, E. Liu, A. Sangiovanni-Vincentelli, Time-domain non-Monte Carlo noise simulation for nonlinear dynamic circuits with arbitrary excitations. *Proceedings of IEEE/ACM Interanational Conference on Computer Aided Design*, pp. 598–603, 1994

351. R. López-Ahumada, R. Rodríguez-Macías, FASTEST: a tool for a complete and efficient statistical evaluation of analog circuits. DC analysis. *Analog Integr Circuits Signal Process*, **29**(3), 201–212 (2001)

352. E. Felt, S. Zanella, C. Guardiani, A. Sangiovanni-Vincentelli, Hierarchical statistical characterization of mixed-signal circuits using behavioral modeling. *Proceedings of IEEE/ACM Interanational Conference on Computer Aided Design*, pp. 374–380, 1996

353. J. Vlach, K. Singhal, *Computer Methods for Circuit Analysis and Design* (Van Nostrand Reinhold, New York, 1983)
354. L.O. Chua, C.A. Desoer, E.S. Kuh, *Linear and Nonlinear Circuits* (Mc Graw-Hill, New York, 1987)
355. L. Arnold, *Stochastic Differential Equations: Theory and Application* (Wiley, New York, 1974)
356. A. Sangiovanni-Vincentelli, Circuit Simulation, in *Computer Design Aids for VLSI Circuits* (Sijthoff & Noordhoff, The Netherlands, 1980)
357. A.S. Hodel, S.T. Hung, Solution and applications of the lyapunov equation for control systems. IEEE Trans. Ind. Electron. **39**(3), 194–202 (1992)
358. R.H. Bartels, G.W. Stewart, Solution of the matrix equation AX+XB=C. Commun Assoc Comput Machin **15**, 820–826 (1972)
359. N.J. Higham, Perturbation theory and backward error for AX−XB=C. BIT Numer. Math. **33**, 124–136 (1993)
360. T. Penzl, Numerical solution of generalized Lyapunov equations. Adv. Comput. Math. **8**, 33–48 (1998)
361. G.H. Golub, C.F. van Loan, *Matrix Computations* (Johns Hopkins University Press, Baltimore, 1996)
362. V. Sima, *Algorithms for Linear-Quadratic Optimization, Vol. 200, Pure and Applied Mathematics* (Marcel Dekker, New York, 1996)
363. P. Benner, E. Quintana-Orti, Solving stable generalized lyapunov equations with the matrix sign function. Numer Algebra **20**, 75–100 (1999)
364. I. Jaimoukha, E. Kasenally, Krylov subspace methods for solving large lyapunov equations. SIAM J. Numer. Anal. **31**, 227–251 (1994)
365. E. Wachspress, Iterative solution of the lyapunov matrix equation. Appl. Math. Lett. **1**, 87–90 (1998)
366. J. Li, F. Wang, J. White, An efficient Lyapunov equation-based approach for generating reduced-order models of interconnect. *Proceedings of IEEE/ACM Design Automation Conference*, pp. 1–6, 1999
367. The Numerics in Control Network, http://www.win.tue.nl/wgs/niconet.htm
368. E. Balestrieri, P. Daponte, S. Rapuano, A state of the art on ADC error compensation methods. IEEE Trans. Instrum. Meas. **54**(4), 1388–1394 (2005)
369. M. Vertregt, P. Scholtens, Scalable high-speed analog circuit design, in *Analog Circuit Design*, ed. by J.H. Huijsing, M. Steyaert, A. van Roermund, vol. 10 (Kluwer, Boston, MA, 2002), pp. 3–11
370. M.F. Toner, G.W. Roberts, A BIST scheme for an SNR test of a sigma-delta ADC. *Proceedings of IEEE International Test Conference*, pp. 805–14, 1993
371. C. Serra, P.S. Girao, Static and dynamic testing of A/D converter using a VXI based system. *Proceedings of IEEE Instrumentation and Measurement Technology Conference*, pp. 903–906, 1994
372. G. Chiorboli, G. Franco, C. Morandi, Analysis of distortion in A/D converters by time-domain and code-density techniques. *IEEE Transaction on Instrumentation and Measurement,* pp. 45–49, 1996
373. K. Arabi, B. Kaminska, J. Rzeszut, A new built-in self-test approach for digital-to-analog and analog-to-digital converters. *Proceedings of IEEE International Conference on Computer Aided Design*, pp. 491–494, 1994
374. K. Arabi, B. Kaminska, Efficient and accurate testing of analog-to-digital converters using oscillation-test method. *Proceedings of IEEE European Design and Test Conference*, pp. 348–352, 1997
375. R. de Vries, T. Zwemstra, E. Bruls, P. Regtien, Built-in self-test methodology for A/D converters. *Proceedings of IEEE European Design and Test Conference*, pp. 353–358, 1997
376. S. K. Sunter, N. Nagi, A simplified polynomial-fitting algorithm for DAC and ADC BIST. *Proceedings of IEEE International Test Conference*, pp. 389–395, 1997

377. F. Azais, S. Bernard, Y. Bertrand, M. Renovell, Towards an ADC BIST scheme using the histogram test technique. *Proceedings of IEEE European Test Workshop*, pp.53–58, 2000

378. A. Charoenrook, M. Soma, Fault diagnosis technique for subranging ADCs. *Proceedings of the IEEE Asian Test Symposium*, pp.367–372, 1994

379. A. Zjajo, M.J. Barragan Asian, J. Pineda de Gyvez, BIST method for die-level process parameter variation monitoring in analog/mixed-signal integrated circuits. *Proceedings of IEEE Design, Automation and Test Europe*, pp.1301–1306, 2006

380. E. Alon, V. Stojanovic, M.A. Horowitz, Circuits and techniques for high-resolution measurement of on-chip power supply noise. IEEE J. Solid-State Circuits **40**, 820–828 (2005)

381. V. Petrescu, M. Pelgrom, H. Veendrick, P. Pavithran, J. Wieling, Monitors for a signal integrity measurement system. *Proceedings of IEEE European Solid-State Circuit Conference*, pp.122–125, 2006

382. E. Sackinger, W. Guggenuhl, A high-swing, high-impedance mos cascode circuit. IEEE J. Solid-State Circuits **25**(1), 89–298 (1990)

383. A. Coban, P. Allen, A 1.75-V rail-to-rail CMOS Opamp. Proc IEEE Int. Symp. Circuits Syst. **5**, 5.497–5.500 (1994)

384. F. Fruett, G.C.M. Meijer, A. Bakker, Minimization of the mechanical-stress-induced inaccuracy in bandgap voltage references. IEEE J. Solid-State Circuits **38**(7), 1288–1291 (2003)

385. M.A.P. Pertijs, G.C.M. Meijer, J.H. Huijsing, Precision temperature measurement using CMOS substrate PNP transistors. IEEE Sens. J. **4**(3), 294–300 (2004)

386. F. Fruett, G. Wang, G.C.M. Meijer, The piezojunction effect in NPN and PNP vertical transistors and its influence on silicon temperature sensors. Sens Actuat A – Phys Sens **85**, 70–74 (2000)

387. R.J. Widlar, New developments in IC voltage regulators. IEEE J. Solid-State Circuits **6**, 2–7 (1971)

388. D. Schinkel, R.P. de Boer, A.J. Annema, A.J.M. van Tuijl, A 1-V 15µW high-precision temperature switch. *Proceedings of IEEE European Solid-State Circuit Conference*, pp. 77–80, 2001

389. C.H. Brown, Asymptotic comparison of missing data procedures for estimating factor loadings. Psychometrika **48**, 269–292 (1983)

390. R.B. Kline, *Principles and Practices of Structural Equation Modeling* (Guilford, New York, 1998)

391. B. Muthen, D. Kaplan, M. Hollis, On structural equation modeling with data that are not missing completely at random. Psychometrika **52**, 431–462 (1987)

392. A.P. Dempster, N.M. Laird, D.B. Rubin, Maximum likelihood from incomplete data via the EM algorithm. J. R. Stat. Soc. **39**, 1–38 (1977)

393. G.J. McLachlan, T. Krishnan, *The EM Algorithm and Extensions* (Wiley-Interscience, New York, 1997)

394. A. Zjajo, S. Krishnan, J. Pineda de Gyvez, Efficient estimation of die-level process parameter variations via the em-algorithm. *Proceedings of IEEE International Symposium on Design and Diagnostic of Electronic Circuits and Systems*, pp. 287–292, 2008

395. C. Cortes, V. Vapnik, Support-vector networks. Mach Learning **20**, 273–297 (1995)

396. V. Franc, V. Hlavac, Multi-class support vector machine. Proc IEEE Int Conf Pattern Recog **2**, 236–239 (2002)

397. R.A. Redner, H.F. Walker, Mixture densities, maximum likelihood and the EM algorithm. Surveys Math. Indus. **26**, 195–239 (1984)

398. S.M. Zabin, H.V. Poor, Efficient estimation of class A noise parameters via the EM algorithm. IEEE Trans. Inf. Theor. **37**(1), 60–72 (1991)

399. Y. Zhao, An EM algorithm for linear distortion channel estimation based on observations from a mixture of gaussian sources. IEEE Trans Speech Audio Process **7**(4), 400–413 (1999)

400. C.F. Wu, On the convergence properties of the EM algorithm. Annu Stat **11**(1), 95–103 (1983)

401. S.P. Lloyd, Least squares quantization in PCM. IEEE Trans. Inf. Theor. **28**(2), 129–137 (1982)

402. D.M. Hummels, F.H. Irons, R. Cook, I. Papantonopoulos, Characterization of ADCs using a non-iterative procedure. Proc. IEEE Int. Symp. Circuits Syst. **2**, 5–8 (1994)

403. D. Hummels, Performance improvement of all-digital wide-bandwidth receivers by linearization of ADCs and DACs. Measurement **31**(1), 35–45 (2002)

404. P. Arpaia, P. Daponte, L. Michaeli, Influence of the architecture on ADC error modeling. IEEE Trans. Instrum. Meas. **48**, 956–966 (1999)

405. K. Noguchi, T. Hashida, M. Nagata, On-chip analog circuit diagnosis in systems-on-chip integration. *Proceedings of IEEE European Solid-State Circuit Conference*, pp. 118–112, 2006

406. A. Charoenrook, M. Soma, A fault diagnosis technique for flash ADC's. IEEE Trans. Circuits Syst. II: Analog Digital Signal Processing **43**, 445–457 (1996)

407. F.H. Irons, D.M. Hummels, I.N. Papantonopoulos, C.A. Zoldi, Analog-to-digital converter error diagnosis. *Proceedings of IEEE Instrumentation and Measurement Technology Conference*, pp. 732–737, 1996

408. A. Zjajo, J. Pineda de Gyvez, Diagnostic analysis of static errors in multi-step analog to digital converters. *Proceedings of IEEE Design, Automation and Test in Europe Conference*, pp. 74–79, 2008

409. U. Eduri, F. Maloberti, Online calibration of a Nyquist-rate analog-to-digital converter using output code-density histograms. IEEE Trans. Circuits Syst. I: Fund. Theor. Appl. **51**(1), 15–24 (2004)

410. J. Tsimbinos, K.V. Lever, Improved error-table compensation of A/D converters. IEE Proc – Circuits, Devices Syst **144**(6), 343–349 (1997)

411. P. Händel, M. Skoglund, M. Pettersson, A calibration scheme for imperfect quantizers. IEEE Trans. Instrum. Meas. **49**(11), 1063–1068 (2000)

412. P. Daponte, R. Holcer, L. Horniak, L. Michaeli, S. Rapuano, Using an interpolation method for noise shaping in A/D converters. *Proceedings of the IEEE European Workshop on ADC Modelling and Testing*, pp. 147–150, 2002

413. B. Widrow, S.D. Stearns, *Adaptive Signal Processing* (Prentice-Hall, Englewood Cliffs, NJ, 1985)

414. M. Vanden Bossche, J. Schoukens, J. Eenneboog, Dynamic testing and diagnostics of A/D converters. IEEE Trans. Circuits Syst. **33**(8), 775–785 (1986)

415. G. Schafer, H. Sapotta, W. Dennerm, Block-oriented test strategy for analog circuits. *Proceedings of IEEE European Solid-State Circuit Conference*, pp. 217–220, 1991

416. A. Zjajo, J. Pineda de Gyvez, DfT for fully accessibility of multi-step analog to digital converters. *Proceedings of IEEE International Symposium on VLSI Design, Automation and Test*, pp. 73–76, 2008

417. A. Osseiran, *Analog and Mixed-Signal Boundary Scan: A Guide to the IEEE 1149.4 Test Standard* (Kluwer, Boston, 1999)

418. M. Karlsson-Rudberg, Calibration of mismatch errors in time interleaved ADCs. Proc. IEEE Int. Conf. Electron, Circuits Syst. **2**, 845–848 (2001)

419. J.-E. Eklund, F. Gustafsson, Digital offset compensation of time-interleaved ADC using random chopper sampling. Proc. IEEE Int. Symp. Circuits Syst. **3**, 447–450 (2000)

420. H. Jin, E.K. Lee, A digital-background calibration technique for minimizing timing-error effects in time-interleaved ADC's. IEEE Trans. Circuits Syst. **47**(7), 603–613 (2000)

421. J. Elbornsson, K. Folkesson, J.-E. Eklund, Measurement verification of estimation method for time errors in a time-interleaved A/D converter system. *Proceedings of IEEE International Symposium on Circuits and Systems*, pp. 129–132, 2002

422. J. Elbornsson, F. Gustafsson, J.-E. Eklund, Blind adaptive equalization of mismatch errors in a time-interleaved A/D converter system. IEEE Trans. Circuits Syst. I **51**(1), 151–158 (2004)

423. A. Zjajo, Diagnostic analysis of bandwidth mismatch in time-interleaved systems. *Proceedings of IEEE International Conference on Electronics, Circuits and Systems*, pp. 105–108, 2008

424. A. Papoulis, *Signal Analysis* (McGraw-Hill, New York, 1977)

425. H. Landau, Necessary density conditions for sampling and interpolation of certain entire functions. Acta Mathematica **117**(1), 37–52 (1967)

426. R.G. Vaughan, N.L. Scott, D.R. White, The theory of bandpass sampling. IEEE Trans. Signal Process. **39**, 1973–1984 (1991)

427. A. Kohlenberg, Exact interpolation of band-limited functions. J. Appl. Phys. **24**, 1432–1435 (1953)

428. C. Herley, P.W. Wong, Minimum rate sampling and reconstruction of signals with arbitrary frequency support. IEEE Trans. Inf. Theor. **45**, 1555–1564 (1999)

429. R. Venkataramani, Y. Bresler, Perfect reconstruction formulas and bounds on aliasing error in sub-Nyquist nonuniform sampling of multiband signals. IEEE Trans. Inf. Theor. **46**, 2173–2183 (2000)

430. J.E. Dennis, R.B. Schnabel, *Numerical Methods for Unconstrained Optimization and Nonlinear Equations* (Prentice-Hall, Englewood Cliffs, NJ, 1983)

431. J. McNeill, M. Coln, B. Larivee, A split-ADC architecture for deterministic digital background calibration of a 16b 1 MS/s ADC. IEEE Int. Solid-State Circuits Conf. Dig. Techn. Pap. **1**, 276–598 (2005)

432. A. Zjajo, J. Pineda de Gyvez, Calibration and debugging of multi-step analog to digital converters. *Proceedings of IEEE International Symposium on Electronic Design, Test and Applications*, pp. 512–515, 2008

433. M.F. Wagdy, S.S. Awad, Determining ADC effective number of bits via histogram testing. IEEE Trans. Instrum. Meas. **40**(4), 770–772 (1991)

Index

Printed in the United States
By Bookmasters